# Adipose Tissue Protocols

# METHODS IN MOLECULAR BIOLOGY™

## John M. Walker, SERIES EDITOR

METHODS IN MOLECULAR BIOLOGY™

# Adipose Tissue Protocols

Edited by

## Gérard Ailhaud

*Université de Nice-Sophia Antipolis,*
*Nice, France*

Humana Press ✳ Totowa, New Jersey

Production Editor: Kim Hoather-Potter.

Cover illustration: From Fig. 4A. in Chapter 2, "Morphologic Techniques for the Study of Brown Adipose Tissue and White Adipose Tissue," by Saverio Cinti, M. Cristina Zingaretti, Raffaella Cancello, Enzo Ceresi, and Plinio Ferrara.

Cover design by Patricia F. Cleary.

For additional copies, pricing for bulk purchases, and/or information about other Humana titles, contact Humana at the above address or at any of the following numbers: Tel.: 973-256-1699; Fax: 973-256-8341; E-mail: humana@humanapr.com; Website: http://humanapress.com

Printed in the United States of America. 10 9 8 7 6 5 4 3 2 1

Library of Congress Cataloging in Publication Data

Main entry under title:Methods in molecular biology™.
Adipose tissue protocols / edited by Gérard Ailhaud.
     p. ; cm. – (Methods in molecular biology ; v. 155)
     Includes bibliographical references and index.
     ISBN 0-89603-747-9 (alk. paper)
          1. Adipose tissue–Laboratory manuals. I. Ailhaud, Gérard. II. Series.
          [DNLM: 1. Adipose Tissue–cytology. 2. Adipose Tissue–microbiology. 3.
          Mycobacterium–isolation & purification. 4. Tissue Culture–methods. QS 532.5.A3
          A2345 2000]
          QP88.15.A35 2000
          571.5'7–dc21
                                                                                                    00-022069

# Preface

Adipose tissue is recognized to be exquisitely sensitive to hormone action, and is also now recognized as a secretory and endocrine organ required for reproduction and good health. Adipocytes are "smart" cells able within the tissue to communicate with surrounding cells, but also with various organs, particularly via leptin acting on the central nervous system.

Brown adipose tissue (BAT) and white adipose tissue (WAT) are known to be distinct tissues, whereas the heterogeneity of WAT depots is well established. Unfortunately, excess WAT leads to obesity, which is the most common health problem in industrialized countries. Therefore, from both a scientific and a technical point of view, the time has come to create a survey of adipose tissues and their neglected adipocytes.

In *Adipose Tissue Protocols*, I have attempted to gather together chapters from all areas of adipose tissue research—from in vivo to in vitro studies—and to provide methods covering a wide variety of techniques, including the choice of adipose tissue depot and of morphological techniques for the study of BAT and WAT; the isolation, subcellular fractionation, and transfection of adipocytes where the low density of these cells must be taken into account; assays of nutrient and ion fluxes and the metabolic aspects of nutrient uptake; assays of lipid-related enzymes; biopsies and quantification of lipid-related mRNAs; cultures of adipose precursor cells from WAT and BAT of various species, including human tissue; measurements of adipose secretory products; and assessment of WAT metabolism in vivo.

Throughout the book, common errors and potential problems, often resulting from the fragility of adipocytes, have been highlighted in the "Notes" section of various chapters.

This book could not have been produced without the efficient cooperation of all the authors, who deserve great thanks for their efforts, and the skillful secretarial assistance of Geneviève Oillaux. I wish to thank Daniel Ricquier, Max LaFontan, and Yannick Le Marchand-Brustel for discussing the content of the book. I wish also to thank John Walker for his advice and efficient help during the preparation of the book.

*Gérard Ailhaud*

# Contents

# Contents

# Contributors

NADA A. ABUMRAD • *Department of Physiology and Biophysics, State University of New York at Stony Brook, NY, USA*

GÉRARD AILHAUD • *Centre de Biochimie, Faculté des Sciences, UMR 6543 CNRS, Parc Valrose, Nice, France*

MARIE-CHRISTINE ALESSI • *Laboratoire d'Hématologie, Faculté de Médecine de la Timone, Marseille, France*

JÉRÔME AUBERT • *Centre de Biochimie, Faculté des Sciences, UMR 6543 CNRS, Parc Valrose, Nice, France*

PIERRE BARBE • *Service d'Endocrinologie-Nutrition and INSERM U-317, Hôpital Rangueil, Toulouse, France*

RAYMOND BAZIN • *Institut Biomédical des Cordeliers, INSERM U-465, Paris, France*

MARTIN BIDLINGMAIER • *Neuroendocrine Unit, Innenstadt University Hospital, Munich, Germany*

JENS BÜLOW • *Department of Clinical Physiology, Bispebjerg Hospital, Copenhagen, Denmark*

DENIS CALISE • *UPRESA-CNRS 5018, CHU Rangueil, Toulouse, France*

RAFFAELLA CANCELLO • *Institute of Normal Human Morphology-Anatomy, School of Medicine, University of Ancona, Ancona, Italy*

BARBARA CANNON • *The Arrhenius Laboratories F3, The Wenner-Gren Institute, Stockholm University, Stockholm, Sweden*

CHRISTIAN CARPÉNÉ • *INSERM U-317, CHU Rangueil, Bât L3, Toulouse, France*

LOUIS CASTEILLA • *UPRESA-CNRS 5018, CHU Rangueil, Toulouse, France*

ENZO CERESI • *Institute of Normal Human Morphology-Anatomy, School of Medicine, University of Ancona, Ancona, Italy*

SAVERIO CINTI • *Institute of Normal Human Morphology-Anatomy, School of Medicine, University of Ancona, Ancona, Italy*

SIMON W. COPPACK • *Academic Medical Unit, Royal London Hospital, London, UK*

MIREILLE CORMONT • *Faculté de Médecine, INSERM E99-11, Nice, France*

BÉATRICE COUSIN • *UPRESA-CNRS 5018, CHU Rangueil, Toulouse, France*

CHRISTIAN DANI • *Centre de Biochimie, UMR 6543 CNRS, Faculté des Sciences, Parc Valrose, Nice, France*

CHRISTIAN DARIMONT • *Nestlé Research Center, Lausanne, Switzerland*

EVA DEGERMAN • *Department of Cell and Molecular Biology, University of Lund, Lund, Sweden*

MARIO DIGIROLAMO • *Department of Medicine, Emory University School of Medicine, Atlanta, GA, USA*

ISABELLE DUGAIL • *Institut Biomédical des Cordeliers, INSERM U-465, Paris, France*

PLINIO FERRARA • *Institute of Normal Human Morphology-Anatomy, School of Medicine, University of Ancona, Ancona, Italy*

PASCAL FERRÉ • *Institut Biomédical des Cordeliers, INSERM U-465, Paris, France*

JACQUELINE B. FINE • *Department of Medicine, Emory University School of Medicine, Atlanta, GA, USA*

KEITH N. FRAYN • *The Oxford Lipid Metabolism Group, University of Oxford, Oxford, UK*

SUSAN K. FRIED • *Department of Nutritional Sciences, Rutgers University, New Brunswick, NJ, USA*

JEAN GALITZKY • *Laboratoire Pharmacologie Médicale et Clinique, INSERM U-317, Toulouse, France*

THIERRY GRÉMEAUX • *Faculté de Médecine, INSERM E99-11, Nice, France*

HANS HAUNER • *Diabetes-Forschungsinstitut an der Heinrich-Heine-Universität Düsseldorf, Düsseldorf, Germany*

CECILIA HOLM • *Department of Cell and Molecular Biology, University of Lund, Lund, Sweden*

HANS-GEORG JOOST • *Institut für Pharmakologie und Toxicologie, Medizinische Fakultät der RWTH Aachen, Germany*

TOVA RAHN LANDSTRÖM • *Department of Cell and Molecular Biology, University of Lund, Lund, Sweden*

YANNICK LE MARCHAND-BRUSTEL • *Faculté de Médecine, INSERM E99-11, Nice, France*

DANIELA MALIDE • *EDMNS/DB/NIDDK/NIH, Bethesda, MD, USA*

VINCENT MANGANIELLO • *Pulmonary/Critical Care Medicine Branch, NHLBI, NIH, Bethesda, MD, USA*

NAIMA MOUSTAID-MOUSSA • *Department for Nutrition, University of Tennessee, Knoxville, TN, USA*

JAN NEDERGAARD • *The Arrhenius Laboratories F3, The Wenner-Gren Institute, Stockholm University, Stockholm, Sweden*

RAYMOND NÉGREL • *Centre de Biochimie, Faculté des Sciences, UMR 6543 CNRS, Parc Valrose, Nice, France*

GUNILLA OLIVECRONA • *Department of Medical Biochemistry and Biophysics, University of Umea, Sweden*

HEIDI K. ORTMEYER • *Obesity and Diabetes Research Center, University of Maryland School of Medicine, Baltimore, MD, USA*

MALIN OTTOSSON • *The Wallenberg Laboratory, Sahlgrenska University Hospital, Göteborg University, Göteborg, Sweden*

LUC PÉNICAUD • *UPRESA-CNRS 5018, CHU Rangueil, Toulouse, France*

SVANTE RESJÖ • *Department of Cell and Molecular Biology, University of Lund, Lund, Sweden*

PERLA SAINT-MARC • *Centre de Biochimie, Faculté des Sciences, UMR 6543 CNRS, Parc Valrose, Nice, France*

ANNETTE SCHÜRMANN • *Institut für Pharmakologie und Toxicologie, Medizinische Fakultät der RWTH Aachen, Germany*

THOMAS SKURK • *Diabetes-Forschungsinstitut an der Heinrich-Heine-Universität Düsseldorf, Düsseldorf, Germany*

JEAN-FRANÇOIS TANTI • *Faculté de Médecine, INSERM E99-11, Nice, France*

MATTHIAS TSCHOEP • *Neuroendocrine Unit, Innenstadt University Hospital, Munich, Germany*

HUBERT VIDAL • *Faculté de Médecine Laennec, INSERM U-449 Lyon, France*

MARTIN WABITSCH • *Department of Pediatrics I, Universitätskinderklinik Ulm, Germany*

M. CRISTINA ZINGARETTI • *Institute of Normal Human Morphology-Anatomy, School of Medicine, University of Ancona, Ancona, Italy*

# 1

## Choosing an Adipose Tissue Depot for Sampling

*Factors in Selection and Depot Specificity*

**Louis Casteilla, Luc Pénicaud, Béatrice Cousin, and Denis Calise**

### 1. Introduction

Adipose tissues (ATs) were long considered as negligible and as simple filling tissues. The increase in knowledge concerning their role in energy balance and the increased occurrence of metabolic disorders, such as obesity and syndrome X, have focused the attention of the scientific community on these tissues. This evolution has been speeded by both the discovery of leptin and the development of transgenic and knockout techniques. The former emphasized the secretory function and the involvement of adipose mass in various general physiological functions, such as reproduction *(1–3)*. The latter made it possible to test hypotheses elaborated from in vitro findings in the organisms, but these approaches also revealed unexpected results and findings concerning the development of AT *(4–8)*. It is true that any changes in fat mass have a rapid effect on body weight (wt), which is quickly and easily identifiable. This chapter aims to give a general picture of the ATs presently available in mammals and to describe their sampling.

### 2. Concept and Principles: How to Classify ATs?

#### 2.1. Brown and White Fat: The Simplest Classification

The simplest definition consists of classifying ATs into two types: brown (BAT) and white adipose tissue (WAT). These tissues are characterized by different anatomical locations, morphological structures, functions, and regulation *(9–13, see also* **Subheadings 2.2–2.4.**). They are called adipose because of the amount of fat stored in both types. BAT is so called because of its char-

From: *Methods in Molecular Biology, vol. 155: Adipose Tissue Protocols*
Edited by: G. Ailhaud © Humana Press Inc., Totowa, NJ

acteristic color, originating mostly from its abundant vascularization and cyto-chromes. Both types of ATs are able to store energy as triglycerides, but, whereas white fat releases this energy according to the needs of the organism, brown fat converts it as heat. WAT is the main store of energy for the organism. BAT plays an important role in the regulation of body temperature in hibernating, as well as in small and newborn mammals *(13)*. The developmental patterns of ATs are different, and are species-dependent (*see* **Subheadings 3.2.–3.4.**). Adipocytes within a pad were long considered to belong to a single phenotype, i.e., either brown or white adipocytes.

Although studies on WAT were always concerned with energy metabolism, studies on BAT were first focused on nonshivering thermogenesis and thermoregulatory purposes *(13)*. Later, the involvement of BAT in diet-induced thermogenesis led researchers also to investigate its role in various conditions associated with changes of energy balance *(14–16)*.

## 2.2. Typical BAT

The main features specific to BAT are summarized in **Table 1** and are compared with WAT. Its thermogenic function is assumed by the numerous mitochondria and by the presence of mitochondrial protein, uncoupling protein 1 (UCP1), in the brown adipocytes. UCP1 is specifically expressed in these cells, and is located in the inner mitochondrial membrane. It is able to uncouple the mitochondria, and enables heat production *(17,18)*. Recently, the homologous proteins, UCP2 and UCP3, have been cloned *(19,20)* and can also be detected in this tissue. Some biochemical or molecular markers, including nuclear factors more or less specific to brown fat, are available, and these are also given in **Table 1** *(21–25)*.

In most mammals, BAT develops during gestation and perinatal life *(26)*. It is prominent in the newborn or in young animals in which nonshivering thermogenesis is necessary to counteract heat loss associated with birth and atmospheric life. It is mostly located around arterial vessels and vital organs. One exception is the piglet, which has no brown fat, and is subject to some disturbances of thermoregulation *(27,28)*. The development and quantity of BAT depend on the degree of nonshivering thermogenesis required by the organism to maintain its body temperature. This need corresponds to the balance between metabolic body mass, which produces heat and heat loss, which is correlated to body surface and the adequacy of insulation. With increasing age, as the rate of heat loss per unit body wt decreases, the tissue becomes indistinguishable from white fat. Nevertheless, in hibernators and in some other small mammals (mice, rats, and so on), it regresses only partially and remains identifiable throughout life. In these species, adipose precursor cells are latent in the tissue, and can be recruited as necessary. This general presentation must be modified according to species and to the developmental stage of the newborn, as summarized in

**Table 1**
**Main Features Differentiating WAT and BAT in Rodents**

|  | WAT | BAT |
|---|---|---|
| Location of main depots | Inguinal, retroperitoneal, gonadal (compare with **Table 2**) | Interscapular, perirenal, axillary, paravertebral |
| Color | Ivory or yellow | Brown |
| Vascular system | ++ | +++ |
| Innervation | Sympathetic (++) | Sympathetic (+++) |
| Adipose cells | Unilocular cells | Multilocular cells |
| Functions | Storage of energy as triglycerides | Storage of energy as triglycerides |
|  | Fatty acids and glycerol release | Heat production |
|  | Secretory tissue | Secretory tissue |
| Mitochondria | + | +++ |
| UCPs | UCP2 (++) | UCP1, UCP2 (+), UCP3 |
| Deiodase type II | + | +++ |
| GMP reductase | – | +++ |
| Leptin | +++ | At birth, not in adult |
| $\alpha,\beta$-Adrenoceptors | $\beta_3$ (++ ), $\alpha_2$ (+) | $\beta_3$ (+++) |
| PGC1 | + | +++ |
| Cig 30 mRNA | – | ++ |

**Table 2 (26).** The most studied typical brown adipose deposits are the inter-scapular (IBAT) and perirenal BATs in rodents and large mammals respectively. IBAT is located subcutaneously between the shoulders, and can easily be dissected (*see* **Subheading 3.**). It is the only fat pad distinguishable at birth in laboratory rodents. Perirenal BAT is brown in large mammals during the perinatal period, its weight is greater, and it is impossible to sample or remove the whole pad without removing the kidney.

One of the strongest inducers of this type of AT is cold exposure. Acute exposure induces marked changes in metabolism and gene regulation, but also stimulates proliferation and differentiation of the precursors into brown adipocytes, leading to development of this tissue in the days or weeks after exposure. Catecholamines or $\beta$-adrenoceptor agonists mimic the majority of these effects *(13,21,22,29)*.

### 2.3. WAT

This is the most abundant tissue of fat mass, and may account for more than half of body wt in severe obesity. It was considered as less vascularized and innervated than brown fat, but various reviews have questioned this opinion

**Table 2**
**Evolution of Typical BAT According to Species**

|  | Immature (hamster) | Altricial (mouse, rat) | Precocial (rabbit, guinea pig, ruminants, primates) |
|---|---|---|---|
| Newborn | Underdeveloped | Nest-dependent | Well-developed |
| Amount at birth | 1–2% body wt | 1–2% body wt | 2–5% body wt |
| Few days after birth | Poorly developed | 11–12% of body wt | Decreasing |
| Few weeks after birth | Developing | Partial regression | Transformation into white fat |
| Adult | Developed | Present | Absent |

*(30–33)*. The importance of white fat in energy balance is well known, and many papers *(84–88)* have dealt with the metabolism of this tissue. Besides this classic view, the wide range of products secreted by adipose cells emphasizes its secretory function, and opens interesting fields for the understanding of the established links between the increase of fat mass and various associated disorders, such as cardiovascular disease *(34–37)*. When white fat is compared to brown fat, it is noteworthy that no specific marker of white fat is now available to positively identify it. In the adult, leptin could be a good marker for positive identification of white fat, but its strong expression at birth in brown fat makes this an open question *(38,39)*. Most of its development occurs after birth, and primarily results from hypertrophy of white adipose cells, which can reach 150 μm in diameter in some species. Nevertheless, a pool of preadipose cells is maintained throughout life in most species, including humans and can participate in this growth *(40,41)*.

## 2.4. Heterogeneity and Plasticity

The above classification must be qualified, because of several findings: the presence of scattered brown adipocytes in white fat; the different properties of WATs according to location; the putative conversion of one AT phenotype to the other.

### 2.4.1. Heterogeneity Within and Between Pads

Brown adipocytes have been observed in non-cold exposed rodents, as well as in several deposits considered as typical white fat in primates. The number of these cells can vary according to the location of fat pads, and are most numerous in the periovarian fat of rodents, which can be compared to a patchwork of brown and white adipocytes *(42,43)*.

It has long been well known that the location of the development of adipose deposits during obesity differs according to gender *(44,45)*. Abdominal obesity is predominant in the male; subcutaneous (sc) fat mass is mostly involved in female obesity. Sex hormones play a major role in these differences *(45)*. Increased intra-abdominal body fat mass is considered as an independent risk factor for health problems linked to obesity, and is positively correlated with increased overall mor-bidity and mortality *(46–48)*. These findings have been the basis for numerous investigations, including genetic approaches to differences of metabolic properties or precursor pools according to location of fat, and they reinforce the concept of heterogeneity, but, in this case, between the fat pads. Taken together, these studies make it possible to distinguish sc from internal fat, and upper from lower body fat. However, this classification is not sufficiently clear and needs further definition: For instance, in women, the round ligament seems to play a particular role *(49,50)*. This heterogeneity exists whatever the species *(51–54)*. For example, in men, omental adipose fat is the most active tissue in lipolysis, as well as in lipogenesis *(55,56)*. Abdominal pads also have higher interleukin 6 or plasminogen activator inhibitor secretion, in vitro differentiation capacity, thiazolidinedione sensitivity, and apoptosis than sc pads *(57–63)*. One exception seems to be leptin expression, which is higher in sc tissue *(64,65)*. It is noteworthy that the regulation of this gene is depot-related *(66)*. These depot-specific properties are partly genetically deter-mined *(67)*. This heterogeneity can also be observed at molecular levels in rodents, as well as in humans *(68–70)*.

## 2.4.2. Plasticity

This term is used to indicate that some deposits are capable of converting from one type of AT to the other. The transformation that has first been described concerns the transformation of BAT into WAT-like AT, which takes place, as previously indicated, during postnatal development *(71)*. The reversibility of this process differs according to species. It does not seem possible in rabbits or rumi-nants, in which even cold exposure or β-adrenoceptor agonist treatment cannot promote reversion *(72)*. In adult rodents, similar transformation has been described after sympathetic denervation of IBAT or during the gestation–lactation cycle. Loncar *(73)* described changes in the inguinal tissue of mice after cold stress. The term of "convertible adipose tissue" was used to describe the corresponding depos-its *(73)*. In fact, the same results can be obtained with all deposits but with different intensity levels *(68)*. Marked development of brown adipocytes occurs among fat considered and studied as typical white fat, i.e., periovarian fat in rats, inguinal fat in mice, and numerous white fat pads in dogs *(72–74)*. The proportion of the two phenotypes of adipose cells changes according to physiological (cold exposure, development, gestation–lactation cycles), pharmacological (β3-adrenergic agonist treatment), or pathophysiological conditions and genetic background *(75–79)*.

Taken together, these findings reveal the heterogeneity within or between pads and the potential for transformation between the two phenotypes of adipose tissues, for which we have proposed the term plasticity *(73,78)*.

In any event, the investigator must be cautious and take into account this aspect of AT biology when: mice or rats are used as a model for humans; fat pads have to be pooled to obtain sufficient sample quantity; or it is only possible to remove an aliquot of AT to interpret the results as the index of the whole fat pad or the whole fat mass.

## 3. Methods

### *3.1. Choice of Species*

The criteria of choice are numerous, and are grouped here into three levels (**Table 3**). The first is the scientific aim, and the choice of species will be strictly dependent on it. In other cases, the decision is less clear, and each aspect may need discussion. Nevertheless, an aid to decision can be suggested as illustrated in **Table 4**, which shows that rodents are valuable and convenient models in most cases, except for human studies. The chief reasons are given in **Table 5**. From these data, it is clear that classic laboratory rodents are not a good model for humans in metabolic or developmental studies. When metabolic features are considered, no important difference exists between adipocytes from nonhuman and human primates *(80)*. Therefore, the only physiological models available as human models are primates. For developmental studies, large animals, and also rabbits, seem to present the same features, and can be used at least until weaning *(81)*. After this time, the great difference in metabolism excludes the use of these animal species as human models.

### *3.2. Choice of Pad*

When the species has been decided upon, the location of the fat pad to be studied must be chosen. Two aspects must be considered: once again, the scientific aim and the amount of tissue needed. Both aspects are summarized in **Tables 6** and **7** for rats or mice. The coarse ratio between IBAT and the three other sites described in **Table 7** is quite different in these two closely related rodent species, which suggests that IBAT is relatively more important in the mouse, the species used for transgenic studies, than in the rat.

Whatever the species, because of heterogeneity between pads, they should not be pooled, or, if pooling is necessary to obtain sufficient quantity, then only pads with similar metabolic and cellular features should be pooled.

Whatever the AT, fine dissection is required, because of the developed vascular system and numerous lymph nodes.

**Table 3**
**Parameters to be Taken into Account in the Study of AT**

|  | Parameters |
|---|---|
| Questions | Model for human physiology or pathology; BAT or WAT studies; physiological investigations; in vivo developmental studies; plasticity studies; transgenic model; photoperiod effect; in vitro studies. |
| Extrinsic | Species; breed; age; sex ; diet; room temperature; photoperiod; gestation–lactation cycles |
| Intrinsic | Importance of metabolic pathways; BAT vs WAT phenotype; plasticity (*see* **Subheading 2.4.2.**); sc vs internal; accessibility of innervation or vascular system (depending on studies) |

**Table 4**
**Advisable Species According to the Investigations**

| Investigation fields | Advisable species |
|---|---|
| Model for humans | Primates, not rodents |
| BAT or WAT studies | All species |
| Physiological investigations | Rats |
| In vivo developmental studies | All species |
| Plasticity studies | Mouse, rat, rabbit |
| Transgenic and knockout models | Mouse |
| Photoperiodic effect | Sensitive species: hamster |
| Diet-obesity | Rats according to breed; dogs with high-fat diet; primates |

## *3.3. Sampling IBAT in Mice or Rats (Fig. 1)*

1. After euthanasia *(82)*, the animals are placed on the abdomen, the head toward the investigator.
2. The shoulder region is abundantly rinsed with diluted EtOH (70%) to wet the coat, and to avoid having hairs on the samples.
3. The skin just behind the head is grasped with tongs, lifted, and incised with scissors.
4. The skin is widely incised from this point to the middle of the back, and the field is opened.
5. The butterfly-shaped IBAT is revealed.
6. There are then two possibilities: Rub the fat pad with a paper tissue to discard the white part just above the IBAT, then carefully dissect the pad. A binocular microscope can be used, but, with some practice, this is not necessary; or remove the fat pad, and afterwards carefully dissect the butterfly of brown fat. In all cases, care must be taken to avoid the muscle closely associated with the brown fat.

Table 5
**Main Features Differentiating ATs in Rats, Mice, and Humans**[a]

|                                          | Rats and mice              | Humans                     |
| ---------------------------------------- | -------------------------- | -------------------------- |
| Location of fat pad:                     |                            |                            |
|    Interscapular          | +                          | –                          |
|    Periovarian            | +                          | –                          |
|    Epididymal             | +                          | –                          |
| Persistence of brown fat in adults       | +++                        | +/–                        |
| Convertible features                     | Mice > rats                |                            |
|    BAT -> WAT             | +                          | +++                        |
|    WAT -> BAT             | +++                        | ?                          |
|                                          | PO > RP > Ep               |                            |
| Main site of lipogenesis                 | AT                         | Liver                      |
| Glucose transport sensitive to insulin   | +++                        | +                          |
| Catecholamine-stimulated lipolysis       | $\beta_1, \beta_2, \beta_3$ (+) | $\beta_1, \beta_2$ (+) $\alpha_2$ (–) |

[a]PO, periovarian; RP, retroperitoneal; Ep, epididymal.

Table 6
**Choice of Fat Pads in the Rat or Mouse According to the Aim of Investigation**[a]

| Aim                 | IBAT | PO  | RP  | Inguinal | Ep  |
| ------------------- | ---- | --- | --- | -------- | --- |
| sc vs abdominal     |      |     |     | +        | +   |
| Plasticity          | +    | Rat |     | Mouse    |     |
| Denervation         | +    | +   | +   |          |     |
| Vascular system     | +    |     |     | +        | +   |
| Isolated adipocytes | +    |     |     | +        | +   |
| Primary culture     | +    |     |     | +        | +   |

[a]IBAT, interscapular brown adipose tissue; PO, periovarian adipose tissue; RP, retroperitoneal adipose tissue; Inguinal, Inguinal adipose tissue; Ep, epididymal adipose tissue.

7. The sample is ready and the parts of the pad can be separated as required. If mRNA is to be extracted, freeze the tissue by immersion in liquid nitrogen, put it into a box or other container, and store at –80°C. It is better to freeze it at once than to freeze it after putting it into a container, to prevent it sticking to the walls.

### 3.4. General Considerations for Sampling White Fat

WATs are organized in lobules, and the various pads can be found together within connective tissue, particularly in obese animals. So, before cutting with

**Table 7**
**Weights of Major Sites of AT in Young Adult Rats (9–10 wk old)
and Mice (7–8 wk old)**[a]

| | IBAT (g) | Inguinal (g) | Gonadal (g) | Retroperitoneal (g) | IBAT/WAT |
|---|---|---|---|---|---|
| Mice | 0.14 | 0.35 | 0.2–0.4 | Negligible | Approx 20% |
| Rats | 0.3 | 2.5 | 1.2 | 1 | Approx 7% |

[a]IBAT, interscapular brown adipose tissue; WAT, white adipose tissue.

scissors, it is sometimes better to separate the different parts by hand, taking care to remove the whole fat pad of interest, and this alone. AT can be frozen by immersion in liquid nitrogen, as for brown fat. After freezing, the tissue can be reduced to powder to facilitate and homogenize the sample.

### 3.4.1. Sampling Inguinal AT in Mice or Rats *(Fig. 2)*

1. The procedure is the same as that previously described for IBAT, but, in this case, the rodent is placed on its back with the tail toward the investigator.
2. The abdomen is rinsed with EtOH, and the skin is widely incised.
3. After removing the pad, dissect and discard the lymph nodes present among the fat. For females, take care not to confuse the fat pad and the mammary gland, which is involuted.
4. If sampling is done to study gene expression, depending on the size of the pad, it may be preferable to reduce the pad to powder, in order to use only the amount required for the study.

### 3.4.2. Sampling Gonadal AT in Mice or Rats *(Fig. 2)*

1. Open the abdominal wall.
2. Extract the genital parts (ovaries or testes, according to the sex) from the abdominal cavity.
3. Remove carefully, by dissecting the fat tissue or handling the gonadal tract with one hand, separate fat from other tissues by gently pulling them with the other hand.

## 3.5. Denervation Studies (Fig. 3)

### 3.5.1. Example of IBAT *(Fig. 3A)*

1. Proceed as in **Subheading 3.3.** to reach the IBAT.
2. Carefully separate AT from muscle above the shoulders.
3. Carefully start to raise the IBAT; nerve fibers can now be seen arising from under each shoulder muscle.
4. Cut them at two points, and remove the fragment to block regeneration.
5. Suture skin.

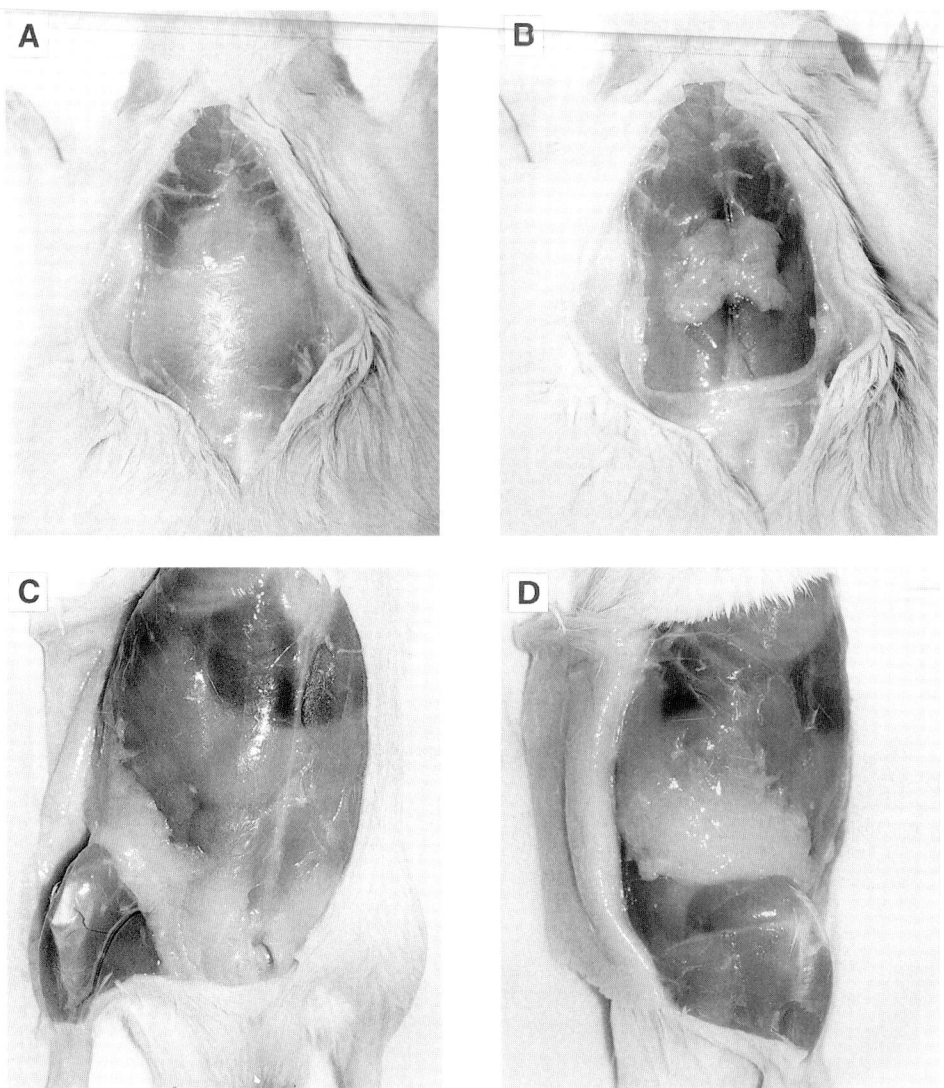

Fig. 1. IBAT and inguinal adipose tissue in rats. (**A**) aspect of IBAT before removing the white part. (**B**) The white part of the pad has been carefully dissected. Brown fat appears as a butterfly between shoulders. (**C,D**) Front and side views of inguinal fat.

### 3.5.2. Example of Retroperitoneal AT *(Fig. 3B)*

1. The retroperitoneal fat pad is innervated by three nerve fibers, and the contralateral pad can be used as control. In this case, use the left or right one at random.

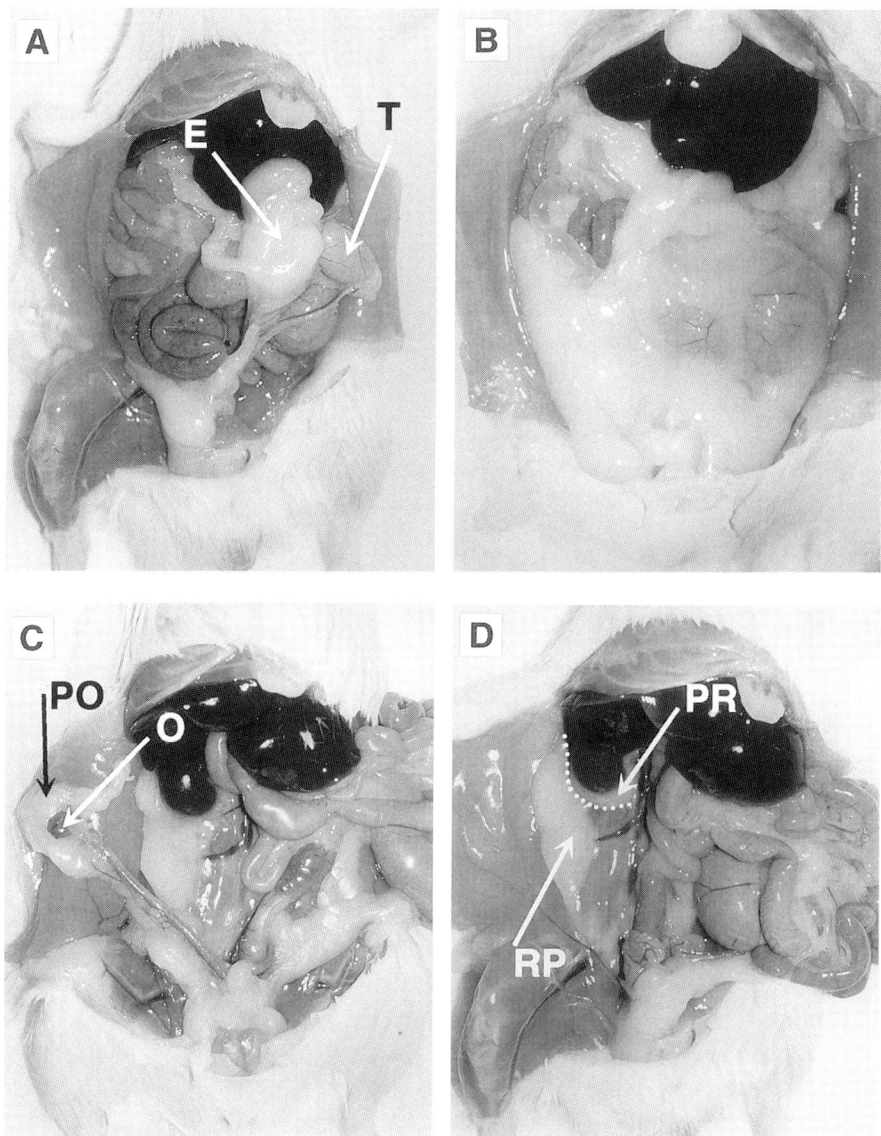

Fig. 2. Internal ATs in male or female older rats. **(A)** The abdominal cavity of a 200-g wt male rat is opened. The left part corresponds to the natural appearance of abdominal cavity. In the right part, the testis (T) and the associated epididymal adipose pad (E) have been put in a prominent position. **(B)** Appearance of abdominal cavity of an older rat (body wt: 400 g). **(C)** Intestines are put on the right side and the left ovary (O) and periovarian pad (PO) placed on the opened abdominal wall. **(D)** The genital part was removed, and it is then possible to better distinguish the retroperitoneal (RP) and the perirenal fat (PR).

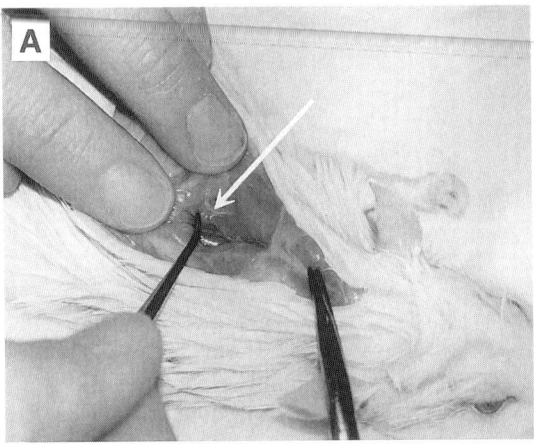

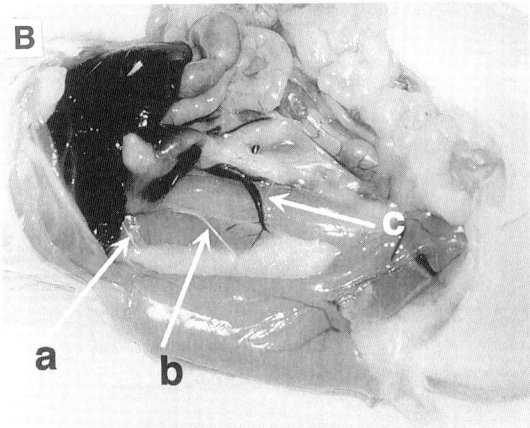

Fig. 3. **(A)** The IBAT is lifted by fingers and the nervous fibers can be easily identified (arrow). **(B)** a, b, c correspond to the three nervous fibers innervating the retroperitoneal pad.

2. Proceed as in **Subheading 3.6.**, but the opening must be as small as possible. The
   goal is to maintain the animal alive after surgery.
3. Cut the three nerves at two points and remove the fragments.
4. Close the abdominal wall, then the skin.

### 3.6. Investigations via Vascular System (Fig. 4)

Almost all fat pads are individually vascularized, and it is possible to use such a
feature to investigate some parameters (blood flow, arteriovenous differences, and
so on) via the vascularization system or to inject particles (e.g., virus) with good
preservation of the anatomy and the cellular interactions of the pad.

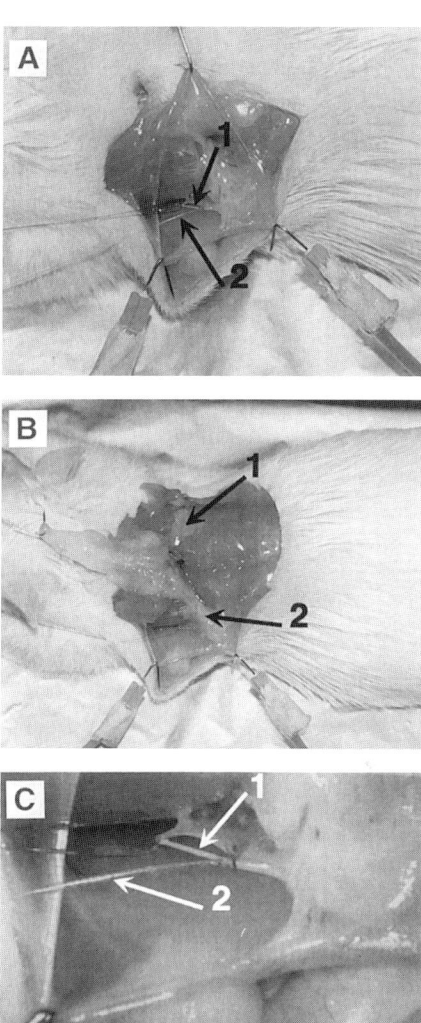

Fig. 4. (**A**) The rat was opened on the side, and, after careful dissection, the arterial vessel of IBAT can be visualized (*arrow 1*). A catheter (*arrow 2*) was introduced into the vessel, and a surgical silk positioned to secure it in place. (**B**) After catheterization, the IBAT was washed with physiological buffer, and the IBAT was dissected and exposed. The washed part of the pad (*arrow 2*) appears clearer than the other part (*arrow 1*). (**C**) Enlargement of the catheterization. 1, arterial vessel; 2, catheter.

### 3.6.1. Example of IBAT

1. To easily obtain cannula with different diameters, use yellow tips for pipetman. Bring the center of the tip, which you hold at each end, near a flame. When the tip

begins to melt and becomes translucent, rapidly stretch it to obtain catheters. The diameter depends on length of stretch. Cut one end obliquely.

2. Lay the anesthezied animal on one side, and carefully dissect the arterial vessel (*see* **Fig. 4A**)
3. Put in position the surgical silk, before doing a small incision in the artery.
4. After introducing the catheter into the artery, one can then inject anything (drugs, viral particles, and so on).

### 3.6.2. Example of White Pad

A very good description of surgical and technical procedure was made by Scow *(83)* for periovarian pad. Similar experiments can be performed with most fat pads. In the case of the human species, this aspect will be developed in Chapter 22.

## References

1. Zhang, Y., Proenca, R., Maffei, M., Barone, M., Leopold, L., and Friedman, J. M. (1994) Positional cloning of the mouse obese gene and its human homologue. *Nature* **372,** 425–432.
2. Flier, J. S. (1998) Clinical Review 94. What's in a name? In search of leptin's physiologic role. *J. Clin. Endocrinol. Metab.* **83,** 1407–1413.
3. Frühbeck, G., Jebb, S. A., and Prentice, A. M. (1998) Leptin: physiology and pathophysiology. *Clin. Physiol.* **18,** 399–419.
4. Bosch, F., Pujol., A., and Valera, A. (1998) Transgenic mice in the analysis of metabolic regulation. *Ann. Rev. Nutr.* **18,** 207–232.
5. Levine, A. S. and Billington, C. J. (1998) Obesity: progress through genetic manipulation. *Curr. Biol.* **8,** R251, R252.
6. Lowell, B. B. (1998) Using gene knockout and transgenic techniques to study the physiology and pharmacology of beta3-adrenergic receptors. *Endocrinol. J.* **(Suppl. 6),** S9–S13.
7. Beattie, J. H., Wood, A. M., Newman, A. M., Bremner, I., Choo, K. H., Michalska, A. E., Duncan, J. S., and Trayhurn, P. Obesity and hyperleptinemia in metallothionein (-I and -II) null mice. *Proc. Natl. Acad. Sci. USA* **95,** 358–363.
8. Zennaro, M. C., Le Menuet, D., Viengchareun, S., Walker, R., Ricquier, D., and Lombes, M. (1998) Hibernoma development in transgenic mice identifies brown adipose tissue as a novel target of aldosterone action. *J. Clin. Invest.* **101,** 1254–1260.
9. Vague, J. and Fenasse, R. (1965) Comparative anatomy of adipose tissue, in *Adipose Tissue, (Handbook of Physiology)* (Renold, A. E. and Cahill, G. F., eds.), American Physiological Society, Washington, DC, pp. 25–36.
10. Napolitano, L. (1965) The fine structure of adipose tissues, in *Adipose Tissue, (Handbook of Physiology)*, (Renold, A. E. and Cahill, G. F., eds.), American Physiological Society, Washington, DC, pp. 109–124.
11. Rodbell, M. (1965) The metabolism of isolated fat cells, in *Adipose Tissue,* (Handbook of Physiology), (Renold, A. E. and Cahill, G. F., eds.), American Physiological Society, Washington, DC, pp. 471–482.

12. Néchad, M. (1986) Structure and development of brown adipose tissue, in *Brown Adipose Tissue* (Trayhurn, P. and Nicholls., D. G., eds.), Arnold E., London, pp. 1–30.
13. Himms-Hagen, J. (1990) Brown adipose tissue thermogenesis: interdisciplinary studies. *FASEB J.* **4,** 2890–2898.
14. Rothwell, N. J. and Stock, M. J. (1986) Brown adipose tissue and diet-induced thermogenesis, in *Brown Adipose Tissue* (Trayhurn, P. and Nicholls., D. G., eds. ), Arnold E., London, pp. 269–298.
15. Trayhurn, P. (1989) Thermogenesis and the energetics of pregnancy and lactation. *Can. J. Physiol. Pharmacol.* **67,** 370–375.
16. Haman, A., Flier, J. S., and Lowell, B. B. (1998) Obesity after genetic ablation of brown adipose tissue. *Z. Ernährungswiss* **37,** 1–7.
17. Nicholls, D. G. and Locke, R. M. (1984) Thermogenic mechanisms in brown fat. *Physiol. Rev.* **64,** 1–64.
18. Ricquier, D., Casteilla, L., and Bouillaud, R. (1991) Molecular studies of the uncoupling protein. *FASEB J.* **5,** 2237–2242.
19. Fleury, C., Neverova, M., Collins, S., Raimbault, S., Champigny, O., Levi-Meyrueis, C., et al. (1997) Uncoupling protein-2: a novel gene linked to obesity and hyperinsulinemia. *Nat. Genet.* **15,** 269–272.
20. Boss, O., Muzzin, P., and Giacobino, J. P. (1998) The uncoupling proteins, a review. *Eur. J. Endocrinol.* **139,** 1–9.
21. Lafontan, M. and Berlan, M. (1993) Fat cell adrenergic receptors and the control of white and brown fat cell function. *J. Lipid Res.* **34,** 1057–1091.
22. Giacobino, J. P. (1995) Beta 3-adrenoceptor: an update. *Eur. J. Endocrinol.* **132,** 377–385.
23. Tvrdik, P., Asadi, A., Kozak, L. P., Nedergaard, J., Cannon, B., and Jacobson, A. (1997) Cig30, a mouse member of a novel membrane protein gene family, is involved in the recruitment of brown adipose tissue. *J. Biol. Chem.* **272,** 31,738–31,746.
24. Puigserver, P., Zhidan, W., Park, C. W., Graves, R., Wright, M., and Spiegelman, B. M. (1998) A cold-inducible coactivator of nuclear receptors linked to adaptive thermogenesis. *Cell* **92,** 829–839.
25. Salvatore, D., Bartha, T., and Larsen, P. R. (1998). The guanosine monophosphate reductase gene is conserved in rats and its expression increases rapidly in brown adipose tissue during cold exposure. *J. Biol. Chem.* **273,** 31,092–31,096.
26. Nedergaard, J., Connolly, E., and Cannon, B. (1986) Brown adipose tissue in the mammalian neonate, in *Brown Adipose Tissue* (Trayhurn, P. and Nicholls., D. G., eds.), Arnold E., London, pp. 152–213.
27. Herpin, P., Bertin, R., Le Dividich, J., and Portet, R. (1987) Some regulatory aspects of thermogenesis in cold-exposed piglets. *Comp. Biochem. Physiol.* **87,** 1073–1081.
28. Trayhurn, P., Temple, N. J., and Van Aerde, J. (1989) Evidence from immunoblotting studies on uncoupling protein that brown adipose tissue is not present in the domestic pig. *Can. J. Physiol. Pharmacol.* **67,** 1480–1485.
29. Girardier, L. and Seydoux, J. (1986) Neural control of brown adipose tissue, in *Brown Adipose Tissue* (Trayhurn, P. and Nicholls., D. G., eds.), Arnold E., London, pp. 122–151.

30. Wirsen, C. (1965) Distribution of adrenergic nerve fibers in brown and white adipose tissue, in *Adipose Tissue, (Handbook of Physiology)*, (Renold, A. E. and Cahill, G. F., eds.), American Physiological Society, Washington, DC, pp. 197–200.

31. Rosell, S. and Belfrage, E. (1979) Blood circulation in adipose tissue. *Physiol. Rev.* **59,** 1078–1104.

32. Crandall, D. L., Hausman, G. J., and Kral, J. G. (1997) A review of the microcirculation of adipose tissue: anatomic, metabolic and angiogenic perspectives. *Microcirculation* **4,** 211–232.

33. Bartness, T. and Bamshad, M. (1998) Innervation of mammalian white adipose tissue: implications for the regulation of total body fat. *Am. J. Physiol.*, **275,** R1399–R1411.

34. Siiteri, P. K. (1987) Adipose tissue as a source of hormones. *Am. J. Clin. Nutr.* **45,** 277–282.

35. Ailhaud, G. (1997) Secretory function of the adipocyte. *J. Ann. Diabetol. Hôtel Dieu* 125–129.

36. Weigle, D. S. (1997) Leptin and other secretory products of adipocytes modulate multiple physiological functions. *Ann. Endocrinol.* **58,** 132–136.

37. Mohamed-Ali, V., Pinkney, J. H., and Coppack, S. W. (1998) Adipose tissue as an endocrine and paracrine organ. *Int. J. Obesity* **22,** 1145–1158.

38. Dessolin, S., Schalling, M., Champigny, O., Lönnqvist, F., Ailhaud, G., Dani, C., and Ricquier, D. (1997) Leptin gene is expressed in rat brown adipose tissue at birth. *FASEB J.* **11,** 382–387.

39. Cancello, R., Zingaretti, M. C., Sarzani, R., Ricquier, D., and Cinti, S. (1998) Leptin and UCP1 genes are reciprocally regulated in brown adipose tissue. *Endocrinology* **139,** 4747–4750.

40. Ailhaud, G., Grimaldi, P., and Négrel, R. (1992) Cellular and molecular aspects of adipose tissue development. *Ann. Rev. Nutr.* **12,** 207–233.

41. Prins, J. B. and O'Rahilly, S. (1997) Regulation of adipose cell number in man. *Clin. Sci.* **92,** 2–11.

42. Cousin, B., Cinti, S., Morroni, M., Raimbault, S., Ricquier, D., Pénicaud, L., and Casteilla, L. (1992) Occurrence of brown adipocytes in rat white adipose tissue: molecular and morphological characterization. *J. Cell Sci.* **103,** 931–942.

43. Viguerie-Bascands, N., Bousquet-Mélou, A., Galitzky, J., Larrouy, D., Ricquier, D., Berlan, M., and Casteilla, L. (1996) Evidence for numerous brown adipocytes lacking functional β3-adrenoceptors in fat pads from nonhuman primates. *J. Clin. Endocrinol. Metab.* **81,** 368–375.

44. Vague, J., Rubin, P., Jubelin, J., and Vague, P. (1974) Various forms of obesity *Triangle* **13,** 41–50.

45. Björntorp, P. (1996) Regulation of adipose tissue distribution in humans. *Int. J. Obesity* **20,** 291–302.

46. Björntorp, P. (1991) Metabolic implications of body fat distribution. *Diabetes Care* **14,** 1132–1143.

47. Kissebah, A. H., and Krakower, G. R. (1994) Physiological review: regional adiposity and morbidity. *Am. Physiol. Soc.* **74,** 761–798.

48. Abate, N. and Carg, A. (1995) Heterogeneity in adipose tissue metabolism: causes, implications and management of regional adiposity. *Prog. Lipid Res.* **34,** 53–70.
49. Marette, A. Mauriège, P., Marcotte, B., Atgié, C., Bouchard, C., Theriault, G., et al. (1997) Regional variations in adipose tissue insulin action and GLUT4 glucose transporter expression in severely obese premenopausal women. *Diabetologia* **40,** 590–598.
50. Mauriège, P., Prud'homme, D., Marcotte, M., Yoshioka, M., Tremblay, A., and Després, J. P. (1997) Regional differences in adipose tissue metabolism between sedentary and endurance-trained women. *Am. J. Physiol.* **273,** E497–E506.
51. Bouchard, C., Després, J. P., and Mauriège, P., (1993) Genetic and nongenetic determinants of regional fat distribution. *Endocr. Rev.* **14,** 72–93.
52. Tavernier, G., Galitzky, J., Valet, P., Remaury, A., Bouloumié, A., Lafontan, M., and Langin, D. (1995) Molecular mechanisms underlying regional variations of catecholamine-induced lipolysis in rat adipocytes. *Am. J. Physiol.* **268,** E1135–E1142.
53. Bartness, T. J., Hamilton, J. M., Wade, G. N., and Goldman, B. D. (1989) Regional differences in fat pad responses to short days in Siberian hamsters. *Am. J. Physiol.* **257,** R1533–R1540.
54. Fried, K., Lavau, M., and Pi-Sunyer X. (1982) Variations in glucose metabolism by fat cells from three adipose depots of the rat. *Metabolism* **31,** 876–883.
55. Jensen, M. D. (1997) Lipolysis: contribution from regional fat. *Ann. Rev. Nutr.* **17,** 127–139.
56. Arner, P. (1995) Differences in lipolysis between human subcutaneous and omental adipose tissues. *Ann. Med.* **27,** 435–438.
57. Fried, S. K., Bunkin, D. A., and Grenberg, A. S. (1998) Omental and subcutaneous adipose tissues of obese subjects release interleukin-6: depot difference and regulation by glucocorticoid. *J. Clin. Endocrinol. Metab.* **83,** 847–850.
58. Hauner, H., Wabitsch, M., and Pfeiffer, E. F. (1988) Differentiation of adipocyte precursor cells from obese and nonobese women and from different adipose tissue sites. *Horm. Metab. Res. Suppl.* **19,** 35–39.
59. Hauner, H. and Entenmann, G. (1991) Regional variation of adipose differentiation in cultured stromal-vascular cells from the abdominal and femoral adipose tissue of obese women. *Int. J. Obesity* **15,** 121–126.
60. Shimomura, I., Funahashi, T., Takahashi, M., Maeda, K., Konati, K., Nakamura, T., et al. (1996) Enhanced expression of PAI-1 in visceral fat: possible contributor to vascular disease in obesity. *Nat. Med.* **2,** 800–803.
61. Adams, M., Montague, C. T., Prins, J. B., Holder, J. C., Smith, S. A., Sanders, L., et al. (1997) Activators of peroxisome proliferator-activated receptor gamma have depot-specific effects on human preadipocyte differentiation. *J. Clin. Invest.* **100,** 3149–3153.
62. Niesler, C. U., Siddle, K. and Johannes, B. P. (1998) Human preadipocytes display a depot-specific susceptibility to apoptosis. *Diabetes* **47,** 1365–1368.
63. Carola, U. N., Siddle, K., and Prins, J. B. (1998) Human preadipocytes display a depot-specific susceptibility to apoptosis. *Diabetes* **47,** 1365–1368.
64. Masuzaki, H., Ogawa, Y., Isse, N., Satoh, N., Okazaki, T., Shigemoto, M., et al. (1995) Human obese gene expression. Adipocyte-specific expression and regional differences in the adipose tissue. *Diabetes* **44,** 855–858.

65. Montague, C. T., Prins, J. B., Sanders, L., Digby, J., and O'Rahilly, S. (1997) Depot-and sex-specific differences in human leptin mRNA expression: implications for the control of regional fat distribution. *Diabetes* **46,** 342–347.

66. Russell, C. D., Petersen, R. N., Rao, S. P., Ricci, M. R., Prasad, A., Zhang, Y., Brolin, R. E., and Fried, S. K. (1998) Leptin expression in adipose tissue from obese humans: depot-specific regulation by insulin and dexamethasone. *Am. J. Physiol.* **275,** E507–E515.

67. Bouchard, C. (1997) Genetic determinants of regional fat pad distribution. *Human Reprod.* **12,** 1–5.

68. Cousin, B., Casteilla, L., Dani, C., Muzzin, P., Revelli, J. P., and Pénicaud, L. (1993) Adipose tissues from various anatomical sites are characterized by different patterns of gene expression and regulation. *Biochem. J.* **292,** 873–876.

69. Lefebvre, A. M., Laville, M., Vega, N., Riou, J. P., Van Gaal, L., Auwerx, J., and Vidal, H. (1998) Depot-specific differences in adipose tissue gene expression in lean and obese subjects. *Diabetes* **47,** 98–103.

70. Montague, C. T., Prins, J. B., Sanders, L., Zhang, J., Sewter, C. P., Digby, J., Byrne, C. D., and O'Rahilly, S. (1998) Depot-related gene expression in human subcutaneous and omental adipocytes. *Diabetes* **47,** 1384–1391.

71. Gemmell, R. T., Bell, A. W., and Alexander, G. (1972) Morphology of adipose cells in lambs at birth and during subsequent transition of brown to white adipose tissue in cold and in warm conditions. *Am. J. Anat.* **133,** 143–164.

72. Nouguès, J., Reyne, Y., Champigny, O., Holloway, B., Casteilla, L., and Ricquier, D. (1993) Beta 3-adrenoceptor agonist ICI-D7114 is not as efficient on reinduction of uncoupling protein mRNA in sheep as it is in dogs and smaller species. *J. Anim. Sci.* **71,** 2388–2394.

73. Loncar, D. (1991) Convertible adipose tissue in mice. *Cell Tissue Res.* **266,** 149–161.

74. Champigny, O., Ricquier, D., Blondel, O., Mayers, R. M., Briscoe, M. G., and Holloway, B. R. (1991) Beta 3-adrenergic receptor stimulation restores message and expression of brown-fat mitochondrial uncoupling protein in adult dogs. *Proc. Natl. Acad. Sci. USA* **88,** 10,774–10,777.

75. Ashwell, M. (1985) Is there a continuous spectrum of the adipose tissues in animals and man?, in *Metabolic Complications of Human Obesities* (Cague J., et al., eds.), Elsevier, New York, pp. 265–274.

76. Villarroya, F., Felipe, A., and Mampel, T. (1986) Sequential changes in brown adipose tissue composition, cytochrome oxidase activity and GDP binding throughout pregnancy and lactation in the rat. *Biochim. Biophys. Acta* **882,** 187–191.

77. Casteilla, L., Forest, C., Robelin, J., Ricquier, D., Lombet, A., and Ailhaud, G. (1987) Characterization of mitochondrial uncoupling protein in bovine fetus and newborn calf. Disappearance in lambs during aging. *Am. J. Physiol.* **252,** E627–E636.

78. Casteilla, L., Cousin, B., Vigueries-Bascands, N., and Pénicaud, L. (1994) Hétérogénéité et plasticité cellulaires des tissus adipeux. *Méd. Sci.* **11,** 1099–1106.

79. Guerra, C., Koza, R. A., Yamashita, H., Walsh, K., and Kozak, L. P. (1998) Emergence of brown adipocytes in white fat in mice is under genetic control: effect on body weight and adiposity. *J. Clin. Invest.* **102,** 412–420.

80. Lafontan, M., Bousquet-Melou, A., Galitzky, J., Barbe, P., Carpéné, C., Langin, D., et al. (1995) Adrenergic receptors and fat cells: differential recruitment by physiological amines and homologous regulation. *Obesity Res.* **3**, 507S–514S.

81. Cambon, B., Reyne, Y., and Nougues, J. (1998) In vitro induction of UCP1 mRNA in preadipocytes from rabbit considered as a model of large mammals brown adipose tissue development: importance of PPARgamma agonists for cells isolated in the postnatal period. *Mol. Cell Endocrinol.* **146**, 49–58.

82. Peeters, L. L. H., Martensson, L., Gilbert, M., and Pénicaud, L. (1984) The pregnant guinea pig, rabbit and rat as unstressed catheterized models, in *Animal Models in Fetal Medicine* (Nathanielz, P. W., ed.), Perinatology, Los Angeles, CA, pp. 74–108.

83. Scow R. O. (1965) Perfusion of isolated adipose tissue: FFA release and blood flow in rat parametrial fat body, in *Adipose Tissue, (Handbook of Physiology),* (Renold, A. E. and Cahill, G. F., eds.), American Physiological Society, Washington, DC, pp. 435–454.

84. Rothwell, N. J. and Stock, M. J. (1981) Regulation of energy balance. *Ann. Rev. Nutr.* **1**, 235–256.

85. Flatt, J. P. (1995) Use and storage of carbohydrate and fat. *Am. J. Clin. Nutr.* **61**, 952S–959S.

86. Hirsch, J. (1999) Human obesity: a sufficient cause. *Curr. Opin. Clin. Nutr. Metab. Care* **2**, 101–104.

87. Björntorp, P. (1997) Hormonal control of regional fat distribution. *Human Reprod.* **12**, 21–25.

88. Lafontan, M. and Berlan, M. (1993) Fat cell adrenergic recptors and the control of white and brown fat cell funtion. *J. Lipid Res.* **34**, 1057–1091.

# 2

## Morphologic Techniques for the Study of Brown Adipose Tissue and White Adipose Tissue

**Saverio Cinti, M. Cristina Zingaretti, Raffaella Cancello, Enzo Ceresi, and Plinio Ferrara**

## 1 Introduction

### 1.1. Light Microscopy

Light microscopy (LM) allows study of the anatomy of the adipose organ. For instance, it permits investigation of its microscopic organization (possible lobular subdivision) and the arrangement in the tissue of vessels, nerves, and main cell types *(1)*. The unilocular or multilocular organization of adipocytes is already clearly visible at low magnification (×2.5), but the identification of other cell types, such as mast cells (particularly frequent in brown tissue), histiocytes (particularly frequent in white tissue, especially in fasting animals and in animals under diet restrictions), and inflammatory cells, requires magnification of at least ×40–100. Other cell types such as fibroblasts, preadipocytes and pericytes, are not easily recognized at LM. This technique also provides data on the functional state of the adipose tissue (AT): the size of white adipose tissue (WAT) adipocytes is indicative of the animal's nutritional state and areas of capillary bed expansion appear in fasting animals. In brown adipose tissue (BAT), multilocularity is observed in various degrees, depending on the functional state of the organ. When this is stimulated by cold exposure or diet, lipid vacuoles are small and numerous, at rest (i.e., at warm environmental temperature or in fasting animals), they increase in size and gradually merge into fewer vacuoles, and eventually into a single vacuole. The functional activation of mitochondriogenesis in BAT also determines increased eosinophily of adipocyte cytoplasm. LM observation requires specimen processing by the following steps: fixation, dehydration, embedding, sectioning, and staining *(2)*.

From: *Methods in Molecular Biology, vol. 155: Adipose Tissue Protocols*
Edited by: G. Ailhaud © Humana Press Inc., Totowa, NJ

## 1.2. Immunohistochemistry

Immunohistochemistry (IHC) is the technique of choice for the localization of intra- and extracellular proteins at LM level. The goal of IHC investigations is to evidence a tissue protein (antigen [Ag]) by means of specific purified antibodies (Abs) and Abs labeled with markers which, at the end of the immunologic sequence, appear colored and can be seen at the LM. This procedure allows protein Ags to be localized in the tissue, with the cytological detail of LM. Successful Ag localization will depend on the state of preservation of the tissue and of the protein's antigenicity following processing. The technique described below, routinely used in this laboratory, provides the best morphologic detail, and at the same time allowing several protein Ags to be evidenced in the AT *(3)*.

## 1.3. In Situ *Hybridization*

*In situ* hybridization (ISH) is a laboratory method developed to localize specific nucleic acid (DNA or RNA) sequences in chromosomes, cells, and histologic sections. In combination with IHC analysis, ISH allows one to relate microscopic topology information to transcriptional activity. It has therefore proved to be a valuable tool to analyze the differences in the biosynthetic activity of single cells. The principle on which this technique rests is that of the complementarity of the sequences of nucleic acids, which, in particular conditions, can pair up and form hybrids (DNA–DNA, DNA–RNA, RNA–RNA). By labeling one of the two sequences, therefore, called a probe, with specific (radioactive or nonradioactive) molecules conjugated to the nucleotides (nt), the target sequence with which it has paired up can be localized *in situ*, i.e., at the site where it lies. Morphologic analysis will then allow identification of the cell compartment where the signal is produced. There is no need for expert knowledge of molecular biology techniques to perform ISH experiments, but familiarity with cloning and plasmid preparation techniques, digestion with restriction enzymes, and in vitro transcription is helpful *(4)*.

Several ISH protocols are described in the literature *(5–7)*, all based on a series of common procedures: probe and tissue preparation, hybridization, rinsing, and signal visualization. These protocols are distinguished by the type of target nucleic acid and type of probe (DNA or RNA) utilized. In the authors' experience, RNA probes, also termed riboprobes, are specific, but require extreme care during processing steps because of the high instability of RNA itself. One limiting factor with this technique is the level of background, nonspecific signal, which depends on probe specificity: if the target nucleic acid is not abundant, probe specificity needs to be very high. By contrast, signal specificity depends on stringency, i.e., the degree of rigorousness of the hybridization conditions, which determines the proportion of specific hybrids out of all the hybrids that have formed.

The detection of a labeled hybrid depends on the type of probe labeling: for a hybrid labeled radioactively with phosphorus, sulfur, or iodine detection is performed with the classic methods for the detection of radioisotopes. For a hybrid labeled nonradioactively, detection may be direct, e.g., with fluorochromes, or indirect, with digoxigenin (DIG) or biotin: for example, reporter molecules inserted in the probe chemically or enzymatically and detectable with immunocytochemical methods (Abs). Regarding nonradio-active labeling, the biotin (detected with streptavidin) or the DIG (detected with specific anti-DIG Abs) methods are the more common. Obviously, these molecules must not interfere with the hybridization reaction or destabilize the hybrids that have formed. ISH is especially interesting for the study of AT, because it allows localization at the morphological level of the nucleic acids (mRNA) connected with the transcriptional activity of adipocytes in different morphologic and functional conditions, as well as molecules identified too recently for specific Abs to be available for IHC tests. In addition, in serial sections, i.e., consecutive sections containing the same cells, IHC experiments can be performed for comparison with ISH tests, to study the presence of messenger RNA or proteins.

### *1.4. Transmission Electron Microscopy*

The resolution obtained with transmission electron microscopy (TEM) is approx 100× that of LM (*see* **Fig. 1**). Whereas the latter allows observation only gross intracellular features, such as lipid vacuoles and the nucleus, the superior resolution of TEM allows visualization of cells and tissues, not only of all cytoplasmic organelles, but even of some important structures that are found inside them, such as mitochondrial cristae *(1)*.

TEM also allows the functional status of the adipocytes, observable at the LM, to be more easily interpreted: for instance, the slimming cells in various stages of lipolysis observed in the white tissue of fasting animals exhibit several ultrastructural features that distinguish them from every other tissue element, such as microvillous cytoplasmic projections rich in pinocytotic vesicles. With a particular method (tannic acid, explained in **Subheading 3.4.2.**), masses of various sizes of myelinated material (interpreted as fatty acids) can be seen leaving these cells during the lipolytic process. Also, in BAT, TEM allows detection of further indications of the cells' functional status besides those observable by LM. In functionally activated adipocytes, mitochondria are larger and richer in cristae, and lipid vacuoles are surrounded by cisternae of smooth endoplasmic reticulum. In nonactivated adipocytes, besides the features that can be observed at LM level, mitochondria become smaller and have a smaller number of cristae. Glycogen deposits can also be observed *(8–11)*.

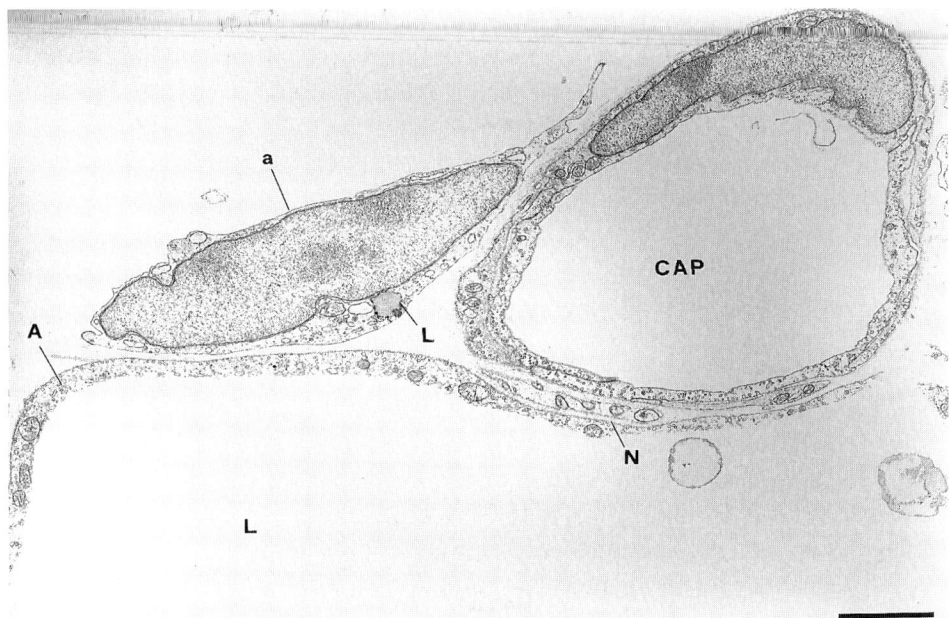

Fig. 1. Mouse retroperitoneal AT. A mature adipocyte (A) and an adipoblast (a) can be seen near a cross sectioned capillary (CAP). N, small nerve; L, lipid droplet. TEM. Bar, 2.2 μm.

## 1.5. Immunoelectron Microscopy

This technique combines the aims of electron microscopy (EM) and IHC. It is therefore a powerful technique that allows Ags to be localized at the level of organelles, but it is also very complex to perform. Tissues must be processed in such a way as to preserve both their ultrastructure and their antigenicity. Because these are conflicting goals, a compromise must be reached. Among the several different techniques reported in the literature, the one described in this chapter is commonly adopted in this laboratory. This technique allows one to address most of the problems connected with ultrastructural immuno-localization in a comparatively simple way, and especially obviates the need to upgrade the existing EM equipment. Different techniques may however be better suited to address different specific problems.

This laboratory uses the immunogold technique on ultrathin sections of AT embedded in metacrylate resins (postembedding) (*see* **Fig. 2**). Gold particles are to be preferred for immunoelectron microscopy, because of their electron density under the beam (*11–14*).

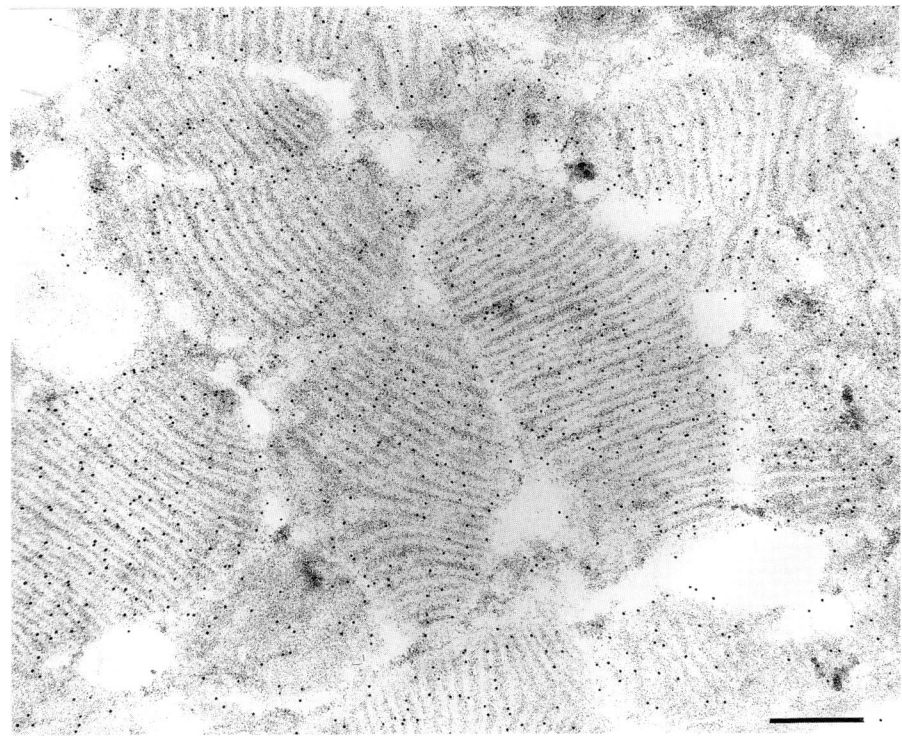

Fig. 2. BAT of cold-acclimated mouse. Immunogold staining with UCP1 antise-rum. UCP1-gold particles are lined on mitochondrial cristae. TEM. Bar, 0.3 μm.

## 1.6. Scanning Electron Microscopy

Scanning electron microscopy (SEM) allows study of the surface of tissue samples with a resolution similar to that of TEM, and visualization of the organization of tissues three-dimensionally. The chief limitation of conventional SEM is that the structures contained inside cells cannot be visualized. A newly available processing technique ($OsO_4$ maceration: *see* **Subheading 3.6.2.2.**), however, combined with the use of a special microscope (emission field microscope, or other high-resolution scanning microscope), allows study of the internal structures of cells and organelles (*see* **Fig. 3**). The description that follows applies specifically to conventional SEM, but reference is also be made to the emission field method (*14*).

## 1.7. Freeze-Fracturing

This technique is used to study cell membrane surfaces in vivo. Freeze-frac-turing is based on the principle that a cooled knife cutting through a frozen

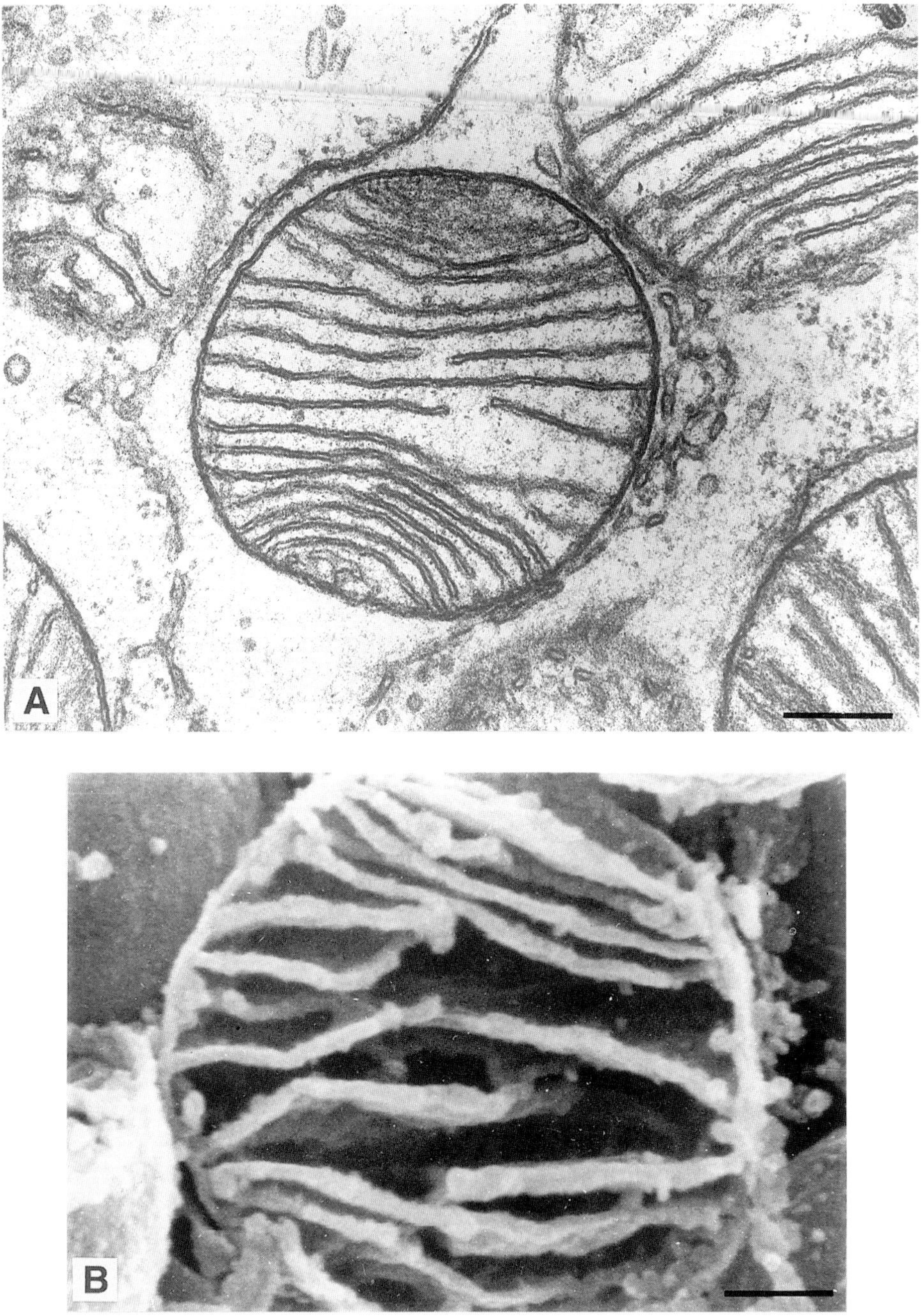

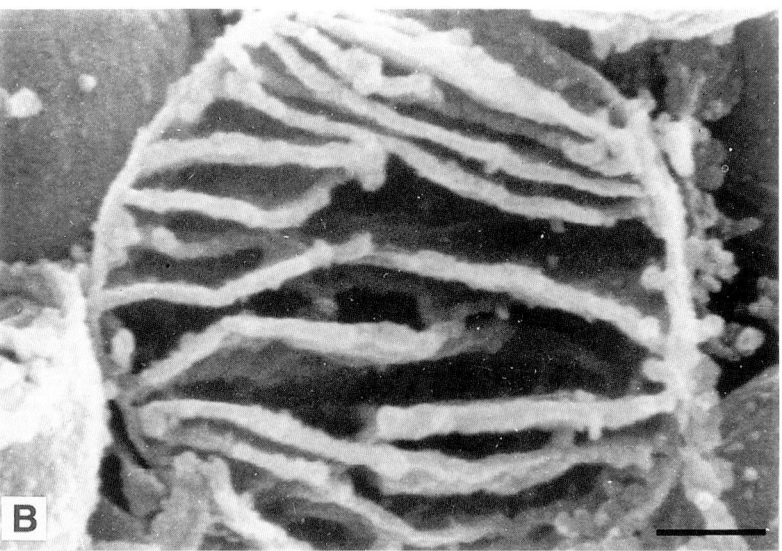

Fig. 3. **(A)** Rat BAT mitochondrion. Numerous cristae extending across the full width of the organelle are visible. TEM. Bar, 0.2 μm. **(B)** Three-dimentional image of a rat BAT mitochondrion similar to that shown in A, can be obtained with the $OsO_4$ maceration technique. SEM (emission field scanning electron microscope). Bar, 0.2 μm.

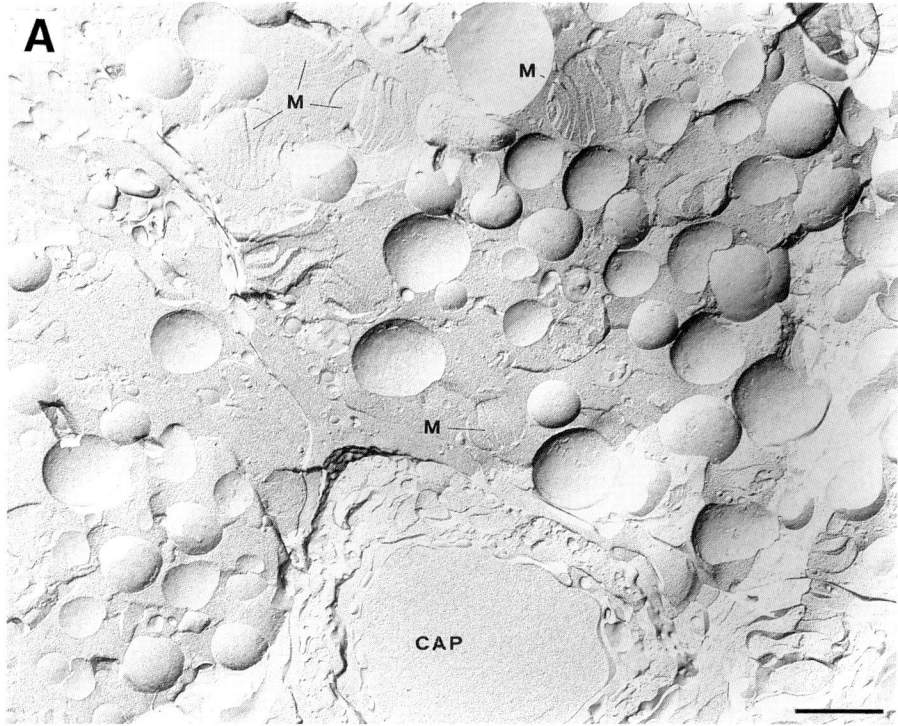

Fig. 4. (**A**) Rat IBAT after acute cold stimulation. Many spherical particles corresponding at least partially to mitochondria are visible. Some are sectioned (M) and show the typical cristae. A cross-sectioned capillary (CAP) is also visible. TEM. Freeze-fracturing technique. Bar, 1.1 µm.

specimen produces a fracture that passes through structurally weak paths within the frozen membranes, exposing their hydrophobic faces. The fracture thus exposes varying amounts of membrane hemisurfaces on which protein particles, such as proteins organized into gap junctions *(15)* and the organization of the membranes themselves, such as evaginations and invaginations, can be recognized. All the structures that can be observed by conventional TEM can also be studied with this technique *(1)* (*see* **Fig. 4A,B**).

## 1.8. Electron Microscopy of Cells in Culture

TEM and SEM can also be applied to the study of cells in culture *(1)*. EM is useful in the study of fine cellular detail in the various stages of controlled growth (**Fig. 5**). Different types of cell cultures have been used to study the development and reactivity of white and brown adipocytes. A widely used method of cell isolation from AT is that which allows isolation

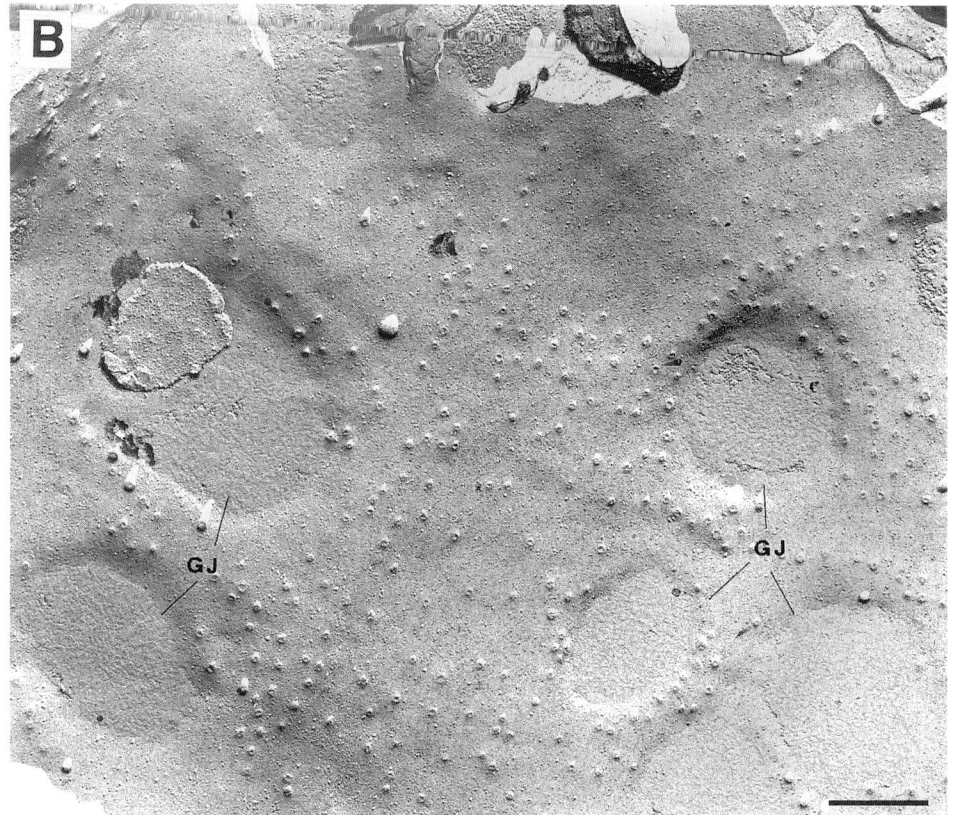

Fig. 4. **(B)** BAT of cold-acclimated rat. Gaps junctions (GJ) can be seen in the brown tissue membrane. TEM. Freeze-fracturing technique. Bar, 0.5 μm.

of the stromal-vascular fraction from the ATs in which white and brown adipocyte precursors, which proliferate, then develop in primary cultures, are believed to originate *(16,17)*.

### 1.9. Morphometry

The utilization of morphometric techniques for the study of the cell elements of AT has become an indispensable tool for the correct interpretation of morphologic data. Morphometric studies are time-consuming and, if EM observation is also required, expensive.

The basic technique rests on established stereologic principles, which allow recovery of a considerable amount of the information lost with sectioning *(19,20)*. In practice, from sections processed for LM and EM, it is possible to

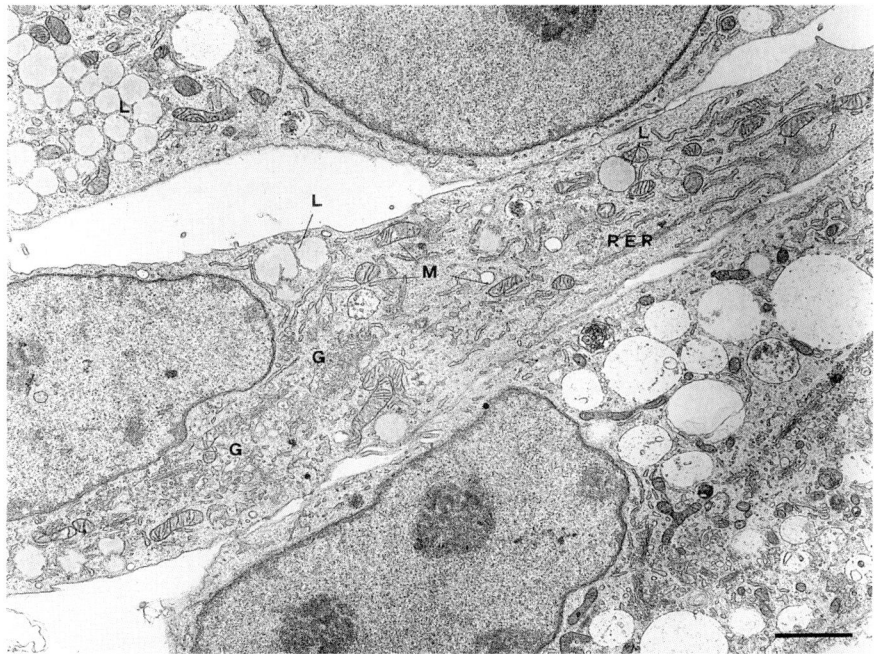

Fig. 5. White adipocyte precursors from the epididymal AT of adult rat after 4 d into primary culture. Many small lipid droplets (L), elongated mitochondria (M), rough endoplasmic reticulum cisternae (RER), and hypertrophic Golgi complex (G) are visible. TEM. Bar, 1.9 μm.

establish the size and relative (sometimes even the absolute) number of cell elements, as well as data on intracellular organelles. Several software packages are available to make morphometric analyses easier, or sometimes even possible. However, in our experience, the majority of morphometric problems require expert interpretation by the operator.

Our morphometric apparatus consists of a computer equipped with a Matrox MGA Millenium Power Desk and a KS100 image-analyzing program (Kontron, Germany). This is a manual system that allows one to work freely on the AT and its distinctive morphologic features. Images can be acquired from a LM fitted with a color video camera, which shows the selected field directly on the computer screen; from a scanner through which photographs and slides can be acquired; and from a graphic tablet, which allows acquisition of measurements and data from any two-dimensional object. Through the video camera fitted onto the EM, images can be acquired directly or stored in an electronic medium (CD-ROM). The last option is the least expensive, and therefore the most cost-effective for morphometric purposes.

The quantitation of morphological data allows comparison of control animals with animals subjected to pharmacologic treatment, to exposure to different temperatures, or to diet modifications. For studies to achieve statistical significance, data should always refer to the highest possible number of treated and control subjects, although this aspect needs to be balanced against the duration and overall cost of the study. It is also important to remember that measurements are aimed at obtaining both absolute and relative determinations in different experimental conditions.

## 2. Materials

1. Fixatives: 4% paraformaldehyde in 0.1 $M$ phosphate buffer (PB), pH 7.4; 2% glutaraldehyde–2% paraformaldehyde in 0.1 $M$ PB, pH 7.4; 3% glutaraldehyde in 0.2 $M$ Na-cacodylate–HCl buffer, pH 7.4; 1% tannic acid in 0.2 $M$ Na-cacodylate–HCl buffer pH 7.4; 1% osmium tetroxide in 0.1 $M$ PB, pH 7.4; 4% paraformaldehyde in 0.1 $M$ PB, pH 7.4; 0.5% glutaraldehyde in 0.1 $M$ Na-cacodylate–HCl buffer, pH 7.4; 2% glutaraldehyde in 0.1 $M$ PB, pH 7.2; ketamine and xylazine are commercially available (ketamine: Ketavet, Gellini, Aprillia, Italy; xylazine: Rompun, Bayer, Leverkusen, Germany).
2. Buffers: 0.1 $M$ PB, pH 7.4, is composed of 0.2 $M$ $Na_2HPO_4$ and 0.2 $M$ $NaH_2PO_4$ mixed with bidistilled water (v:v; 1:1); phosphate buffered saline (PBS) is composed of 0.008 $M$ $Na_2HPO_4$, 0.0015 $M$ $KH_2PO_4$, 0.027 M KCl, and 0.137 $M$ NaCl; 0.2 $M$ Na-cacodylate–HCl buffer, pH 7.4; sterile water pretreated with 0.1% diethylpyrocarbonate (DEPC).
3. Reagents for embedding: propylene oxide (1,2-epoxypropane); paraffin, xylol, epon-araldite mixture (dodecenyl succinic anhydride [DDSA]): 6 parts; EPON 812: 2 parts; Durcupan A/M resin: 2 parts and 2% 2,4,6-tridimethylamino methyl phenol (DMP 30) (Multilab Supplies, Fetcham, UK); London Resin White (LR) (London Resin Company, Basingstoke, Hampshire, UK).
4. Reagents for immunohistochemistry (IHC): avidin–biotin complex (ABC) (Vector Labs, Burlingame, CA); 0.3% $H_2O_2$ in methanol; 3,3'-diaminobenzidine hydrochloride chromogen (DAB) (Sigma, St. Louis, MO).
5. Reagents for ISH: sterile water pretreated with 0.1% DEPC; acetone; denatured salmon sperm DNA; 50% formamide; standard sodium citrate (SSC): 0.15 $M$ NaCl, 0.015 Na-citrate, nitrobblueditetrazolium/bromochloroindolilphosphate (NTB/BCIP) (Roche Diagnostics, Milan, Italy).
6. Staining: hematoxylin, eosin, and toluidine blue are commercially available products of Merck, Darmstadt, Germany.
7. Counterstaining: uranylacetate (Agar Scientific, Stansted, UK) 0.05 mCi/mg is a saturated solution used with methanol (1:1; v:v); lead citrate (Merck) is a 0.25% solution in bidistilled water.
8. Cryoprotection: 20% glycerol in 0.15 $M$ PB, pH 7.4.

# 3. Methods

## 3.1. Light Microscopy

### 3.1.1. Fixation

Fixation can be considered the most important processing step, because specimen observation and interpretation depend crucially on this procedure. AT, especially WAT, has a high lipid content that must be extracted progressively by dehydration. The tissue must then be suitably fixed to preserve the integrity of the thin rim of cytoplasm of each cell. For this reason, accurate fixing by perfusion should be performed whenever possible. Transcardiac perfusion allows the complete replacement of blood with fixative, which can, at least theoretically, reach the capillaries and, finally, the cells. In a well-perfused animal, it is easier to dissect all the adipose depots and to recognize their anatomic connections with the other organs and structures.

With the animal deeply anesthetized (for mice, we use an intraperitoneal injection of 100 mg/kg ketamine in combination with 10 mg/kg xylazine), or after sacrifice, the thoracic cavity is opened and a blunt-tip needle of a size appropriate for the animal is inserted at the level of the apex of the left ventricle into the aorta. Immediately after beginning the perfusion, the right atrium must be excised to allow the blood, and subsequently the excess fixative, to flow out. Based on the weight of the animal, the optimal fixation time can be established: it is normally 10–15 min/100 g animal weight *(20)*.

As fixative, we use 4% paraformaldehyde in 0.1 *M* PB, pH 7.4, which allows one to obtain excellent histologic sections (also for morphometry purposes) and, at the same time, preservation of sufficient antigenicity for IHC reactions.

After perfusion, tissue samples are collected and the fixation procedure continues by immersion in a suitable amount of the same fixative (which should allow tissue fragments to float freely in the vial) to a final fixation time of 12–15 h at 4°C. After fixation, the tissue must be rinsed and washed in PB to remove residual fixative.

### 3.1.2. Sample Collection

The size of AT samples collected from human and animal subjects for LM study must be sufficient to preserve the greatest possible morphologic detail without being excessive, to avoid embedding problems. In our experience with AT, optimal sample size is about 2 cm$^3$. Because of the texture of this tissue and to its high lipid content, all the parts exhibiting mechanical artifacts (compression and deformation caused by incorrect handling of forceps and knives) must be discarded. To perform this operation, we make a clean cut with a razorblade.

### 3.1.3. Dehydration and Embedding

The final aim of these procedures is to obtain thin sections that can be stained and observed with a LM. To this end, after fixation, specimens must be embedded in a medium (paraffin) that will allow sectioning. Because paraffin is not miscible with water, and since specimens cannot be transferred directly into melted paraffin immediately after fixation, tissue fragments must first be dehydrated with increasing concentrations of EtOH, which gradually replaces water, then cleared in a solvent (xylol) miscible with paraffin, before impregnation.

1. Fix in 4% paraformaldehyde in 0.1 *M* PB, pH7.4 (for both perfusion and immersion), for 12–15 h, at 4°C.
2. Rinse in PB, and store in the same buffer at 4°C until embedding.
3. Dehydrate with gradual steps of EtOH as follows: 30 min in 75%, twice for 75 min in 95%, and 3× for 60 min in 100%, at room temperature (RT), stirring.
4. Clear specimens in xylol, 2 × 60 min.
5. Impregnate with paraffin at 58°C overnight.
6. Change the paraffin twice.
7. Orientate samples in embedding moulds.

### 3.1.4. Sectioning

AT is cut into sections of 3–4 μm with a normal sliding or rotary microtome, which are then placed on a glass, and dried overnight at 40°C.

### 3.1.5. Staining

Routine processing consists of dehydration and staining first with hematoxylin, which stains the nucleus, and then with eosin, which is an acid cytoplasm stain *(21)*.

1. Deparaffinize sections with xylol (2 × 10 min), and hydrate to water with decreasing alcohol concentrations.
2. Mayer's hematoxylin for 2 min.
3. Rinse in tap water.
4. Stain with 1% eosin (in dH$_2$O) for 15 s to 1 min.
5. Dehydrate in upgraded alcohols, until the excess eosin is removed, to xylol. Check the staining under the microscope.
6. Mount sections in Entellan (Merck) or other resin.

Results: nuclei, blue; cytoplasm, various shades of pink—identifying of different tissue components.

## 3.2. Immunohistochemistry
### 3.2.1. Fixation

To perform an IHC investigation of paraffin-embedded tissue, the animal should be perfused transcardially (as described under LM, **Subheading 3.1.1.**).

Fixation must allow maintenance of the best tissue morphology consistent with the preservation of antigenicity. In our experience with AT, paraformaldehyde, with its low crosslinking properties, is usually to be preferred for these purposes; the 4% concentration is also used for LM. Fixation must not exceed 12–15 h to avoid blocking Ag activity.

### 3.2.2. Tissue Embedding

For paraffin-embedding, the procedure is the same as for LM, except that the melting temperature of the paraffin used must not exceed 58°C, to avoid protein denaturation.

### 3.2.3. Sample Sectioning

Sections are obtained as for LM. Serial sectioning is useful for performing comparisons, when required, among different reactions for the same Ag or among reactions for different Ags. Given that sections are 3–4 μm in thickness, that the thickness of brown adipocytes is 30–50 μm, and that of white adipocytes is 80–150 μm, it is theoretically possible to perform different reactions on the same cell in different sections.

### 3.2.4. Incubation (see **Note 1**)

Our immunologic sequence is indirect. The specific Ab (primary antibody), raised against the Ag to be localized, binds directly to the tissue Ag in the section. Another Ab against the primary Ab (called "bridge" serum), conjugated to one biotin molecule, represents the second step of the sequence. As detection system, we use the ABC *(22)*, which is based on the capacity of avidin (glycoprotein of egg albumin) to bind to four molecules of biotin (vitamin H). The incubation step with ABC allows nonimmunologic crosslinking between the biotin conjugated to the bridge serum, itself bound to the primary Ab, and the avidin present in the complex, starting a chain reaction that progressively amplifies the signal. The peroxidase in the ABC is localized with a solution containing diaminobenzidine (DAB) as the substrate and hydrogen peroxide; the reaction product is brown-colored and easy to localize at LM *(23–30)*.

1. Place 3-μm paraffin sections in clean glasses, and let them dry in a slide-warming table overnight.
2. Deparaffinize sections, and hydrate to water.
3. Dewaxed sections are then processed through the following 12 incubation steps. Sections are covered with aliquots of serum solution in a humid chamber; for the washing steps and the enzymatic reaction, the glasses are immersed in Hallendal jars.
4. 0.3% $H_2O_2$ in methanol for 30 min, to block endogenous peroxidase activity.
5. Wash in PBS, pH 7.3, 2 × 15 min, gently stirring.

6. Normal serum (raised from the same animal that produced the secondary Ab) 1:75 in PBS, to reduce nonspecific background staining.
7. Primary Ab (poly- or monoclonal) raised against the Ag to be localized (the Ab should be tested to find the best working dilution), diluted in PBS overnight at 4°C in a humid chamber.
8. Wash in PBS 2 × 15 min.
9. Biotinylated secondary Ab immunoglobulin G against the primary Ab, raised in the same animal as the normal serum in step 2, 1:200 in PBS for 30 min at RT.
10. Wash in PBS 2 × 15 min.
11. ABC complex for 1 h at RT.
12. Wash in PBS 2 × 15 min.
13. Histoenzymatic visualization of peroxidase using 3',3'DAB and $H_2O_2$ for 4–5 min in a dark chamber (0.075% DAB in 0.05 $M$ Tris-HCl buffer, pH 7.6, and 0.03% $H_2O_2$).
14. Wash in tap water.
15. Weakly stain nuclei with hematoxylin, and mount sections in Entellan or other syntethic resin.

### 3.3. In Situ *Hybridization*
### 3.3.1. Sample Collection

This technique does not differ from that described for LM.

### 3.3.2. Tissue Fixation

Tissue is fixed by transcardial perfusion with a 4% paraformaldehyde-1X PBS solution. Perfusion time and volume of fixative depend on the animal's weight (*see* **Subheading 3.1.1.**). Following perfusion, the tissue is dissected and placed in the solution used for perfusion for 12–15 h at 4°C. The fixative is removed the next day, and samples are stored in 1X PBS at 4°C until embedding.

### 3.3.3. Embedding

Tissue specimens are embedded in paraffin previously melted and filtered (paraffin is solid at ambient temperature and melts over 50°C). The embedding procedure requires sequential washes with saline solution to remove the fixative and tissue dehydration by multiple passages in ascending alcohol concentrations. In brief, the sequence (performed at RT unless otherwise indicated) envisages: 1X PBS at 4°C, 0.85% saline at 4°C, 0.85% saline/100% EtOH 1:1 v/v, 70% EtOH, 85% EtOH, 95% EtOH, and 100% EtOH, with occasional shaking. Samples are then immersed in xylol–paraffin, 1:1, at 58°C and, in paraffin at 58°C: this step is performed 3×. The last time, the sample is oriented in the paraffin block. After solidification, the sample is ready for sectioning with the microtome.

### 3.3.4. Sample Sectioning

The slides on which sections are placed need to be clean and free of fat. To accomplish this, they are usually cleaned by immersing them in a 2 $N$ HCl solution, in sterile H$_2$O pretreated with 0.1% DEPC and finally in acetone. Because of the numerous rinses and high processing temperatures of the hybridization protocol, tissue sections may become detached from the slides. To prevent this, the slides should be pretreated with an adhesive substance, such as poly-L-lysine *(4)*.

To avoid contamination with RNase while sectioning the blocks, it is recommended that one wear latex gloves and lay the sections on DEPC-treated water before placing them on the slides. After overnight drying at 38–40°C, slides are stored in slide boxes at 4°C.

### 3.3.5. Reaction

The ISH reaction consists of the following phases: preparation of the labeled probe, prehybridization, hybridization, posthybridization and immunologic signal detection.

#### 3.3.5.1. PREPARATION OF LABELED PROBE (*SEE* **NOTE 2**)

As mentioned in **Subheading 1.3.**, different types of nucleic acid probes, such as DNA, RNA, and oligonucleotides, can be used, and they can be labeled in different ways. One important parameter is probe length, since too short a probe may to produce a weak or nonspecific signal; excessively long probes may have steric difficulties in penetrating the tissue and require tissue permeability treatments with protease or detergents. We primarily use RNA probes (riboprobes) *(31)*. To obtain this type of probe RNA must be synthesized in vitro. A plasmid containing an RNA polymerase enzyme promoter (e.g., SP6, T3, T7,) must also be available, as well as a cDNA encoding a specific target mRNA inserted in the region between two promoters. Two probes can thus be polymerized, an antisense probe (complementary to the target sequence) and a sense probe for use as negative control (identical to the target sequence).

The plasmid vector must prelimarily be linearized at one end with a specific restriction enzyme, and subsequently purified. The success of in vitro synthesis greatly depends on the pureness, not the amount, of plasmid DNA. Indeed, about 1 µg linearized and adequately purified plasmid can produce, through the polymerase reaction, 10 µg labeled RNA. During polymerization, a labeled nucleotide (nt) is inserted whenever the sequence requires it: with DIG, the labeled nt will be uridine triphosphate (UTP), which is complementary to the adenine molecules in the sequence. The in vitro synthesis reaction is performed at 37°C by incubating linearized plasmid DNA with a mixture containing RNA polymerase, triphosphate nts, including the nt with the labeling molecule (adenosine, guanosine, or cytosine triphosphates, or, UTP + UTP–DIG), and the transcription buffer for 30 min.

When the probe is too long (>300 bp), it must be reduced by alkaline degradation, which cuts it into smaller fragments, ensuring better tissue penetration.

After in vitro synthesis, RNA must be purified, for instance, by precipitation in 4 $M$ LiCl/EtOH, resuspended in a suitable volume of sterile and 0.1% DEPC-treated water, and tested for the degree of incorporation of the labeled nt. The latter (spot test) is a qualitative test that is performed by comparison with an equal volume of a labeled standard solution provided with the labeling kit on nylon membranes.

The labeled probe, which can be employed in several ISH reactions, is then stored at –20°C until use. Probe concentration depends on the type of reaction and of tissue.

### 3.3.5.2. PREHYBRIDIZATION

Before performing a hybridization reaction, paraffin needs to be removed (by two passages in xylol, each of at least 10 min). Sections are rehydrated in decreasing EtOH concentrations (100, 95, 75, 50, and 30%), then in $H_2O$ DEPC. They are subsequently incubated in 1X PBS, then in 1X PBS containing 0.1% DEPC, and finally in 5X SSC (0.75 NaCl, 0.075 $M$ Na-citrate).

Sections can be prehybridized for 2 h at 58°C in the hybridization mixture of 50% formamide, 5X SSC, and denatured salmon sperm DNA *(32)*.

### 3.3.5.3. HYBRIDIZATION

After denaturation for 5 min at 80°C, the probe is added to the prehybridization mixture, and incubated for 4 h (sufficient for abundant transcripts) or overnight. The hybridization solution is carefully placed, with a micropipet, directly on sections, which are then covered with a specific hydrophobic cover slip. Incubation is performed at 58°C in a humid chamber, to prevent evaporation caused by high temperature.

### 3.3.5.4. POSTHYBRIDIZATION AND IMMUNOLOGIC SIGNAL DETECTION

Following incubation, sections are washed in different stringency conditions, depending on temperature and salt concentrations: 2X SSC for 30 min at RT, 2X SSC for 1 h at 65°C, and 0.1X SSC for 1 h at 65°C. They are then rinsed in 100 m$M$ Tris-HCl, 150 m$M$ NaCl for 5 min *(32)*.

At this stage, the immunologic step of the reaction can be performed: sections are incubated with the Ab that recognizes as Ag the DIG molecule present in the previously formed hybrids. This Ab is directly conjugated to the alkaline phosphatase enzyme, which reacts with NBT/BCIP, producing a purple-blue precipitate. Sections are incubated with the Ab for 2 h at RT, washed to remove excess Ab and incubated in the dark in a solution containing NBT/BCIP. The colorimetric reaction requires monitoring and incubation from a minimum of 2 h to a maximum of 3 d, depending on the amount of transcript.

When the color has developed, the reaction is blocked in Tris-HCl buffer, pH 8.0, containing 1 m*M* EDTA, and slides are mounted with an aqueous medium (e.g., glycerol gelatin) and cover-slipped.

## *3.4. Transmission Electron Microscopy*

### *3.4.1. Sample Collection*

Sample collection for TEM studies is critical, because of the artifacts that may be introduced in this phase. This is primarily because samples for fixation must be very small (approx 1 mm$^3$), and are therefore difficult to manipulate. For optimal fixation, tissue should be collected in samples of about 0.5 cm$^3$, and then reduced with a razorblade to fragments not exceeding 1 mm$^3$. When collecting a tissue sample, particular care must be exercised in identifying with precision the sampling area. The macroscopic features of the zone of collection must be kept in mind, because, unlike LM analysis, TEM examination is performed on tissue fragments that are much too small to allow subsequent orientation, given the inherent absence of microscopic topographic references. The rationale for this small sample size is twofold: first, fixatives for TEM processing have a bulky steric configuration, and thus have reduced penetration properties; second, the tissue sections that can be observed at the TEM are squares of about 200 μm to a side.

### *3.4.2. Fixation*

For TEM studies, a fixative based on buffered aldehyde solution is preferred. We recommend a final solution composed of 2% glutaraldehyde–2% paraformaldehyde in 0.1 *M* PB, pH 7.4, for 4 h, preferably at 4°C. Tissues fixed in this way are chiefly or absolutely intended for TEM analysis. Following primary fixation in aldehyde, tissues are washed in PB. AT (especially white) fragments float in the fixative, and are so small that they risk being lost during washing.

Postfixation in OsO$_4$ is a usual step in TEM processing (osmium has a high mol wt which makes the tissue very electron-dense and visible at EM level), but, in the case of ATs, it is probably the most important. Osmium is the main fixative of lipids, which, in this case, are not extracted during dehydration, allowing adipocytes to maintain their original morphology *(20)*.

For ultrastructural demonstration of lipolytic events (*see* **Subheading 1.4.**), the fixative is 3% glutaraldehyde in 0.2 *M* cacodylate buffer pH 7.4, for 1 h at 4°C, followed by washing in the same buffer, by a incubation with 1% tannic acid in cacodylate buffer for 45 min at 24–27°C, and by postfixation in 1% OsO$_4$ in 0.1 *M* PB, pH 7.4 for 2 h, at 4°C *(33)*.

### 3.4.3. Embedding

There are several important differences between sections prepared for LM and for EM. For the latter, the material to be sectioned must be able to withstand the heat generated by the electron beam; therefore, specimens must be embedded in a hard epoxy resin.

After fixation, specimens are dehydrated in increasing concentrations of dehydrating agents, such as alcohol. This is necessary because the embedding medium is not water soluble. After dehydration, specimens are infiltrated with a liquid embedding medium (a plastic monomer), which is then polymerized to produce a solid block using the epon-araldite mixture *(20)*.

1. Fix small fragments of tissue (1 mm$^3$) in 2% glutaraldehyde–2% paraformaldehyde in 0.1 $M$ PB, pH 7.4, for 4 h at 4°C.
2. Wash overnight in the same buffer at 4°C.
3. Postfix in 1% $OsO_4$ in PB for 1 h at 4°C.
4. Dehydrate in upgraded EtOH series: 25, 50, 75% for 15 min at 4°C, 95% 2 × 15 min, 100% 3 × 15 min at RT.
5. Propylene oxide 2 × 15 min.
6. Infiltration at RT with propylene oxide: resin, 1:1, for 30 min.
7. Propylene oxide: resin, 1:3, for 1 h and 30 min.
8. Absolute resin overnight.
9. Embed in silicone rubber moulds or in gelatin capsules.
10. Polymerize at 60°C overnight.

### 3.4.4. Sample Sectioning

For TEM examination at suitable resolution, samples must be very thin. Polymerization yields hard resin blocks in which a tissue fragment (blackened by osmium) is embedded at one end. The sectioning instruments, called ultramicrotomes (highly sophisticated versions of common paraffin microtomes), are fitted with glass or diamond knives. Sections must be extremely small, because they will be placed on grids 3 mm in diameter, and they must not exceed 50–60 nm in thickness, because the electron beam passing through them should undergo the least possible deviation (and must thus meet the smallest possible number of atoms). Given the very small field of observation of EM, an area of greatest interest is selected in the best sections. Therefore, before preparing ultrathin sections, thicker sections (0.5–2 μm) are usually obtained from resin blocks with glass knives for LM observation. Semithin sections are observed following staining with toluidine blue, a basic monochromatic stain. The study of these sections following histochemical staining (periodic acid Schiff staining for carbohydrates and mucoproteins) and other stains typical of LM (paraffin sections) is made more difficult by the network polymerization of epoxy resin.

Once the area of interest for TEM observation has been selected, 50-nm sections can be obtained with the ultramicrotome (in this laboratory, an MT-X ultratome [RCM, Tuscon, AZ] and a Ultracut [Reichert-Jung, Vienna, Austria]), fitted with a diamond knife, and set on supporting metal grids. Sections are usually stained directly on the grid with uranyl acetate and lead citrate to increase contrast, and therefore electron density *(34)*.

### 3.4.5. Observation (Photos or Digital Images)

This institute has a CM10 TEM and an EM 208 TEM (Philips, Eindhoven, Netherlands). TEM images are virtual images on a phosphorescent screen. It is therefore recommended to capture the more significant images of each observation session on photographic plates to obtain prints, or, through a video camera fitted to the microscope, transform them into digital images that can be stored for later use.

## 3.5. Immunoelectron Microscopy

### 3.5.1. Fixation

The procedure is the same as in sample processing for LM. Neither glutaraldehyde, which induces crosslinks, nor $OsO_4$, which drastically reduces tissue antigenicity, can be employed. The price to be paid for the preservation of antigenicity is decreased ultrastructural quality. This laboratory uses tissues fixed in 4% paraformaldehyde in 0.1 $M$ PB, pH 7.4, overnight at 4°C.

### 3.5.2. Dehydration and Embedding

After perfusion, small fragments (2 mm³) of AT are excised from the main sample and stored in small glass bottles. After washing in PB, specimens are dehydrated in EtOH and infiltrated with LR white, a hydrophylic resin with low viscosity, which, besides preserving tissue antigenicity, is also stable to the electron beam.

1. Fix small fragments of AT directly or after perfusion in 4% paraformaldehyde in 0.1 $M$ PB, pH 7.4, overnight at 4°C.
2. Wash in PB.
3. Dehydrate in graded EtOH series: 75% 15 min, 95% 15 min, and 100% 2 × 15 min, stirring, at RT. Infiltrate with LR white (in the authors' experience with AT, the medium grade) at RT, 2–3 changes, 1 h each, then leave overnight.
4. Polymerization: the authors use thermal curing at 50°C for 24–48 h. It is important to limit the contact of oxygen with the resin during polymerization. The most convenient way of achieving this is to use gelatin capsules.

### 3.5.3. Tissue Sectioning

Resin blocks are sectioned with an ultramicrotome, and ultrathin sections are placed on 300-mesh nickel grids.

### 3.5.4. Incubation (see **Note 1**)

All the incubation steps are performed with the sections mounted on grids, floating on drops of reagent.

Sections are first exposed to 3% normal serum in PBS for 20 min at RT, then incubated with the primary antiserum, diluted in PBS containing 1% bovine serum albumin, at 10- to 100-fold concentrations with respect to IHC, overnight at 4°C. After several washes in PBS, sections are incubated with the secondary Ab conjugated to 10–15 nm gold particles diluted 1:50 in PBS for 30 min at RT. Sections are then washed in PBS and $dH_2O$ and counterstained with uranyl acetate and lead citrate solutions.

## 3.6. Scanning Electron Miscroscopy (see **Note 3**)

### 3.6.1. Sample Collection

The principal aim of SEM observation is the study of tissue surfaces. Processing methods are therefore of crucial importance. In each phase of the procedure, careful handling will ensure that specimens are not subjected to microtraumas, which may entail modifications of surface structures *(35)*.

For sample collection, the procedure is the same as that described for LM examination. Note that the tissue surface that can be observed at the microscope is approx 1 $cm^2$, and that only one side of the tissue can be examined, since sections are glued to the stub.

### 3.6.2. Fixation

#### 3.6.2.1. METHOD 1 (*SEE* NOTES 4,5)

1. The fixation procedure is identical to that described for the study of TEM samples: 2% glutaraldehyde–2% paraformaldehyde in 0.1 *M* PB, pH 7.4, 4 h at 4°C.
2. Wash overnight in PB.
3. Postfix with 1% $OsO_4$ in 0.1 *M* PB, pH 7.4, 1 h at 4°C.
4. Wash in 1X PBS.

#### 3.6.2.2. METHOD 2: $OsO_4$ MACERATION TECHNIQUE

1. Quick perfusion (when this is possible) with the fixative indicated at **item 3**.
2. Reduction of the tissue into thin strips (1 × 1 × 5 mm).
3. Fixation by immersion in 0.5% glutaraldehyde–0.5% paraformaldehyde in 0.1 *M* Na-cacodylate-HCl buffer for 15 min at RT in the dark.
4. Rinsing in 0.015 *M* PBS, pH 7.2.
5. Postfixation in 1% $OsO_4$ and 1.25% potassium ferrocyanide for 2 h *(36)*.
6. Embedding in agarose and sectioning (about 150 µm) with a chopper microtome.
7. Washing in PBS.
8. Maceration in 0.1% $OsO_4$ in PBS, for 50 h at 20°C or for 3 h at 45°C.
9. Multiple rinses in PBS.

10. Dehydration in ascending EtOH concentrations and critical-point drying.
11. Coating with a gold-palladium layer (1.5–2 nm) with a sputter coater. This coating is thinner than that used for conventional SEM observation, and allows study of cells in their finest ultrastructural detail, but requires such careful tissue handling that, for instance, specimens cannot even be moved from one laboratory to another.

## 3.7. Freeze-Fracturing (see Note 6)

### 3.7.1. Sample Collection

This operation is performed as for TEM (*see* **Subheading 3.4.1.**).

### 3.7.2. Fixation

We fix ATs in 2% glutaraldehyde in 0.1 *M* PB, pH 7.2, for 2 h at 4°C, followed by washing in the same buffer.

### 3.7.3. Specimen Processing

Small pieces of tissue are cryoprotected with 20% glycerol in 0.15 *M* PB, pH 7.4, for 1–2 h at RT and frozen in freon cooled with liquid nitrogen. Specimens are fractured, and replicated at –112°C in a Balzers freeze-fracturing device (Balzers, Lichenstein). Tissue is then digested with Na hypochlorite, and with a chloroform–methanol mixture and 100% dimethyl formamide, which are used specially for ATs. Replicas are washed with $dH_2O$ and recovered on 200-mesh grids (*37*).

## 3.8. EM of Cells in Culture (see Note 7)

### 3.8.1. Fixation (see **Subheading 3.4.2.**)

Fixation and postfixation are performed for TEM and SEM with the same fixatives and procedures as adopted for tissues, except that these operations last only about 30 min.

### 3.8.2. Sample Collection

Samples are collected following processing of the whole culture. The most significant cells are collected under the microscope.

### 3.8.3. Specimen Processing

#### 3.8.3.1. TEM

After fixation and postfixation, dehydration, and impregnation are performed as described (*see* **Subheadings 3.4.2.** and **3.4.3.**) for tissue samples processed for traditional TEM. The authors perform these procedures on primary cultures of white and brown adipocytes, using the culture support itself as container, in which the various substances envisaged by the protocol are added. For this reason, the ideal support for cultured cells is a glass container, which resists the action of solvents and in which they can be examined under the microscope (e.g., Leighton tubes). After the identification of the areas containing the most significant cells, embedding is performed by tipping gelatin capsules filled with absolute resin onto them, in order as to include only the areas of interest. Leighton tubes containing the gelatin capsules, upside down on cell cultures are then polymerized at 60°C for 12–15 h. Capsules are detached by thermal shock by immersing the tubes in liquid nitrogen for some seconds. The selected culture area is detached from the tube and remains embedded at the flat end of the polymerized resin capsule *(38–41)*.

### 3.8.3.2. SEM

After fixation and postfixation, the tube tops are removed by breaking the tubes. The bottom of the tube, containing the cell monolayer, is then dehydrated and coated as though it were a normal bioptic specimen. Fragments of approx 1 cm$^2$ of the bottom can then be viewed directly under the SEM *(35)*.

### *3.8.4. Sectioning (TEM only)*

The cell monolayer on the capsule-shaped, polymerized, and hardened resin surface (*see* **Subheading 3.4.4.**) is sectioned with an ultramicrotome. Considering that monolayers are, on average 5–10 μm in thickness, that the thickness of a semithin sections is 1 μm, and that of ultrathin sections 50–70 nm, a considerable number of sections can be obtained from a single capsule. Accounting for a certain number of inevitable alignment errors, only one or two ultrathin sections can be obtained from each; for this reason, several areas of the same culture should be encapsulated.

## *3.9. Morphometry*

### *3.9.1. Light Microscopy*

In our studies we have utilized morphometry mostly to determine the size of WAT cells and to count brown adipocytes in BAT and mixed depots.

### 3.9.1.1. White Adipocyte Size

To assess the contribution of adipocyte hyperplasia, compared to hypertrophy, to increased AT mass, we adopted the following procedure *(26,42)*:

1. The tissue sample must consist of sections representing the AT depot. Paraffin-embedded sections of suitable thickness (approx 3 μm), stained with H&E, which allows clear outlining of the cell profile, are ideal to obtain a large number of cells. If only sections processed for EM are available, semithin sections are also suitable, and will in fact provide sharper images, except that the number of available cells will be lower, because of the smaller size of tissue samples.

2. The section areas to be analyzed morphometrically are randomly selected by careful observation under the LM of the morphometric apparatus, studying each section's tissue composition. In these cases, magnification will be in the ×10–20 range.
3. The image analyzer allows calculation of the area and diameter of individual cells after they have been outlined with the mouse or with the pencil of the graphic tablet. Other quantitative and, especially, statistical data can be extrapolated from these measurements, such as mean, standard, and error deviation, and size class distribution.
4. The number of cell elements studied should be as large as possible, and measured on representative sections of the depot. The best solution, when feasible, is to cut serial sections of the whole depot at thickness intervals that prevent repeated observation of the same cells.
5. Unilocular WAT cells are spherical and have a rounded profile. Measuring their real diameter would require the section to cut through the core (and the nucleus) of all the cells lying in the same plane of section. If the cell population is made up of elements of varying size with a normal distribution, it is necessary to account for the loss of smaller profiles. Giger and Riedwyl *(18)* have developed a graphic method by which the correct mean particle diameter, as well as the SD of particle size distribution, can be derived. The principle is to complete and correct the histogram of profile distribution.
6. Not less important is the requirement that all tissues subjected to morphometric analysis be processed in the same way. Fixation and dehydration (and relative lipid extraction) procedures should be compatible with the lowest possible presence of embedding and sectioning artifacts. Sections should be as thin as allowed by paraffin embedding, to visualize clearly cell profiles.
7. The methods described in the literature *(43)* (such as the Hirsh and Gallian method of lipid extraction and $OsO_4$ fixation) do not allow for the presence of brown cells within depots, although the mixed composition of some visceral depots is well established. These methods mainly provide for the calculation of mean values, from which neither frequency data nor the distribution of values in size classes nor the demonstration of variability in the presence of identical means can be derived.

### 3.9.1.2. BROWN ADIPOCYTE COUNT

Brown adipocytes can be counted exclusively at the morphologic level, particularly when tissues are mixed. Because of the presence of several small lipid vesicles, recognizing cell profiles is difficult. One possible solution is to count only cells in which the nucleus is visible: brown adipocyte nuclei are easily distinguished from the others by their round shape, and are often located at the center of the cell.

To demonstrate BAT recruitment resulting from cold acclimation in old rats *(44)*, we have correlated absolute numerical values, adopting the following procedure:

1. After perfusion and fixation in 4% paraformaldehyde in 0.1 *M* PB, pH 7.4, interscapular brown adipose tissue (IBAT) is carefully dissected under the microscope, weighed, and its volume measured by fluid displacement.

2. IBAT is cut into two parts along the middle plane of symmetry; each hemi-IBAT is paraffin-embedded and sectioned along parasagittal planes; 3-μm-thick serial sections are collected every millimeter and stained with H&E.
3. For each section, connective-vascular, unilocular, and multilocular areas are identified with a camera lucida, and measured with the Kontron image analyzer *(45)*.
4. For each animal, 500 profiles of multilocular cells are measured to determine mean cell volume. Unlike unilocular cells, multilocular adipocytes are not spherical but polygonal. In an LM section, most profiles are elliptic, and, given that the largest circular profile has a diameter corresponding to the largest elliptic profile, we consider them as prolate ellipsoids. The major and minor axes of each profile are measured, and the volume of each prolate ellipsoid calculated.
5. The percent volume occupied by multilocular adipocytes ($V_1$) is obtained by applying the principle of Delesse (the volume density of the various components of a tissue can be estimated on random serial sections by measuring the relative areas of their profiles, $A_1/A_0 = V_1/V_0$) *(46)*, where $A_1$ corresponds to the area occupied by multilocular adipocytes, calculated on serial sections, $A_0$ is the total area of all section components, and $V_0$ is the hemi-IBAT volume measured directly by fluid displacement, as described above.
6. The number of multilocular adipocytes is calculated by dividing $V_1$ by the mean volume of multilocular adipocytes.

This procedure can be applied to mouse and rat brown depots, in which the boundary between white and brown areas is sharp, but not to hamster IBAT, in which unilocular and multilocular cells are mixed in the tissue.

### 3.9.1.3. ADIPOCYTE DENSITY

Relative determinations, i.e., counting the number of multilocular cells in mixed white depots, are easier to perform.

We have quantitated the recruitment of brown adipocytes in rat periovarian pads *(13,25)* during cold acclimation, using the following method:

1. 3-μm-thick slices are sectioned every 200 μm and stained with H&E.
2. Each section is projected on a screen by a projecting LM or directly observed by LM at the magnification that allows the best localization of brown adipocytes.
3. Brown fat areas are identified and the number of brown adipocytes is counted.
4. At the same time, the total adipose area of each section is measured with the image analyzer.
5. Brown adipocyte density (number of multilocular adipocytes/$\mu m^2$) is calculated for each section.
6. The distribution of periovarian brown adipocyte density is described by one of the statistical parameters for comparison and significance.

### 3.9.2. Immunohistochemistry

Quantitating morphometrically the results of an IHC reaction is rendered complex by the simultaneous presence of several factors that can affect the results. This may occur when the higher or lower presence of the Ag needs to be demonstrated, based on the color density of the staining reaction. All morphometric programs provide for this analysis through the measurement of greyscale (inexpensive graphic software is also available), but the variables to be kept under control are numerous, and include the different fixation degree attained in different section areas; section thickness, which is difficult to standardize, especially in sections obtained from paraffin-embedded samples (so that a greater amount of Ag and staining may simply be the result of the greater volume of the structure containing it); areas of greater staining density may be nonspecific, because of the nonoptimal adherence of the section to the slide. The literature reports various examples of automatic quantitation of positivity for cytoplasmic antigens *(47)*.

Counting the number of nerve terminals positive for some neuropeptides (e.g., tyrosine hydroxylase [TH], neuropeptide Y [NPY], calcitonin gene related peptide [CGRP], substance P [SP]) in different conditions, and following different treatments, is much easier. Their presence can be expressed as density in a specific tissue area, or as proportion (%) of the positive fibers in the condition of highest expression. An example of the latter situation is described below *(48)*.

### 3.9.2.1. EXAMPLE

Serial sections from the IBAT of rats of all experimental groups were processed for IHC with specific Abs against TH, a specific marker of noradrenergic nerve fibers, and against protein gene product 9.5 (PGP 9.5), an enzyme that is considered a highly sensitive marker for neuronal elements, including axonal projections. TH- and protein gene product (PGP) 9.5-immunoreactive (ir) parenchymal fibers were counted in transverse sections with a LM, at a final magnification of ×630, on 20 random fields for each animal. The overall number of positive fibers was calculated as the mean of the values observed in each animal, and was expressed as proportion of their fibers found in cold-acclimated rats (highest expression). In this condition, the IBAT increase in functional activity is associated with an augmented noradrenaline and TH content. This increase results from the branching of noradrenergic fibers, as demonstrated by the parallel increment in the number of PGP 9.5 ir fibers.

### *3.9.3. Electron Microscopy*

Any structure or organelle can be measured on ultrastructural images. However, morphometric evaluation at the ultrastructural level is rendered expensive by the need for analyzing several areas, and therefore several slides with accompanying prints, to reach statistical significance. With an interactive system, namely a video camera fitted onto the EM, images can be saved in an electronic medium (CD-ROM), reducing cost as well as study duration.

In BAT it is occasionally necessary to study and quantitate mitochondrial morphology at the ultrastructural level, because it varies with the functional activity of the cells and of the adipose depot. There are morphologic features in the development of mitochondrial cristae that allow identification of different cells correlated with different situations, even though the literature reports that this is not always the case *(49)*.

The morphometric evaluation of the multilocular cell mitochondria, found in white or brown depots, is performed at two levels: number and size of mitochondria, and development of cristae with respect to mitochondrial area (density). For each case, several areas on which to make measurements are required. A number of random pictures are generally taken from ultrathin sections in different fields. Magnification must obtain the largest possible area, striking a balance between the exhaustive identification of mitochondria and image resolution.

### 3.9.3.1. MITOCHONDRIAL AREA

The largest possible number of mitochondria (at least 300–500/animal) must be studied. The number of areas (slides or photographs) to be included depends on mitochondrial density in the areas under examination, as well as on the animals available for each treatment. To avoid escalating costs, we work with TEM slides, acquiring images by means of the scanner.

### 3.9.3.2. MITOCHONDRIAL DENSITY

This is the number of mitochondria occupying a given area of the cytoplasm.

### 3.9.3.3. DENSITY OF MITOCHONDRIAL CRISTAE

In this case, magnification of the photograph must allow clear identification of clearly the linear development of cristae. Depending on fixation and final storage, magnification may be satisfactory in the 10–15 K (K = ×1000) range; the zoom of the morphometric apparatus is useful when cristae are very closely packed, for instance, when multilocular cells are strongly activated *(13)*. By a macrosequence specifically programmed for this purpose, the procedure is started by tracing mitochondrial profiles on the screen (to obtain their surfaces in $\mu m^2$). Then, after tracing the cristae from one side to the other, the software adds up the various length measurements (in $\mu m$) and calculates their density (in $\mu m/\mu m^2$).

With a freeze-fracturing system, possible variations in BAT activity can be demonstrated by counting gap junctions (responsible for the electrical coupling among adjacent adipocytes), which change in size following suitable stimulation, and measuring their area *(15)*. Also, in this case, low magnification photographs are taken to identify the largest possible number of membrane areas, and high magnification ones to measure gap junction surface (*see* **Fig. 4B**).

At the ultrastructural level, the number of parenchymal nerve terminals and capillaries can also be counted. In these cases, low magnifications are suitable, and curb at the same time the final cost of the morphometric evaluation, without excessively restricting the sample.

## 4. Notes

1. To obtain reliable results, several control tests (positive and negative) must be performed to check that the sequence is methodologically correct (method controls), and, above all, that the immunostaining is specific for the Ag to be localized (specificity controls) *(3)*. One method control test consists in replacing a component of the sequence (e.g., the primary Ab) with immunoglobulin G raised in the same animal as the first Ab. The reaction must be negative. One specificity test (also negative) consists in incubating the section with the primary Ab adsorbed in a vial with the homologous Ag (the Ag that is sought in the tissue). The excess antigen saturates all Ab binding sites, which will not be ready to begin the immunologic sequence with the Ag present in the section: the reaction must be negative. This test is very important to verify Ag–Ab binding specificity.

    Tissue controls are those performed on sections in which the Ag is known to be present (positive method and specificity-control test) and those, equally important, performed on sections that do not contain the Ag (negative-control test). Sometimes internal controls (positive and negative) are available in sections: these are important, because substrate conditions are identical.

    In our experience, positive controls, where available, are the more important, because they allow interpretation of the positive reaction and distinguish it from the background. One of the fundamental interpreting problems to be addressed when examining an IHC reaction is to understand whether the staining that is observed on a cell or a structure is produced by the Ag–Ab reaction, or is nonspecific (background). In case of uncertainty about the presence of internal controls in sections, it is advisable to add to the series of slides subjected to processing at least one slide containing a structure rich in the Ag that is being investigated. For instance, when studying the noradrenergic fibers of WAT with an antityrosine hydroxylase Ab, a tissue section of an organ containing these fibers should be included among AT sections. If, on the other hand, the AT that is being investigated is adjacent to an organ that contains noradrenergic fibers, that will be the internal control. The staining intensity of the control section is the chief premise of the interpretation of the reaction.
2. Control reactions can be performed with sense riboprobes, which have the same sequence as the target transcript. This provides information on nonspecific background.
3. We use a Philips 505 SEM, which allows study of samples of approx 1 cm$^2$, with a resolution approaching 0.5 nm. Operating voltage is 10–15 kV. Samples can be rotated by 360 degrees to observe the lateral surfaces.

4. Because specimens are examined in a vacuum system, ATs, like all soft tissues, must be dried to avoid the collapse of the structures to be examined. Following dehydration in ascending alcohol concentrations, samples are critical-point dried (CPD010 Balzers).

5. Specimens have to be coated with a thin, conductive metal film (in this case, a 3-µm gold layer), otherwise, charge artifacts will occur when the electron beam hits the specimens, particularly at high accelerating voltage. The gold layer gives a high secondary electron emission coefficient, and also stabilizes the specimen mechanically. It may be applied by a sputter coater (Balzers S150A).

6. Special care must be exercised when placing samples in the microscope's specimen holder. The holder of the authors' Philips CM10 TEM is fitted with a clip that holds the grid in place. Very careful handling is required when clipping the grids to the holder, because replicas are brittle and vulnerable to the smallest trauma. Observation conditions are those adopted for traditional TEM.

7. Observation is conducted as described for conventional TEM and SEM (*see* **Subheading 3.4.5.**).

## References

1. Cinti, S. (1999) *The Adipose Organ*, Editrice Kurtis, Milan, Italy.
2. Bloom, W. and Fawcett, D. W. (1994) *Textbook of Histology*, 12th ed., (Fawcett, D. W., ed.), Chapman and Hill, New York.
3. Polak, J. M. and Van Noorden, S. (1986) *Immunohistochemistry: Modern Method and Applications*, 2nd ed., Wright PGS, Bristol, UK.
4. Nonradioactive *in situ* hybridization. Application Manual. (1996) Boehringher Mannheim, 2nd ed., pp. 44–56, 126–135.
5. Dagerlind, A., Friberg, K., Bean, A. J., and Hokfelt, T. (1992) Sensitive mRNA detection using unfixed tissue: combined radioactive and non radioactive *in situ* hybridization histochemistry. *Histochemistry* **98**, 39–49.
6. Wilkinson, D. G., Bailes, J. A., Champion, J. E., and McMahon, A. P. (1987) Molecular analysis of mouse development from 8 to 10 days post coitum detects changes only in embryonic globin expression. *Development* **99**, 493–501.
7. Sassoon, D. A., Garner, I., and Buckingham, M. (1988) Transcripts of alpha-cardiac and alpha-skeletal actins are early markers for myogenesis in the mouse embryo. *Development* **104**, 155–164.
8. Cinti, S. (1992) Morphological and functional aspect of brown adipose tissue. *Pediatric. Adolesc. Med.* **2**, 125–132.
9. Cinti, S., Cigolini, M., Bosello, O., and Björntorp, P. (1984) Morphological study of the adipocyte precursors. *J. Submicrosc. Cytol.* **16**, 243–151.
10. Rossouw, D. J., Cinti, S., and Dickersin, G. R. (1986) Liposarcoma: an ultrastructural study of 15 cases. *Am. J. Clin. Pathol.* **85**, 649–667.
11. De Mey, J. (1983) Colloidal gold probes in immunocytochemistry, in *Immunocytochemistry, Practical Application in Pathology and Biology* (Polak, J. M. and Van Noorden, S., eds.), Wright-PGS, Bristol, UK.

12. Cinti, S., Zancanaro, C., Sbarbati, A., Cigolini, M., Vogel, P., Ricquier, D., and Fakan S. (1989) Immunoelectron microscopical identification of the uncoupling protein in brown adipose tissue mitochondria. *Biol. Cell* **67,** 359–362.

13. Cousin, B., Cinti, S., Morroni, M., Raimbault, S., Ricquier, D., Pénicaud, L., and Casteilla, L. (1992) Occurence of brown adipocytes in rat white adipose tissue: molecular and morphological characterization. *J. Cell Sci.* **103,** 931–942.

14. Hayat, M. A., ed. (1974). Biological application, in *Principles and Techniques of Scanning Electron Microscopy*, vol. 2. Van Nostrand Reinhold, London.

15. Barbatelli, G., Heinzelmann, M., Ferrara, P., Morroni, M., and Cinti, S. (1994) Quantitative evaluations of gap junctions in old rat brown adipose tissue after cold acclimation: a freeze-fracture and ultrastructural study. *Tiss. Cell* **26,** 667–676.

16. Cinti, S., Cigolini, M., Gazzanelli, G., and Bosello, O. (1985) Ultrastructural study of adipocyte precursors from epididymal fat pads of adult rats in culture. *J. Submicrosc. Cytol.* **14,** 631–636.

17. Cigolini, M., Cinti, S., Bosello, O., Brunetti, L., and Björntorp P. (1986) Isolation and ultrastructural features of brown adipocytes in culture. *J. Anat.* **145,** 207–216.

18. Giger, H. and Riedwyl, H. (1970) Bestimmung der Grössenverteilung von kugeln aus schnittkreisradien. *Biometr. Zschr.* **12,** 156.

19. Williams, M. A. (1977) Quantitative methods in biology, in *Practical Methods in Electron Microscopy*, vol. 6, (Glauert, A. M., ed.), North-Holland, Amsterdam, The Netherlands.

20. Hayat, M. A. (1986) *Basic Techniques for Transmission Electron Microscopy*, Academic Press, New York.

21. Thompson, S. W. (1966) *Selected Histochemical and Histopathological Methods*, Thomas Publisher, Springfield, IL.

22. Hsu, S. M., Raine, L., and Fanger, H. (1981) Use of avidin-biotin-peroxidase complex (ABC) in immunoperoxidase technique: a comparison between ABC and unlabeled antibody (PAP) procedures. *J. Histochem. Cytochem.* **29,** 577–580.

23. Barbatelli, G., Morroni, M., Vinesi, P., Cinti, S., and Michetti, F. (1993) S-100 protein in rat brown adipose tissue under different functional conditions: a morphological, immunocytochemical and immunochemical study. *Exp. Cell Res.* **208,** 226–231.

24. Zancanaro, C., Pelosi, G., Accordini, C., Balercia, G., Sbabo, L. and Cinti, S. (1994) Immunohistochemical identification of the uncoupling protein in human hibernoma. *Biol. Cell* **80,** 75–78.

25. Giordano, A., Morroni, M., Santone, G., Marchesi, G. F., and Cinti, S. (1996) Tyrosine hydroxylase, neuropeptide Y, substance P, calcitonin gene-related peptide and vasoactive intestinal peptide in nerves of rat periovarian adipose tissue: an immunohistochemical and ultrastructural investigation. *J. Neurocytol.* **25,** 125–136.

26. Cinti, S., Frederich, R. C., Zingaretti, M. C., De Matteis, R., Flier, J. S., and Lowell, B. (1997) Immunohistochemical localization of leptin and uncoupling protein in white and brown adipose tissue. *Endocrinology* **38,** 797–804.

27. De Matteis, R., Dashtipour, K., Ognibene, A., and Cinti, S. (1998) Localization of leptin receptor splice variants in mouse peripheral tissues by immunohistochemistry. *Proc. Nutr. Soc.* **57,** 441–448.

28. Giordano, A., Morroni, M., Carle, F., Gesuita, R., Marchesi, G. F., and Cinti, S. (1998) Sensory nerves affect the recruitment and differentiation of rat periovarian brown adipocytes during cold acclimation. *J. Cell Sci.* **111,** 2587–2594.

29. Cancello, R., Zingaretti, M. C., Sarzani, R., Ricquier, D., and Cinti, S. (1998) Leptin and UCP1 genes are reciprocally regulated in brown adipose tissue. *Endocrinology* **139,** 4747–4750.

30. Tonello, C., Giordano, A., Cozzi, V., Cinti, S., Stock, M. J., Carruba, M., and Nisoli, E. (1999) Role of sympathetic activity in controlling the expression of vascular endothelial growth factor in brown fat cells of lean and genetically obese rats. *FEBS Lett.* **442,** 167–172.

31. Shaeren-Wiemers, N. and Gerfin-Moser, A. (1993) Single protocol to detect transcripts of various types and expression levels in neuronal tissue and cultured cells: *in situ* hybridization using digoxigenin-labeled cRNA probes. *Histochemistry* **100,** 431–440.

32. Braissant, O. and Wahli, W. (1998) Simplified *in situ* hybridization protocol using non-radioactively labeled probes to detect abundant and rare mRNAs on tissue sections. *Biochemica* **3,** 10–16.

33. Blanchette-Mackie, E. J. and Scow, R. O. (1981) Membrane continuities within cells and intercellular contacts in white adipose tissue of young rats. *J. Ultrastruct. Res.* **77,** 277–294.

34. Johannessen, J. V., ed. (1978) Instrumentation and techniques, in *Electron Microscopy in Human Medicine,* vol. 1. McGraw-Hill, London, UK.

35. Sbarbati, A., Zancanaro, C., Cigolini, M., and Cinti, S. (1987) Brown adipose tissue: a scanning electron microscopy study of tissue and cultured adipocytes. *Acta Anat.* **128,** 84–88.

36. Riva, A., Congiu, T., and Faa, G. (1993) Application of the OsO$_4$ maceration method to the study of human bioptic material. A procedure avoiding freeze-fracture. *Microsc. Res. Tech.* **26,** 526–527.

37. Willison, J. H. M. and Rowe, A. T. (1980) Replica, shadowing and freeze-etching techniques, in *Practical Methods in Electron Microscopy*, vol. 8. (Glauert, A. M., ed.), North-Holland, Amsterdam, The Netherlands.

38. Cigolini, M., Cinti, S., Brunetti, L., Bosello, O., Osculati, F., and Björntorp P. (1985) Human brown adipose cells in culture. *Exp. Cell Res.* **159,** 261–266.

39. Cinti, S., Cigolini, M., Sbarbati, A., and Zancanaro, C. (1986) Ultrastructure of brown adipocytes mitochondria in cell culture from explants. *J. Submicrosc. Cytol.* **18,** 625–627.

40. Cinti, S., Enzi, G., Cigolini, M., and Bosello, O. (1983) Ultrastructural features of cultured mature adipocytes precursors from adipose tissue in multiple symmetric lipomatosis. *Ultrastruct. Pathol.* **5,** 145–152.

41. Cinti, S., Cigolini, M., Sbarbati, A., Zancanaro, C., and Björntorp, P. (1987) Effects of noradrenaline exposure on rat brown adipocytes in cultures. An ultrastructural study. *Tissue Cell* **19,** 809–816.

42. Cinti, S., Eberbach, S., Castellucci, M., and Accili, D. (1998) Lack of insulin receptors affects the formation of white adipose tissue in mice. A morphometric and ultrastructural analysis. *Diabetologia* **41,** 171–177.

43. Hirsh, J. and Gallian E. (1968) Methods for the determination of adipose cell size in man and animals. *J. Lipid Res.* **9,** 110–119.
44. Morroni, M., Barbatelli, G., Zingaretti, M. C., and Cinti, S. (1995) Immunohistochemical, ultrastructural and morphometric evidence for brown adipose tissue recruitment due to cold acclimation in old rats. *Int. J. Obesity* **19,** 126–131.
45. Sbarbati, A., Morroni, M., Zancanaro, C., and Cinti, S. (1991) Rat interscapular brown adipose tissue at different ages: a morphometric study. *Int. J. Obesity* **15,** 581–588.
46. Weibel, E. R. (1979) Stereological methods in practical methods for biological morphometry. Vol. 1., Academic Press, London.
47. Lehr, H. A., Mankoff, D. A., Corwin, D., Santeusanio, G., and Gown, A. M. (1997) Application of Photoshop-based image analysis to quantification of hormone receptor expression in breast cancer. *J. Histochem. Cytochem.* **45,** 1559–1565.
48. De Matteis, R., Ricquier, D., and Cinti, S. (1998) TH-, NPY-, SP-, and CGRP-immunoreactive nerves in interscapular brown adipose tissue of adult rats acclimated at different temperatures: an immunohistochemical study. *J. Neurocytol.* **27,** 877–886.
49. Loncar, D., Afzelius, B. A., and Cannon, B. (1988) Epididymal white adipose tissue after cold stress in rats. II Mitochondrial changes. *J. Ultrastruct. Mol. Struct. Res.* **101,** 109–122.

# 3

## Confocal Microscopy of Adipocytes

### Daniela Malide

### 1. Introduction

The isolated rat adipose cell experimental system is the principal model for studies of the mechanism of insulin's stimulatory action on glucose transport (reviewed in **refs. *1–4***). Despite the successful use of this preparation in biochemical studies, the unique structure of the adipose cell, with its large central triglyceride storage droplet (80 μm) and thin (1–2 μm) rim of cytoplasm, has caused special problems for morphological approaches. Several methods have been used for studies of protein localization and trafficking by immunocytochemistry (ICC) at both light microscopy (LM) and electron microscopy (EM) levels.

Studies employing sections of the adipose tissue (AT) and of the 3T3-L1 adipocyte cell line remain scarce, because they are technically very difficult to prepare *(5–8)*. Although ICC on ultrathin sections allows precise subcellular localization, the sections give only a limited view of the cell, and the sampling problem can become overwhelming. Thus, converting the information they provide into a three-dimensional (3-D) pattern may be difficult, especially for proteins that are unevenly distributed.

Other studies *(9,10)* have focused on the use of plasma membrane sheets (lawns) as an assay for the translocation of the glucose transporter, GLUT4, in response to insulin by LM and EM ICC. This technique is limited to an examination of structures associated with the inner surface of the plasma membrane.

The specific aim of this chapter is to introduce an alternative morphological approach using confocal microscopy optical sectioning and computer-assisted image reconstruction in the whole adipose cell. As documented here, this approach allows investigators to see the *in situ* localization of proteins, and to trace the changes in response to different stimuli. Furthermore, it overcomes

From: *Methods in Molecular Biology, vol. 155: Adipose Tissue Protocols*
Edited by: G. Ailhaud © Humana Press Inc., Totowa, NJ

some technical difficulties, particularly the sectioning, in studying the rat adipose cell, and it opens a new, more accessible way to investigate the protein trafficking pathways in an insulin-responsive cell of physiological significance (*see* **Note 1**).

## 2. Materials

### 2.1. Adipose Cell Isolation (11)

1. Krebs-Ringer bicarbonate-HEPES buffer supplemented with 1% bovine serum albumin (BSA) (Intergen, Purchase, NY) and 200 n*M* adenosine.
2. Type I and type II collagenase (Worthington, Freehold, NJ); DNase I (Boehringer Mannheim , Indianapolis, IN).
3. A shaking water bath and a low-speed centrifuge.

### 2.2. Immunocytochemistry

1. Fixative: 4% paraformaldehyde (Electron Microscopy Sciences, Ft. Washington, PA) in 0.15 *M* phosphate-buffered saline (PBS), pH 7.4 (prepare fresh before each use).
2. Blocking and permeabilization buffer (B1): 1% BSA, 3% normal goat serum, 0.1% saponin in 0.15 *M* PBS, pH 7.4.
3. Washing buffer (B2): 0.1% saponin in 0.15 *M* PBS, pH 7.4.
4. Primary antibodies (Abs): *see* **Table 1**.
5. Secondary Abs:
   a. Fluorescein isothiocyanate (FITC): lissamine rhodamine sulfonyl chloride and Cy5-conjugated Abs specific for rabbit or mouse immunoglobulins, used at 15 µg/mL from Jackson ImmunoResearch (West Grove, PA).
   b. Rhodamine (Rhd)-conjugated lectin lens culinaris agglutinin (LCA) (50 µg/mL) from Vector (Burlingame, CA).
6. Other materials: Vectashield mounting media from Vector; glass microscopy slides, glass cover slips (no. 1), plastic containers, a rocker platform.

## 3. Methods

The methods to obtain single cell suspensions of white and brown adipose cells from the rat ATs are described in detail in Chapters 5, 10, 15, and 24, and in **refs. *11*** and ***12***. Cells are incubated with insulin or other compounds at 37°C. Isolated rat adipose cells in suspension (floating cells) are used throughout the following staining protocols. ICC can be performed using direct or indirect techniques. Direct methods use a labeled specific (primary) Ab to bind directly (in one step) to the cellular epitope. Indirect methods comprise at least two steps: first an unlabeled primary Ab binds to the specific cellular epitope; in a second step the bound primary Ab is detected by a labeled-secondary Ab.

**Table 1**
**Primary Abs Used in this Work**

| Antigen | Compartment | Name of ab/conc. | Species and type[a] | Source |
|---------|-------------|------------------|---------------------|--------|
| GLUT4 | ? | 0.15 µg/mL | Rabbit pc | Hoffman-La Roche |
| Calnexin | ER | SPA-860/1:100 | Rabbit pc | Stressgen Biotech. |
| Lgp-120 | Lysosomes | Ly1C6/5 µg/mL | Mouse mc | Dr. I. Mellman |
| β-tubulin | Microtubules | JDR.3B8/1:100 | Mouse mc | Sigma |
| Rat MHC-I | ER and pm | OX-18/10 µg/mL | Mouse mc | Serotec |

[a]pc, polyclonal Ab; mc, monoclonal Ab.

### 3.1. Indirect Immunofluorescence Methods Using Fixed and Permeabilized Cells (Figs. 1–3)

1. Fix adipose cells in 4% paraformaldehyde (PFA) in 0.15 $M$ PBS, pH 7.4, for 20 min at room temperature (RT). For ~1 mL packed cells, use 40 mL of fixative solution, with gentle shaking, on a rocker platform (*see* **Note 2**). Use 15- or 50-mL polypropylene tubes (*see* **Note 3**).
2. Wash cells three times by centrifugation (210$g$ for 30 s) with PBS 0.15 $M$, pH 7.4, containing 50 m$M$ glycine, to quench aldehyde-induced nonspecific binding sites.
3. Permeabilize and block nonspecific binding sites with B1 for 45 min at RT. Use 1.5-mL polypropylene screw-cap tubes to incubate ~200 µL packed cells in 1 mL buffer B1.
4. Incubate cells with the primary Ab diluted in B1, for 2 h at RT with gentle shaking (*see* **Note 4**).
5. Aspirate the buffer beneath the floating cells, using a long, fine pipet tip (gel loading tips), and add 100 µL of the Ab solution.
6. Wash cells three times, by centrifugation (210$g$ for 30 s) with buffer B2.
7. Incubate cells with the secondary antibody diluted in B1, for 45 min at RT with gentle shaking.
8. Wash cells three times by centrifugation (210$g$ for 30 s) with buffer B2.
9. Mount cells on glass slides using Vectashield mounting medium. Deposit a droplet of mounting media on a glass microscopy slide, add ~20 µL packed cells, place a cover slip, and seal with nail polish.

Using this protocol, single-, double-, and triple-labeling experiments can be successfully performed in isolated rat adipose cells (*13–17*).

Examine staining with a fluorescence microscope equipped with a confocal laser scanning system. The author used a Nikon Optiphot 2 microscope attached to an MRC-1024 Bio-Rad confocal system controlled by Lasersharp image acquisition and analysis software from Bio-Rad Labs (Hercules, CA) (*see* **Notes 5–7**). For each experimental condition, 10–15 cells were imaged separately by Kalman, averaging 8–10 frames/image, using a planapochromat 60X/1.4NA

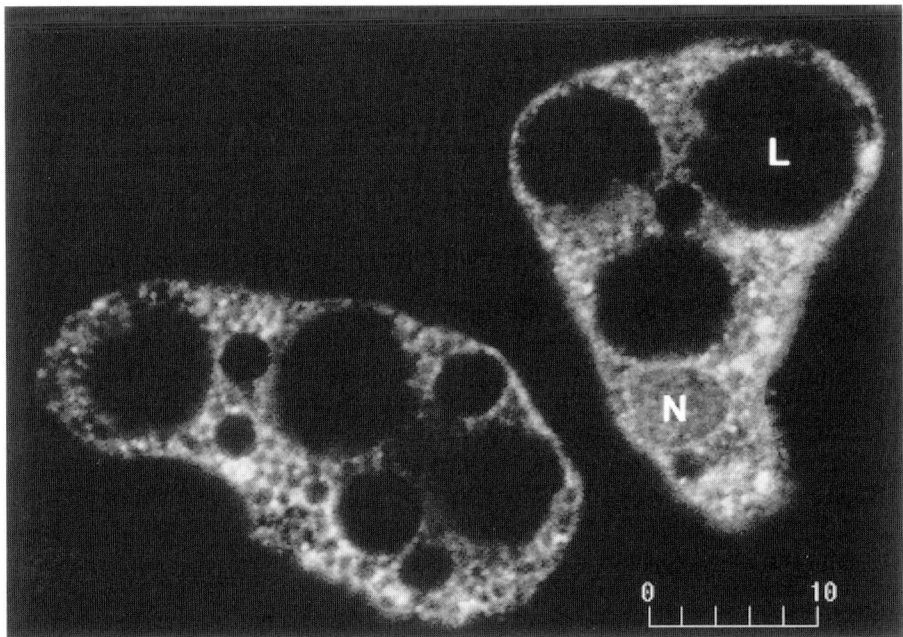

Fig. 1. Localization of the ER-marker calnexin in a single optical section of brown adipose cells. The staining appears intracellularly, displaying a reticular, honeycomb-like pattern throughout the cytoplasm, and outlining the nuclear envelope. N, nucleus; L, lipid droplet.

oil objective at optical zooms between 1 and 2.5. The excitation wavelengths used were 488-nm (FITC), 568-nm (Rhd), and 647-nm (Cy5), from a 15-mW krypton/argon laser. Images were acquired using 522DF32, 605DF32, and 680DF32 bandpass emission filters. For 3-D reconstruction, series of optical sections were collected at 0.5-μm intervals along the *z*-axis. For presentation, digitized images were cropped and assembled using the Adobe Photoshop 5.0 program from Adobe Systems (Mountain View, CA), and printed with a Kodak 8650 PS digital printer (Eastman Kodak, New Haven, CT).

In our experience, careful optimization of fixation and permeabilization conditions allow satisfactory labeling of adipose cells for LM. Confocal imaging allows precise visualization of fluorescent signals within a narrow plane of focus, with exclusion of out-of-focus blur: The technique permits the reconstruction of 3-D structures from serial optical sections *(18,19)*. Several examples of the application of this technique for the localization of proteins used as compartment markers are illustrated: endoplasmic reticulum (ER) (**Fig. 1**), the lysosomes (**Fig. 2**), and the microtubules (**Fig. 3**).

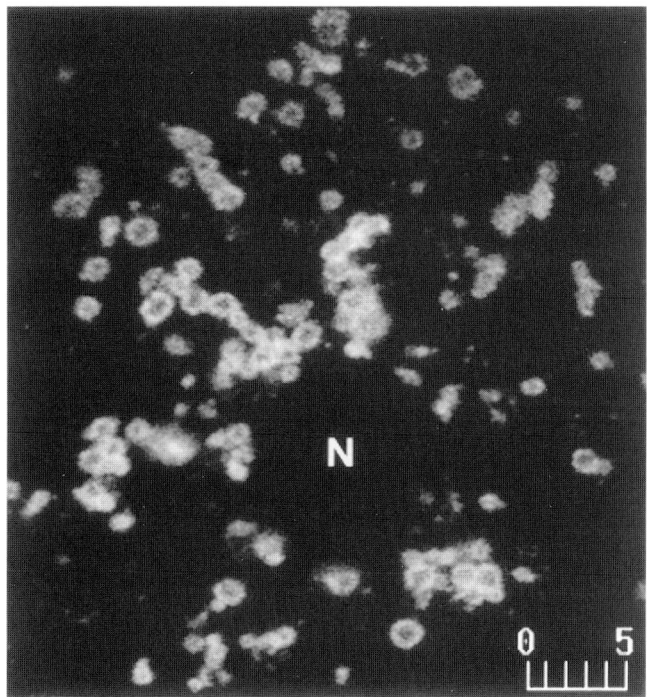

Fig. 2. Localization of the lysosomal membrane protein, lgp-120, in a single optical section of the white adipose cells. Staining of the lysosomes shows large, vacuolar-like structures dispersed throughout the cell. N, nucleus.

## 3.2. Cell Surface Immunofluorescence of Living Cells Stained in Nonpermeabilized State (Figs. 4 and 5)

Living adipose cells can be immunostained using fluorescent-labeled Abs or lectins, in direct or indirect procedures.

### 3.2.1. Analysis of Protein Co-Localization at the Cell Surface by Direct Staining of the Plasma Membrane, Using Fluorescently-Labeled Lectins (**Fig. 4**; see **Note 8**)

1. Rinse adipose cells quickly with ice-cold 0.15 $M$ PBS, pH 7.4, and chill them to 4°C.
2. Incubate cells with the Rhd-labeled lectin LCA for 30 min at 4°C in an ice-water bath, with gentle shaking.
3. Wash cells three times with ice-cold 0.15 M PBS, pH 7.4.
4. Fix cells with 4% PFA in 0.15 $M$ PBS, pH 7.4, for 20 min at RT.
5. Subsequently, the cells can be permeabilized and stained using an Ab to a protein distributed in both intracellular and surface pools (e.g., GLUT4), following the steps described in **Subheading 3.1.**

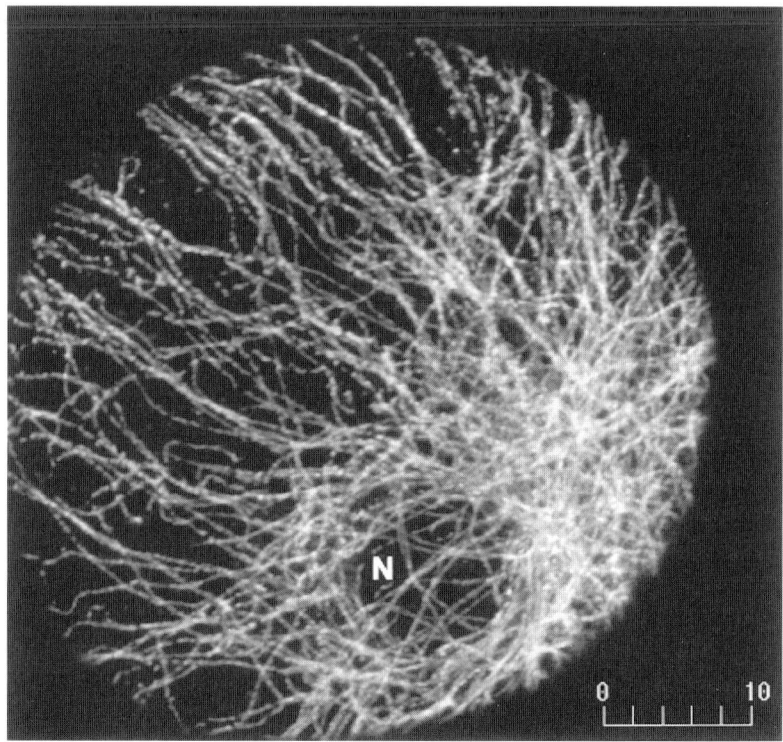

Fig. 3. Localization of the microtubule protein β-tubulin, in a whole white adipose cell. For 3-D reconstruction, series of optical sections are collected at 0.5-μm intervals along the *z*-axis. For visualization purposes, a 2-D projection is generated showing the extensive radial network of the microtubules. N, nucleus.

This allows direct examination (*see* **Fig. 4A**) and quantitative analysis (*see* **Fig. 4B**) of the protein localized at the cell surface.

Co-localization is assessed using Lasersharp 3.1 software which provides a fluorogram image and two (red and green) co-localization coefficients. The pixel fluorogram is a 2-D intensity histogram of the dual-color image, showing the distribution of all pixels within the merged image as a scattergram, similar to the one used in flow cytometry. Pixel values of the green and red channels are displayed along the *x*- and *y*-axes, respectively. The red (or green) co-localization coefficient is the ratio of the sum of the co-localized red (or green) pixel intensities to the sum of all the red (or green) pixel intensities. A high degree of co-localization is revealed by a narrow, diagonal-like distribution (at 45 degrees) of the dual-color pixels on the fluorogram, and the co-localization coefficients are almost equal to 1.0 (*17*).

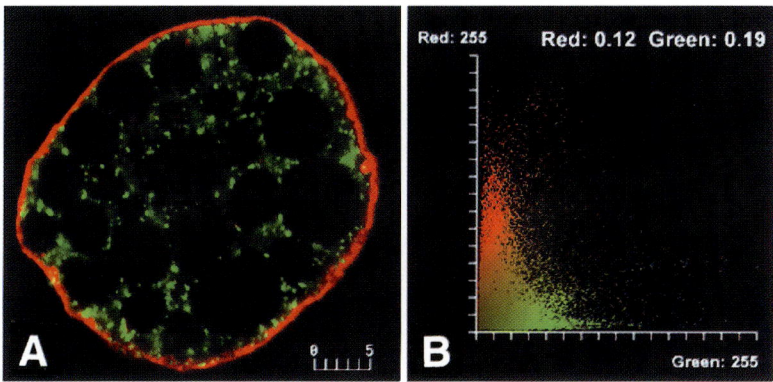

Fig. 4. GLUT4 localization to the cell surface by sequential immunofluorescent-lectin staining of the plasma membrane, followed by GLUT4 immunostaining. **(A)** (merged image) In the absence of insulin, GLUT4 staining (green) shows a punctate pattern beneath and distinct from the plasma membrane identified by lectin staining (red). **(B)** (fluorogram) Consistent with the visual observation (i.e., lack of co-localization), the fluorogram illustrates single-color pixels widely dispersed toward the red and green axis, and the two co-localization coefficients are very low (~0.1).

## 3.2.2. Quantitative Analysis of the Cell Surface Fluorescence, Using Indirect Staining (**Fig. 5**)

1. Rinse adipose cells quickly with ice-cold PBS, and chill them to 4°C to stop protein trafficking.
2. Incubate living cells with the primary Abs diluted in B1 without saponin, for 30 min, in an ice-water bath, with gentle shaking.
3. Wash cells three times with ice-cold PBS.
4. Incubate cells with the secondary Abs for 45 min in an ice-water bath.
5. Wash cells three times with ice-cold PBS.
6. Fix cells with 4% PFA in PBS for at least 20 min at RT.
7. Wash cells with PBS, and mount them on glass microscopy slides, only shortly before examination.

In order to obtain valuable quantitative data, images must be acquired during the same day for all experimental conditions, using identical settings of the instrument that avoid saturation of the brightest pixels. Series of images along $z$-axis are obtained for whole cellular surface labeling, and processed using Lasersharp 2.1 software. The integrated sum of total fluorescence through all optical sections in the 3-D series is calculated and compared among cells of similar size from different experimental conditions (*17*). **Figure 5** illustrates a quantitative analysis of cell-surface major histocompatibility complex-I (MHC-I) immunofluorescence in brown adipose cells.

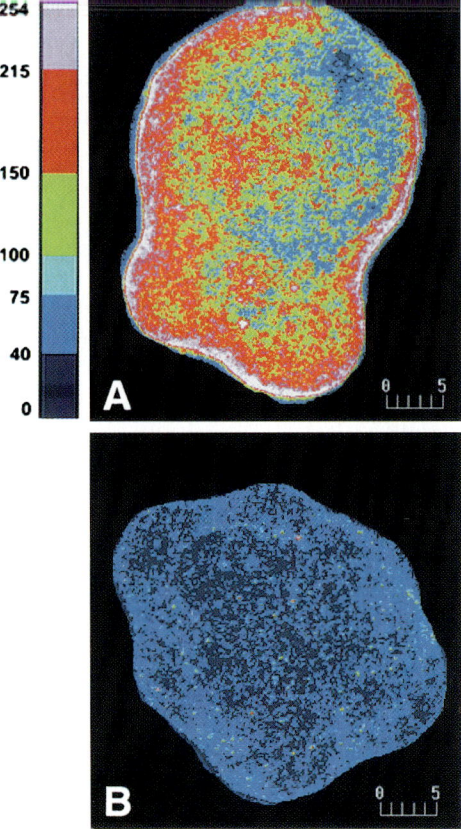

Fig. 5. Quantitative analysis of cell-surface MHC-I immunofluorescence. 2-D projections (using a maximum-pixel-intensity algorithm) of the 3-D fluorescence are generated. Fluorescence intensities are color-coded using a look-up table, with a 0–255 range, to create pseudocolor-mapped images. These show a threefold increase in cell-surface associated MHC-I molecules in response to insulin treatment (**A**), compared to untreated (**B**) brown adipose cells.

### 3.3. Simultaneous Direct Green Fluorescent Protein Detection, and Indirect Ab Immunostaining (Fig. 6)

Green fluorescent protein (GFP) expression and localization can be visualized directly, and can be used to "paint" particular cells or cellular processes. Thus, GFP can be used in conjunction with the immunostaining of other proteins on fixed cells, or for high-resolution imaging of subcellular events in living cells. Although recent studies provide evidence that the latter possibility, the real-time imaging of GFP-tagged proteins in 3T3-L1 adipocytes, is a

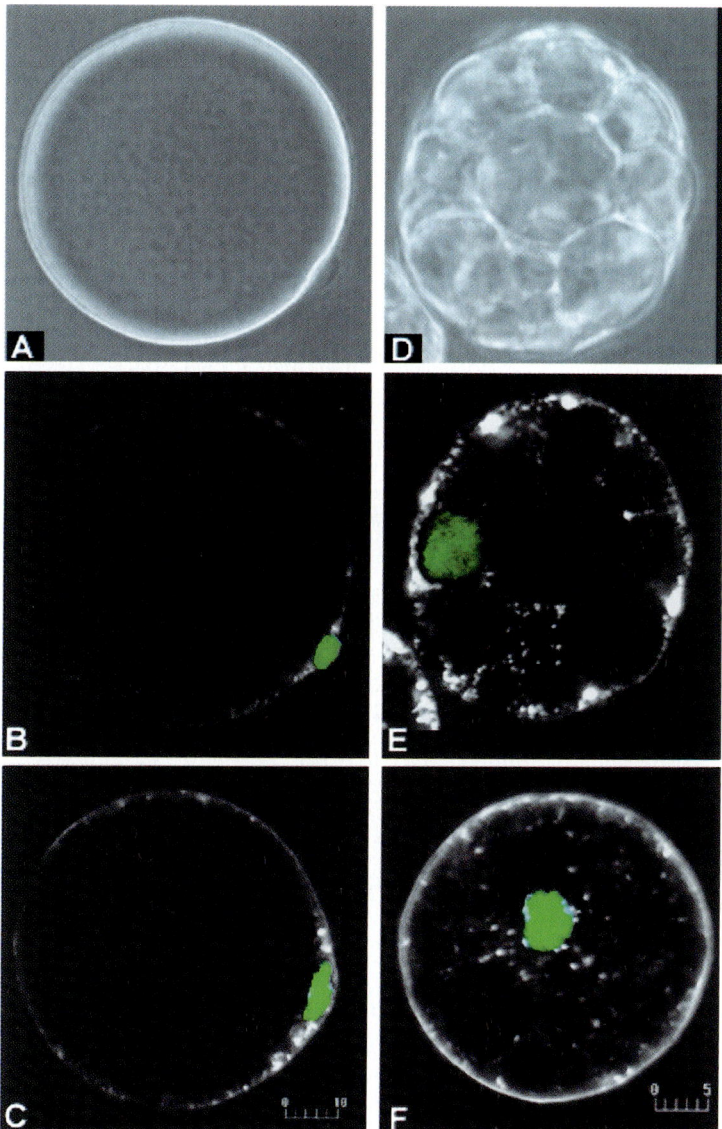

Fig. 6. Insulin induced GLUT4 translocation in adipose cells expressing a GFP-tagged influenza virus nuclear protein (GFP-NP). Under phase-contrast microscopy (using the transmitted-light setting of the confocal system), the white adipose cell (**A**) contains only one large lipid droplet surrounded by a thin cytoplasmic rim (~1 µm). In cells incubated without insulin (**B**), GLUT4 shows a punctate intracellular pattern. In cells incubated with insulin (**C**), GLUT4 displays a rim-like pattern at the cell periphery. The brown adipose cells (**D**) is smaller and contains multiple lipid droplets under phase microscopy. In these cells, GLUT4 shows an intracellular punctate staining in the absence of insulin (**E**); GLUT4 staining is outlining the cell periphery after insulin treatment (**F**). GFP-NP is directly visualized (using a FITC setting) in the nuclei of the adipose cells by GFP (green) fluorescence (B,C,E,F).

realistic goal *(20)*, technical problems remain to be solved in isolated rat adipose cells *(21,22)*. We illustrates the first possibility in **Fig. 6**, showing insulin-induced GLUT4 translocation in cells expressing a GFP-tagged influenza virus nuclear protein (GFP-NP). For gene delivery, a recombinant vaccinia virus system is used to express GFP-NP in isolated white and brown adipose cells. After 5 h infection time, cells are incubated in the absence or presence of 67 n*M* insulin at 37°C for 30 min. Cells are then aldehyde-fixed, permeabilized, and stained for endogenous GLUT4, following the indirect protocol described in **Subheading 2.1.** *(17)*.

## 4. Notes

1. We use confocal microscopy to investigate insulin action on isolated rat adipose cells in this laboratory. These techniques can be easily adapted for studies of the cultured 3T3-L1 adipocytes *(22)*. These methods are only guidelines for confocal microscopy of adipocytes; alternative protocols remain to be designed for specific applications.

2. Other fixative and permeabilization procedures, such as –20°C methanol, can be used when working with isolated rat adipose cells or 3T3-L1 adipocyte cell line *(22–24)*.

3. In working with primary rat adipose cells, compared to cell lines, the major challenge derives from the fact that isolated adipose cells float in suspension throughout the staining and examination steps. Adipose cells in suspension are large, fragile cells, and should be manipulated gently, in order to avoid damage. These cells tend to lyse when contacting charged nonbiological surfaces, particularly glass. Thus, use of polypropylene and polyethylene containers, and the continued presence of BSA, are recommended throughout the procedures.

4. Adequate amounts of cells (~100–200 μL packed cells) are required for these methods because the multiple washing steps will probably result in a significant loss of cells. In relation to the high number of cells in suspension, the Abs should also be applied in excess.

5. Adipose cells float, even when mounted on the microscopy slides, and thus working with an upright microscope is preferred. This is in contrast to cultured cells (attached to surfaces) when an inverted microscope is more appropriate.

6. A common problem with thick samples is light scattering, interfering with accurate imaging beyond ~60 μm depth, when using a conventional confocal microscope. This phenomenon often results in a progressive decrease in the fluorescent signal in the optical sections taken further away from the cover slip. Thus a multiphoton system may be considered to image the large size adipose cells, and to perform accurately quantitative immunofluorescence analyses.

7. For reasons that are unclear, long-term storage of mounted cells on slides results in deterioration of the overall cell morphology and a loss of the lipid droplet(s). When possible, images should be acquired within a short time (days) following immunostaining.

8. When using the direct immunofluorescence of living cells, it is important to ensure that the lectin conjugate binds to carbohydrate moieties exposed to the

outer surface of the plasma membrane and is not allowed to enter the cytoplasm. If the cells to be stained are kept at 4°C, to avoid endocytosis of the fluorochrome-conjugated lectin, the immunolabeling can be performed when cells are still alive. In our experience, this procedure results in much more reliable data, preventing possible fixative-induced redistribution of molecules within different domains of the plasma membrane. This method should be used for staining of aldehyde-sensitive epitopes exposed at the cell surface.

## Acknowledgments

I thank Dr. Samuel W. Cushman for his generous support, helpful advice, and encouragement throughout this work, and Drs. Ian A. Simpson and Evelyn Ralston for critical reading of the manuscript and many helpful discussions.

## References

1. Simpson, I. A. and Cushman, S. W. (1986) Hormonal regulation of mammalian glucose transport. *Annu. Rev. Biochem.* **55,** 1059–1089.
2. Birnbaum, M. J. (1992) The insulin-sensitive glucose transporter. *Int. Rev. Cyto.* **137A,** 239–297.
3. Holman, G. D. and Cushman, S. W. (1994) Subcellular localization and trafficking of the GLUT4 glucose transporter isoform in insulin-responsive cells. *Bioessays* **10,** 753–759.
4. Stephens, J. M. and Pilch, P. F. (1995) The metabolic regulation and vesicular transport of GLUT4, the major insulin-responsive glucose transporter. *Endocr. Rev.* **16,** 529–546.
5. Slot, J. W., Geuze, H. J., Gigengack, S., Lienhard, G. E., and James, D. E. (1991) Immuno-localization of the insulin regulatable glucose transporter in brown adipose tissue of the rat. *J. Cell Biol.* **113,** 123–135.
6. Smith, R. M., Charron, M. J., Shah, N., Lodish, H. F., and Jarett, L. (1991) Immunoelectron microscopic demonstration of insulin-stimulated translocation of glucose transporters to the plasma membrane of isolated rat adipocytes and masking of the carboxyl-terminal epitope of intracellular GLUT4. *Proc. Natl. Acad. Sci. USA* **88,** 6893–6897.
7. Blok, J., Gibbs, E. M., Lienhard, G. E., Slot, J. W., and Geuze, H. J. (1988) Insulin-induced translocation of glucose transporters from post-Golgi compartments to the plasma membrane of 3T3-L1 adipocytes. *J. Cell Biol.* **106,** 69–76.
8. Martin, S., Slot, J. W., and James, D. E. (1999) GLUT4 trafficking in insulin-sensitive cells. A morphological review. *Cell Biochem. Biophys.* **30,** 89–113.
9. Robinson, L. J., Pang, S., Harris, D. S., Heuser, J., and James, D. E. (1992) Translocation of the glucose transporter (GLUT4) to the cell surface in permeabilized 3T3-L1 adipocytes: effects of ATP, insulin, and GTPγS and localization of GLUT4 to clathrin lattices. *J. Cell Biol.* **117,** 1181–1196.
10. Voldstedlund, M., Tranum-Jensen, J., and Vinten, J. (1993) Quantitation of Na+/ K+-ATPase and glucose transporter isoform in rat adipocyte plasma membrane by immunogold labeling. *J. Membrane Biol.* **136,** 63–73.

11. Weber, T. M., Joost, H. G., Simpson, I. A., and Cushman, S. W. (1988) Methods for assessment of glucose transport activity and the number of glucose transporters in isolated rat adipose cells and membrane fractions, in *The Insulin Receptor Part B*, (Kahn, C. R. and Harrison, L. C., eds.), Alan R Liss, New York, NY, pp. 171–187.

12. Omatsu-Kanbe, M., Zarnowski, M. J., and Cushman, S. W. (1996) Hormonal regulation of glucose transport in a brown adipose cell preparation isolated from rats that show a large response to insulin. *Biochem. J.* **315,** 25–31.

13. Malide, D., Dwyer, N. K., Blanchette-Mackie, E. J., and Cushman, S. W. (1997) Immunocytochemical evidence that GLUT4 resides in a specialized translocation post-endosomal, VAMP-positive compartment in rat adipose cells in the absence of insulin. *J. Histochem. Cytochem.* **8,** 1083–1096.

14. Malide, D. and Cushman, S. W. (1997) Morphological effects of wortmannin on the endosomal system and GLUT4-containing compartments in rat adipose cells. *J. Cell Sci.* **110,** 2795–2806.

15. Malide, D., St-Denis, J-F., Keller, S. R., and Cushman S. W. (1997) Vp165 and GLUT4 share similar vesicle pools along their trafficking pathways in rat adipose cells. *FEBS Lett.* **409,** 461–468.

16. Barr, V. A., Malide, D., Zarnowski, M. J., Taylor, S. I., and Cushman, S. W. (1997) Insulin stimulates both leptin secretion and production by rat white adipose tissue. *Endocrinology* **138,** 4463–4472.

17. Malide, D., Yewdell J. W., Bennink, J. R., and Cushman, S. W. MHC class I and GLUT4 define two distinct insulin-responsive compartments and exocytic pathways in rat adipose cells. In review.

18. Matsumoto, B., ed. (1993) *Cell Biological Applications of Confocal Microscopy*: *Methods in Cell Biology*, vol. 38, Academic Press, San Diego, CA.

19. Pawley, J., ed. (1995) *The Handbook of Biological Confocal Microscopy*. IMR Press, Madison, WI.

20. Oatey, P. B., Van Weering, D. H., Dobson, S. P., Gould, G. W., and Tavare, J. M. (1997) GLUT4 vesicle dynamics in living 3T3-L1 adipocytes visualized with green-fluorescent protein. *Biochem. J.* **327,** 637–642.

21. Venkateswarlu, K., Oatey, P. B., Tavare, J. M., and Cullen, P. J. (1998) Insulin-dependent translocation of ARNO to the plasma membrane of adipocytes requires phosphatidylinositol 3-kinase. *Curr. Biol.* **8,** 463–466.

22. Blanchette-Mackie, E. J., Dwyer, N. K., Barber, T., Coxey, R. A., Takeda, T., Rondione, C. M., Theodorakis, J. L., Greenberg, A. S., and Londos, C. (1995) Perilipin is located on the surface layer of intracellular lipid droplets in adipocytes. *J. Lipid Res.* **36,** 1211–1226.

23. Martin, S., Reaves, B., Banting, G., and Gould, G. W. (1994) Analysis of the colocalization of the insulin-responsive glucose transporter (GLUT4) and the trans-Golgi network marker TGN38 within 3T3-L1 adipocytes. *Biochem. J.* **300,** 743–749.

24. Chakrabarti, R., Buxton, J., Joly, M., and Corvera, S. (1994) Insulin-sensitive association of GLUT-4 with endocytic clathrin-coated vesicles revealed with the use of Brefeldin-A. *J. Biol. Chem.* **269,** 7926–7933.

# 4

## Cellularity Measurements

### Mario DiGirolamo and Jacqueline B. Fine

## 1. Introduction

Adipose tissue (AT) makes up 10–30% of body weight in normal-weight individuals, but its mass can vary from as little as 3–5% of body wt in lean, athletic subjects to 70% of body wt in very obese subjects.

The main cellular components of AT are adipocytes, spherical cells with a small rim of cytoplasm and a nucleus, often called "signet-ring" cells because of their appearance. The lipid contained in the cells is mostly triglyceride (TG). In addition to adipocytes, AT contains stromal-vascular cells *(1)*.

Because of the unique ability of adipose tissue to expand and contract under different conditions of age, energy balance and endocrine status, units of reference for metabolic activity of the tissue, such as wet weight, lipid, protein, and DNA content, have not provided suitable means of comparison. The search has been ongoing for simple and reproducible ways to estimate the number of adipocytes present in a fragment of AT or in a suspension of isolated fat cells.

Two methods are described here: (1) an optical method, based on direct microscopic determination of the diameter of fat cells isolated by collagenase incubations; and an (2) electronic method, based on counting adipocytes, fixed with osmium tetroxide ($OsO_4$), in a Coulter counter. Pros and cons of these two widely used methods are presented to assess their ability to accurately and reproducibly define fat cell size and fat cell number in AT (*see* **Note 1**).

## 2. Materials
### 2.1. Optical Method

The optical method *(2–4)* is relatively simple and economic, and requires a microscope with a caliper attachment which allows consecutive measurements of the diameter of 200–300 isolated adipocytes floating on the surface of an

From: *Methods in Molecular Biology, vol. 155: Adipose Tissue Protocols*
Edited by: G. Ailhaud © Humana Press Inc., Totowa, NJ

aqueous medium It can be used to measure cell diameters from 11 μm to maximum size (140–180 μm). It provides information on the frequency distribution for diameter, surface area and volume of the cell population. Its chief limitation is that it requires isolation of the adipocytes by collagenase in a 45-min incubation, and that it takes approx 10 min for each determination, thus limiting the number of samples that can be examined during an experiment (usually 8–10).

## 2.1.1. Equipment

A microscope of good quality; we have used a Zeiss microscope with a focusing eye piece. A microscope caliper is inserted into the eye piece to produce a projected caliper scale. When magnified at ×190–200, the caliper scale should be calibrated to give unit marks with a constant interval of 7 μm. A Zeiss calibration slide (1 mm) is used to calibrate the microscope. By doubling the magnification, larger populations of fat cells can be measured with caliper constant intervals of 14 μm. A Dubnoff metabolic shaker and a standard centrifuge are also needed.

## 2.1.2. Chemicals and Reagents

1. Incubation buffer: Krebs-Ringer bicarbonate (KRB) with 4% bovine serum albumin (BSA) should be prepared on the morning of the experiment. Reagents for KRB are prepared ahead of time, and can be stored at 4°C for 1 mo.
     To make up one batch of KRB solution (total 130 mL), add 100 mL 0.9% NaCl (0.154 $M$) to 4 mL 1.15% KCl (0.154 $M$), 3 mL 1.22% $CaCl_2$ (0.11 $M$), 1 mL 2.11 $KH_2 PO_4$ (0.154 $M$), 1 mL 3.82% $MgSO_4$·7 $H_2O$ (0.154 $M$), and 21 mL 1.3% $NaHCO_3$ (0.154 $M$). Once assembled, the KRB solution is gassed with 95% $O_2$ 5% $CO_2$ (3 L/min) for 20–30 min. Glucose is added at 3 mM (3 μmol/mL).
2. BSA (Sigma A-4503) is added to KRB solution at a concentration of 4% (40 mg/mL). After stirring the albumin in solution, pH of 7.4 is achieved by addition of 10 $N$ NaOH.
3. Collagenase from *Clostridium Histolyticum* (Worthington, no. 4197 C/SI) is added to KRB-BSA buffer at a concentration of 1.5 mg/mL to digest the tissue and isolate the adipocytes. Three to 5 g minced tissue are incubated in 20 mL buffer plus collagenase.
4. Dole's extraction solution: 40 mL isopropyl alcohol, 10 mL heptane, 1 mL 1 $N$ $H_2 SO_4$ *(5)*.
5. Methylene blue: Dissolve 20 mg in 1 mL water. Store at room temperature (RT).

## 2.1.3. Miscellaneous Supplies

Siliconized glass microscope slides, cover slips, silicone grease, plastic pipets, and centrifuge tubes.

## 2.2. Electronic Method

The electronic method involves more time and financial commitment than the optical method, and requires a Coulter counter (particle characterization) machine that is used to count fat cells, which have been incubated and fixed with $OsO_4$, washed, and suspended in normal saline solution. The literature that accompanies the machine provides detailed information about the theory behind electronic particle characterization, and Hirsch and Gallian (6) and Cartwright (7) are valuable reference resources. This method provides information on the total number of fat cells and size-frequency distribution in a given sample of AT, and, when used in conjunction with the Folch lipid extraction assay, can be used to determine mean fat cell size and volume. The main advantage of this method is that it allows collection of multiple AT samples per experiment, because determinations are not dependent on viability of the adipocyte preparation. Thus, analysis of a sample does not have to occur at the time of collection (*see* **Notes 1** and **2**).

### 2.2.1. Equipment

Coulter counter (Beckman Coulter; we use a Model ZM machine, but newer models are available); fume hood (very important); vacuum unit within the fume hood; 37°C water bath, with racks to hold scintillation vials.

### 2.2.2. Chemicals and Reagents

**CAUTION:** Some of the chemicals used are toxic and should be handled with appropriate safety precautions. Those include collidine buffer kit (collidine + HCl; Polysciences), and $OsO_4$ (crystalline form, sold in 1-g ampules; Stevens Metallurgical, New York, NY). Also needed are generous amounts of 0.9% (w/v) saline solution that have been filtered through a Whatman P8 paper filter; and corn oil to neutralize osmium liquid waste (*see* **Note 3** and **ref. 8**).

### 2.2.3. Glassware and Miscellaneous Supplies

Clear glass, heavy-duty stoppered flasks (preferably plastic-coated); amber storage bottles or flasks (variable size, dependent on desired volumes of buffer and $OsO_4$ solutions); graduated cylinders; 150-mL beaker; 100- and 200-mL volumetric flasks; 400-mL glass beakers (one per sample to be counted), which have been weighed and have the tare weight recorded; borosilicate glass scintillation vials; Pasteur pipets and bulbs; a rubber spatula (or a pipet bulb slipped over a rod small enough to fit into a scintillation vial); aluminum foil; stir plates and stir bars; indelible lab marker; 250-μm and 25-μm Nitex mesh (Sefar America) cut into 4 x 4 in. squares; Parafilm (Fisher Scientific); squirt bottles to be filled with either deionized water or 0.9% saline solution; wet-strengthened 18.5-cm pleated paper filter (Fisher Scientific, no. 113V).

## 2.2.4. Cell-Washing Apparatus (To Be Constructed)

1. 4-L vacuum flask and a rubber stopper with a hole in the middle.
2. 25-mL glass pipet with the tip cut. Insert the pipet through the hole in the stopper until 4–5 in. of the top of the pipet remains outside the stopper.
3. An ~14 in. piece of thick-walled tubing to fit over the top of the pipet.
4. An ~8 in. piece of thick-walled tubing to connect the flask to the vacuum nozzle or pump.
5. Washing and waste-capture unit made from two disposable vacuum filtration units (Nalgene), with the upper portions detached from the lower portions, and the membranes removed. The bottom piece of one of the units will serve as the cup to capture waste liquid and will have a small piece of tubing inserted that will remove the waste to the vacuum flask. The upper portion of the unit will snap onto the lower portion, to stabilize a piece of the 25-μm Nitex mesh. A piece of the 250-μm Nitex mesh is placed over the top, and is secured and made taut with the upper portion of the other unit. The tapered bottom must be cut to achieve a snug fit. The top cup can be held in place with a clamp on a ringstand.
6. Connect the small piece of tubing coming out of the waste cup to the tubing coming out of the top of the vacuum flask.

## 3. Methods

### 3.1. Optical Method

### 3.1.1. AT Preparation

1. Fragments of AT (usually 50–100 mg in weight) are removed from various sites (6–8) of a given adipose depot, and placed in 5 mL warm KRB medium with glucose and collagenase.
2. After 45–60 min of gentle shaking at 37°C in a Dubnoff metabolic shaker (60–80 strokes/min), the fat cells are isolated by three consecutive washings in KRB–4% BSA medium without collagenase. Letting the cells float by gravity or by gentle centrifugation (1 min at 1000$g$) allows for separation from the medium and easy removal of infranatant prior to adding fresh medium. After removal of the infranatant, and addition of fresh buffer, it is important to cap the vial and gently invert it two to three times to produce an even suspension.

### 3.1.2. Determination of Fat Cell Diameter

1. A 0.5-mL aliquot of the isolated fat cell suspension is added to 2–3 mL warm medium in a plastic scintillation vial, to which about 1 mg methylene blue (about one drop) is added. After staining at 37°C for 3–5 min, aliquots of the stirred suspended cells are removed with a plastic pipetter and placed on a siliconized glass slide in a silicone well and covered with a glass coverslip.
2. After bringing the cells into focus, the fat cells are recognized by their spherical shape, the stained nucleus, and cytoplasm (*see* **Note 4**). The cells are aligned on the caliper scale with systematic motion of the stage control knobs (*see* **Notes 5**

and **6**). The transverse diameter of 200–300 fat cells is measured and recorded in successive 7-μm multiples, to provide a frequency distribution of diameter in 9–15 categories of size *(2)*.

### 3.1.3. Calculation of Mean Diameter and Cell Volume (see **Note 7**)

Data for diameter are converted to mean diameter and standard deviation (SD). The heterogeneity of the cell population is expressed as coefficient of variation (CV = SD/mean). From the diameter data, the frequency distribution can also be calculated for fat cell surface area ($SA = \Pi D^2$) and cell volume ($[V = \Pi D^3]/6$), taking into consideration the nonlinear transformation *(2)*.

Cell diameter ($D$) is a normally distributed variable, but the volume is skewed. To calculate mean volume, Goldrick *(9)* has suggested a formula: $E$ (cell volume) = $(\Pi/\sigma) (3\sigma^2 + X^2)X$, where $X$ and $\sigma^2$ are, respectively, the mean and the variance.

### 3.1.4. Estimation of the Number of Fat Cells in Suspension

The number of fat cells in an isolated fat cell preparation can be estimated after performing three key steps: determination of the lipid content of the sample; determination of the mean cell volume, as described in **Subheading 3.1.3.**; and determination of the lipid density, which the authors have found to be 0.915 (*see* **Note 8**).

The number of fat cells per milliliter of fat cell suspension is calculated as:

$$\frac{\text{TG content of cell suspension/TG content of mean fat cell}}{(\text{mean cell volume} \times \text{density})}$$

The lipid content of the fat cell suspension is determined by the Dole's lipid extraction assay *(5)*, which can be summarized as follows: Add 0.5 mL fat cell suspension to 5 mL Dole's mixture in a 15-mL glass centrifuge tube. Shake well. Add 3 mL heptane and 3 mL dH$_2$O. Shake well, then centrifuge for 5 min in a desktop centrifuge at 1000–2000$g$. Measure upper (heptane) phase, usually 4.3 mL. Remove 2 mL upper phase into a pretared glass vial for gravimetric measurement of lipid, after overnight drying in a hood.

Calculate the amounts of TG in 0.5 mL fat cell suspension by multiplying the weighed TG × 4.3/2.

## 3.2. Electronic Method

### 3.2.1. Fixation of AT with OsO$_4$

**CAUTION:** These five steps should be performed under the fume hood.

1. Make stock collidine buffer (0.2 $M$) according to manufacturer directions, and store refrigerated in an amber bottle (stable at 4°C for several months).
2. Working collidine buffer (0.05 $M$) is made by adding one part of the stock solution to three parts of 0.9% NaCl.

3. Make a 2% solution of $OsO_4$. The ratio of $OnO_4$ to working buffer is 1 g:50 mL. One 1-g ampule osmium is needed per 10 fat samples to be fixed. Open the ampules by dropping them into the heavy-duty stoppered flask, and shaking the flask until the ampules break into small pieces and spill their contents. Add the appropriate amount of working buffer and a stir bar to the flask, and stir for 30–45 min. The flask should be shielded with aluminum foil, because collidine is light-sensitive. Use a Whatman 113V filter paper and a funnel to filter the solution into a dark storage bottle or flask.

4. Distribute 5 mL $OsO_4$ solution into glass scintillation vials (1/sample) cap and warm to 37°C in a water bath under the fume hood. Refrigerate and store unused solution.

5. Fix samples of freshly harvested AT in $OsO_4$. Mince the tissue, rinse it with warm saline to remove visible oil droplets, blot it and weigh out representative pieces (total of ~40–50 mg). Place the sample in one of the glass scintillation vials containing 5 mL 2% $OsO_4$ solution (cap vial tightly). Incubate the samples in the water bath under the fume hood for 72–96 h, remove from the water bath, and store in a refrigerator or cold room until they can be processed for counting (*see* **Note 9**).

An analogous fat sample (about 100 mg in duplicate or triplicate) is extracted using the Folch assay (*see* **Note 10** and **ref. 10**), so that mean fat cell size of the sample can be determined. Briefly, place each of the duplicate or triplicate samples to be processed in a disposable centrifuge tube containing 5 mL extraction cocktail (2:1 chloroform:methanol; **CAUTION:** this mixture is volatile, and should be dispensed under fume hood). Shake for 8 h. Squeeze the remaining tissue dry, and place it in an identically labeled centrifuge tube with 5 mL fresh extraction cocktail, and shake for 2 h. Remove the tissue fragments and dry in an oven to determine depot composition (*see* **Note 10**). Combine the solution from the original tube with that in the duplicate tube and add 2.4 mL 0.74% KCl. Vortex the tubes briefly, and spin them in a tabletop centrifuge for 10 min at top speed. Remove the aqueous phase (top layer). We recommend adding 1 mL deionized $H_2O$ to each tube, vortexing, and centrifuging briefly (up and down) before transferring the infranatant to a tared and labeled glass scintillation vial. Place vials under the fume hood and weigh them when dry. Percent lipid is calculated for each of the duplicate or triplicate samples as lipid weight/ sample wet weight. Use the average percent value for each set of duplicates or triplicates in the calculations below.

### 3.2.2. Processing of Fixed Samples for Electronic Counting

**CAUTION:** These four steps should be performed under the fume hood.

1. Prepare the filtration unit for sample processing as described in **Subheading 2.2.4.** Place the 4 × 4 in. square of 25-μm mesh on the bottom piece of the unit, mount flush the larger of the two cups on top of the mesh, place the 4 × 4 in. square of 250-μm mesh over the top of this cup, then place the smaller cup on top

of the mesh and secure it, so that the mesh is very taut. Place ~2 in. water in the bottom of the 4-L vacuum flask with ~1 in. of corn oil on top, before sealing flask with the rubber stopper. Turn on the vacuum system, and verify that it is working properly.

2. Pour the sample to be washed into the filtration unit. Gently swirl the vial to get the sample into suspension, and carefully pour the contents of the vial into the top of the filtration unit. Rinse the vial with saline to capture bits of sample left behind. Use the rubber spatula to swab the sides and bottom of the vial (be sure to rinse the pieces of sample that have stuck to the rubber spatula into the filtration unit), and continue rinsing until the entire sample has been transferred into the filtration unit.

3. Push the sample through the 250-μm mesh. Use the rubber spatula to gently push the sample through the mesh, rinse with deionized $H_2O$, and continue this process until as much of the sample has been pushed through as possible. When the water looks clear rather than gray, this has been accomplished.

4. Rinse the fixed cells with deionized $H_2O$ and saline. Remove the top cup from the filtration unit, and dispose of the 250-μm mesh. Then, using squirt bottles, gently squirt deionized $H_2O$, then saline at the cells on the 25-μm mesh. It is critical that the last rinse be with saline.

Carefully remove the bottom detachable cup from the waste cup, while supporting the mesh with your fingers, and check the lower rim for stray cells: These should be rinsed with saline onto the mesh. Place the mesh against the side of the beaker (cell side up), and gently rinse the cells off the mesh with saline. Bring the volume of saline in the beaker up to ~50–75 mL, and cover with Parafilm. Place beakers with washed cells in refrigerator or cold room for storage until they can be counted.

### 3.2.3. Electronic Counting of Washed Samples

1. Prepare washed sample for counting. Remove sample(s) to be counted from refrigerated storage, and leave at RT until counting is performed (*see* **Note 11**). For each sample to be counted, the tare weight of the beaker should be recorded, and the beaker should be filled with 0.9% saline, to just below the lip, and weighed. Both the tare weight and the filled weight are important for subsequent calculations.

2. Prepare Coulter counter for sample counting. We routinely use a 450-μm aperture to count fixed adipocytes from several rodent models. The machine should be calibrated accordingly, following manufacturer's instructions. The manometer volume setting should be 2000 μL for this aperture. We set the Full Scale to 1, the alarm to "Off," Preset Gain to 1, and the polarity to auto. The other settings are varied according to the measurements desired. Several blank counts should be taken with saline alone, to assure that the aperture is clean, that there is no electrical noise (*see* **Note 12**), and that the machine is functioning properly.

3. Count the samples. We use an older model ZM machine that allows analysis of the total number of cells per sample, as well as a frequency distribution of cell diameter by performing multiple counts with manually size-adjusted windows (*see* **Note 13**). The new models determine both the total and frequency distribution counts at one time.

### 3.2.4. Determination of Total Number of Fat Cells per Sample

To count the total number of cells in the sample, the lower threshold should be no smaller than the diameter of the lower mesh onto which the cells were captured during washing (25-μm), and the upper threshold should be wide open. On our machine, the settings for total count are: current = 700; attenuation = 2; lower threshold = 20.9; and upper threshold = 99.9. At these settings, the machine counts all cells with a diameter of at least 25.02-μm. Turn on the stirrer, and make sure that the suspension is homogeneous before counting. At minimum, triplicate counts should be taken, and the counts should have a CV of less than 5%. Determine the average count. To calculate the total number of cells in the sample, it is necessary to know the volume of saline in which the cells were suspended. Thus, subtract the tare weight of the beaker from its filled weight. Total cells in the sample = (average count/2) × volume of saline. To extrapolate to the total number of cells in a fat depot, multiply the total counts by the weight (in mg) of the depot and divide by the weight (in mg) of the sample that was fixed in $OsO_4$.

### 3.2.5. Determination of Size Frequency Distribution

To determine the frequency distribution of cell diameter in the sample, the lower and upper threshold settings and the attenuation setting must be varied to achieve the desired window sizes. The authors use 10 window settings that give a size range, in increments, from 25–240 μm (*see* **Note 13**).

### 3.2.6. Determination of Mean Fat Cell Volume

Determine the total micrograms of lipid in the sample counted:

$$\text{Sample weight (in mg)} \times \text{Percent lipid (Folch assay)} \times 1000$$

To determine micrograms of lipid per cell, divide total lipid by total number of cells in sample (*see* **Subheading 3.2.4.**). To convert to ng of lipid per cell, multiply by 1000. To convert ng lipid to picoliters, divide by 0.915 (the density of TG).

## 4. Notes

1. Both methods described here (the optical method and the electronic method) are widely used. In a few instances, the two methods have been compared, and have given similar results (*4,11*), although basic differences exist. It needs to be emphasized that the optical method provides a direct observation of the cells and the measurement of the diameter of a representative sample of the isolated adipocyte preparation. This offers accurate information not only on the mean cell size, but also on other characteristics of the cell population, such as heterogeneity in size and frequency distribution. In this method, the estimate of cell number is derived from knowledge of lipid content and conversion of cell volume to mean cell weight.

In the electronic method, the estimate of fat cell number in a given fragment of AT is directly assessed by the Coulter counter; mean cell volume is derived from knowledge of lipid content and estimate of cell number. It needs to be further emphasized that a precise assessment of the cell number in a fragment of tissue is still not possible, because of the inability of either method to measure all the cells present in the tissue. The optical method usually includes cells with a diameter above 11–15 μm; the electronic method excludes cells smaller than 25 μm in diameter.

2. A simplified electronic determination of size and number of fat cells has been described *(11)* in an isolated, unfixed adipocyte population, but this method is only limited to small cells (<60 μm in diameter), because it loses reliability for cells larger than 60 μm.

3. A 2% solution of $OsO_4$ can be neutralized by twice the volume of oil, according to Cooper *(8)*. Corn oil is usually used, because of the high percentage of unsaturated bonds. The authors always keep a good supply of corn oil near the fume hood, in case of an accidental osmium spill, as well as to neutralize the liquid waste generated during the processing of osmium-fixed samples. The waste can be fully neutralized in our 4-L capture flask by adding 2 in. of water to the flask, with an inch of corn oil on top of the water prior to capturing the used osmium. Because much water and saline are used to wash the fixed samples, there is a great dilution of the original 2% $OsO_4$ solution. When the flask is full, the vacuum tubing is removed and sealed with a solid rubber stopper. The waste is fully neutralized after 36–48 h. **CAUTION:** Test for neutralization under the fume hood by suspending a piece of filter paper that has been soaked in corn oil over the solution. If the solution is completely neutralized, there should be no blackening of the filter paper. The authors' chemical safety department retrieves and disposes of the neutralized osmium waste. All lab personnel preparing or using $OsO_4$ or processing fixed samples should use appropriate safety precautions, and have a plan to deal with an accidental spill. We cannot stress strongly enough that $OsO_4$ in its crystalline form, or in 2% solution, should be handled only under a fume hood.

4. For an inexperienced observer, at times, the microscope field shows spherical lipid droplets that may be confused with very small fat cells. Staining of cytoplasm and nucleus, and systematic focusing on the horizon of the fat cell at its maximal diameter, is frequently sufficient to avoid measurement error. Occasionally, bilocular cells are seen; they are noted, but usually do not exceed 1–2% of the population.

5. Usually two silicon wells are filled with the stained isolated fat cell suspension. After placing the glass cover and positioning the microscope slide on the stage, the fat cell layer is brought into focus. A systematic sweep (down, then lateral, then up) brings the cells in alignment with the caliper scale, and the transverse diameter is read for each individual cell focused in its horizontal plane. About one-half of the cells (100–150) are measured in the first well, and the rest in the second well.

6. In our experience, class intervals of 7 or 14 μm (at a higher magnification) have given suitable intervals for a frequency distribution. Depending on the microscope, the lenses, and the caliper available, class intervals of 5, 6, or 8 are equally satisfactory, and the calculations should reflect class interval, mid-class diameter, and frequency distribution.

7, We have developed a computer program that gives final calculation of mean and SD for fat cell diameter and volume in a few seconds, after logging in class size and frequency distribution. This is based on computing mean cell diameter by multiplying each class by the frequency, and dividing cumulative diameter by the number of cells measured. The mean volume is calculated by the Goldrick formula already described. The reader is referred to the original description of the optical method *(2)* and subsequent papers *(12,13)*.

8. More recently, this laboratory has published a simple and reliable method to estimate cell density and promote cell incubations at preordained cell concentration. The reader is referred to this publication for details *(14)*.

9. The traditional protocol instructs that 5 mL 2% $OsO_4$ be used to fix AT samples weighing 40–60 mg *(7)*. We have used 3 mL the solution to fix samples in the 40–60 mg range, and find no difference in the final result, if they are incubated at 37°C for 72–96 h. This modification will not only save money, but will reduce the amount of liquid osmium waste that is disposed of in toxic waste dumps.

10. Knowledge of the relative and absolute tissue content of lipid, water, and defatted dry residue (DDR) (a measure of protein content) is often useful. The Folch assay provides absolute values of sample lipid content, which are converted to percentage values for calculation of mean cell size. The absolute DDR can be calculated by drying and weighing the tissue fragments removed from the duplicate centrifuge tubes. The tissue water is determined by subtracting the absolute weights of the lipid and DDR from the original wet weight of the sample. The proportion that each constituent contributed to the wet weight also can be determined.

11. We have found that the most consistent counts are achieved when the samples are at RT. This is probably the result of temperature-specific differences in electrolyte conductance.

12. The aperture is cleaned and blank counts are taken before each new sample is counted. This reduces the incidence of erroneous counts that can occur because of interference by cells from a previous sample (shadowing effect; *see* machine manual for details), and can alert the user to electrical noise that will prevent attainment of correct counts. Optimally, the Coulter counter should be isolated from other electrical equipment in the lab, especially centrifuges, homogenizers, and so on. It is futile to count samples when electrical interference is present.

13. We use the following 10 size ranges (in μm) to determine a size frequency distribution for each sample: 25–30, 31–40, 41–50, 51–60, 61–70, 71–80, 81–100, 101–140, 141–180, and 181–240. Remember that there is a limited volume of saline in the beaker, and that the volume decreases with each count, so the number of size ranges must be limited. When the volume in the beaker drops by ~2/3, the stir paddle will generate bubbles that cannot be distinguished from cells by the machine. Hint: it is extremely important to remember to weigh the filled beaker just prior to counting the sample, because the volume decreases by more than the 2000 μL that is used to determine each count. Therefore, it is impossible to reconstruct this measure after the fact.

# References

1. Gurr, M. L. and Kirtland, J. (1978) Adipose tissue cellularity: a review. 1. Techniques for studying cellularity. *Int. J. Obes.* **2,** 401–427.
2. DiGirolamo, M., Mendlinger, S., and Fertig, J. W. (1971) A simple method to determine fat cell size and number in four mammalian species. *Am. J. Physiol.* **221,** 850–858.
3. Cushman, S. W. and Salans, L. B. (1978) Determination of adipose cell size and number in suspensions of isolated rat and human adipose cells. *J. Lipid Res.* **19,** 269–273.
4. Smith, U., Sjöström, L., and Björntorp, P. (1972) Comparison of two methods for determining human adipose cell size. *J. Lipid Res.* **13,** 822–824.
5. Dole, V. P. (1955) A relation between non-esterified fatty acids in plasma and the metabolism of glucose. *J. Clin. Invest.* **35,** 150–154.
6. Hirsch, J. and Gallian, E. (1968) Methods for the determination of adipose cell size in man and animals. *J. Lipid Res.* **9,** 110–119.
7. Cartwright, A. L. (1987) Determination of adipose tissue cellularity, in *Biology of the Adipocyte: Research Approaches.* (Hausman, G. H. and Martin, R. J., eds.), Van Nostrand Reinhold, pp. 229–254.
8. Cooper, K. (1988) Neutralization of osmium tetroxide in case of spillage and for disposal. *Bull. Microscop. Soc. Canada* **8,** 24–28.
9. Goldrick, R. B. (1967) Morphological changes in the adipocyte during fat deposition and mobilization. *Am. J. Physiol.* 212, 777–782.
10. Folch, J., Lees, M., and Stanley, G. H. S. (1957) A simple method for the isolation and purification of total lipids from animal tissues. *J. Biol. Chem.* **226,** 497–509.
11. Maroni, B. J., Haesemeyer, R., Wilson, L. K., and DiGirolamo, M. (1990) Electronic determination of size and number in isolated unfixed adipocyte populations. *J. Lipid Res.* **31,** 1703–1709.
12. DiGirolamo, M. and Mendlinger, S. (1971) Role of fat cell size and number in enlargement of epididymal fat pads in three species. *Am. J. Physiol.* **221,** 859–864.
13. DiGirolamo, M. (1972) Fat cell size, metabolic capacities and hormonal responsiveness of adipose tissue in spontaneous obesity in the rat. *Israel Jr. Med. Sci.* **8,** 807.
14. Fine, J. B. and DiGirolamo, M. (1997) A simple method to predict cellular density in adipocyte metabolic incubations. *Int. J. Obes.* **21,** 764–768.

# 5

# Subcellular Fractionation of Adipocytes and 3T3-L1 Cells

## Hans-Georg Joost and Annette Schürmann

## 1. Introduction

For more than 30 yr, adipocytes have been an almost ideal model for the study of insulin action, because they can be prepared easily as a homogenous population of cells, they exhibit a marked hormonal response, and they can be fractionated into relatively pure membrane preparations. This subcellular fractionation has provided the breakthrough finding that insulin stimulates a translocation of glucose transporters (GLUTs) from an intracellular pool into the plasma membrane *(1)*. A second model for the study of adipose tissue (AT) metabolism is the 3T3-L1 cell line, which differentiates to an adipocyte-like phenotype. In these cultured cells, long-term effects of hormones and agents can be investigated. As in fat cells, the effects of insulin on the subcellular distribution of GLUTs can be demonstrated in 3T3-L1 cells after fractionation *(2)*. In addition, 3T3-L1 cells allow the preparation of membrane sheets still adhering to the culture dish, e.g., for immunofluorescence *(3)*.

Yield and purity of fat cell membrane fractions depend in large part on the quality of the fat cell preparation. Fat cells are prepared in a simple and straightforward procedure by digestion of AT with collagenase, and are separated from stromal cells by centrifugation. However, it is not a trivial task to prepare cells responding well to insulin (for an earlier description of the method, *see* **ref. 4**). The most important variables are the animals (strain, age, housing conditions), the plastic materials, and the batches of collagenase and albumin used in the procedure. All materials must be tested according to the criteria of insulin responsiveness and fragility of cells.

From: *Methods in Molecular Biology, vol. 155: Adipose Tissue Protocols*
Edited by: G. Ailhaud © Humana Press Inc., Totowa, NJ

For the fractionation of adipocytes, the quality of the cells and the conditions of the homogenization are the major (and possibly only) determinants of yield and purity of the membrane fractions, as judged from the response to insulin and from marker enzymes. Variables are the type of grinder, its clearance, and the homogenization temperature. In our experience, best results are obtained with a Potter-Elvehjem homogenizer (Thomas Scientific, Philadelphia, PA) with a teflon pestle and maximal clearance.

The fractionation of 3T3-L1 cells follows a protocol that is similar to that of the fractionation of fat cells. In fact, the centrifugation steps are essentially identical. However, the homogenization conditions must be much gentler (use of hand strokes, instead of a motor-driven grinder). A considerable crosscontamination is inevitable, even under the best conditions.

## 2. Materials

### 2.1. Preparation of Fat Cells

1. Male Wistar rats (160–180 g).
2. Narrow-mouth bottles, 30 mL (Nalgene, cat. no. 2003-0001).
3. Nylon mesh (pore size 300 mesh).
4. 100 mL polypropylene urine sample containers (Sarstedt, cat. no. 563).
5. Collagenase type I (Boehringer, Indianapolis, IN).
6. Bovine serum albumin (BSA), fraction V (Sigma, St. Louis, MO).
7. Crystalline pork zinc insulin (Sigma).
8. Krebs Ringer HEPES (KRH) buffer: Stock solution I (g/L): NaCl, 70.08; $KH_2PO_4$, 5.46; $MgSO_4x7H_2O$, 2.46; $CaCl_2$, 1.1. Stock solution II (g/L): $NaHCO_3$, 8.4. Stock solution III (g/L): HEPES, 71.5. The stocks are stored at 4°C for up to 3 mo. 30 mL of each stock solution are mixed, and *aqua bidest.* is added to a final volume of 300 mL. D-glucose (final concentration 2.5 m$M$), adenosine (200 n$M$; stock solution 200 m$M$, store at –20°C) and albumin (12 g) are added, and the buffer is vigorously stirred until the albumin is dissolved. Prewarm the buffer to 37°C and adjust the pH to 7.4.

### 2.2. Fractionation of Adipocytes and 3T3-L1 Cells

1. Potter-Elvehjem grinder 60 mL, specific clearance 0.15 mm (Thomas Scientific, cat. no. 34310-E25).
2. Tris-EDTA-sucrose (TES) buffer: 20 m$M$ Tris-HCL, 1 m$M$ EDTA, 8.7% sucrose; pH 7.4, at 4°C.
3. Sucrose cushion: 38.5% (1.12 $M$) sucrose, 20 m$M$ Tris-HCL, 1 m$M$ EDTA, pH 7.4, at 4°C.

For 3T3-L1 cells, the same materials as in the fractionation of fat cells are used, except that phenylmethylsulfonic fluoride (0.2 m$M$) is added to the TES.

# 3. Methods

## 3.1. Preparation of Fat Cells

1. Dissect epididymal fat pads from 10–12 male Wistar rats. Carefully remove blood vessels and epididymal tissue, and transfer the fat pads to 2 narrow-mouth Nalgene bottles (approx 6 g tissue/bottle), which contain 5.5 mL prewarmed KRH-buffer (*see* **Note 1**).
2. Cut the fat pads manually for 1 min with sharp dissecting scissors, and add 15 mg collagenase in 0.5 mL buffer. Incubate in a shaking water bath with rapid shaking (120 cycles/min; stroke length, 5 cm) for 60 min at 37°C (*see* **Notes 2** and **3**).
3. Fill bottle with KRH buffer and squeeze cells gently through nylon screen (pore size 300 mesh), which is attached to the mouth of the bottle with a rubber band. Centrifuge cells at 400*g* and room temperature for 1 min, withdraw infranatant, and resuspend with 30 mL fresh buffer (37°C). Repeat centrifugation and washing four times. Resuspend cells in 30 mL KRH (approx $2.5 \times 10^6$ cells/mL), and incubate in polypropylene containers under the desired conditions (e.g., with or without insulin) for 30 min at 37°C with moderate shaking (40 cycles/min) (*see* **Note 4**).

## 3.2. Fractionation of Adipocytes

1. Wash isolated cells (approx $7 \times 10^7$ cells or 5 mL packed cells are isolated from 10 rats) once with 20 mL TES buffer (18°C), and resuspend cells in 28 mL TES buffer (*see* **Note 8**).
2. Homogenize adipocytes in the precooled (4°C) motor-driven Potter-Elvehjem grinder with 10 strokes at 1400 rpm (*see* **Note 9**).
3. Centrifuge homogenized cells at 12,000 rpm in a fixed angle JA20 rotor (Beckman) for 1 min at room temperature, then for 15 min at –3°C.
4. Remove and discard the solidified fat. Remove and spin the supernatant at 20,000 rpm (JA20 rotor) for 30 min at 4°C. The pellet contains the high-density microsomes. Resuspend the pellet, and centrifuge again at 20,000 rpm.
5. Centrifuge the supernatant for 75 min at 200,000*g* (4°C) to pellet the low-density microsomes. Resuspend the pellet, and centrifuge again at 200,000*g*.
6. The supernatant of the first 200,000*g* spin contains the cytosol, and can be concentrated by centrifugation in Centricon tubes.
7. The pellet of the first centrifugation (12,000 rpm, **step 3**) is resuspended in TES buffer (2 mL Potter; 20 strokes at 1400 rpm), loaded on the sucrose cushion (38.5%), and centrifuged for 60 min at 100,000*g* (4°C) in a swing-out rotor (e.g., SW27, Beckman). The resulting pellet contains nuclei and mitochondria. Plasma membranes are collected from the top of the sucrose cushion, resuspended in TES buffer, and repelleted by centrifugation at 31,000*g* (rotor JA20, 4°C) for 60 min (*see* **Note 10**).
8. Resuspend and homogenize all pellets (**steps 4, 5,** and **7**) in 0.4–1 mL TES in a 2-mL Potter with 10–20 strokes at 1400 rpm (*see* **Notes 11–13**).

## 3.3. Fractionation of 3T3-L1 Cells

1. Wash differentiated 3T3-L1 adipocytes (six culture dishes, diameter 8.5 cm) twice with TES (4°C), and scrape cells off the carrier with a rubber policeman.

Homogenize cells in TES buffer in a total volume of 2.3 mL (390 µL/dish) in a 30-mL Potter-Elvehjem grinder with 15 handstrokes (see Notes **14** and **15**).

2. Centrifuge the homogenized cells at 8000*g* in a JA20 rotor (Beckman) for 15 min.
3. All subsequent centrifugations are identical to the steps performed in the fat cell fractionation (*see* **Notes 16** and **17**).

## 4. Notes

1. Removal of blood vessels and epididymal tissues from the fat pad is essential for the quality of the fat cells; these seem to contain factors lysing the adipocytes.
2. Amplitude and frequency of shaking of the water bath is very important in the collagenase digestion, and should be optimized by testing. During incubation with insulin, slow shaking may cause the cells to float to the surface. When cells are incubated with lipolytic agents, cells must be shaken at a higher rate, in order to avoid intracellular accumulation of fatty acids.
3. For each batch of collagenase, the digestion conditions (concentration, time) have to be optimized. Thereafter, keep the ratio of collagenase per tissue weight constant. In each fat cell preparation, the digestion time can be adjusted (± 5 min) after inspection of the flask, shortly before the end of the routine digestion time.
4. It is important to work fast during washing of cells, because these washes remove lysing activities. Try to keep cells always at 37°C, because cooling causes loss of specific functions, e.g., hormone-stimulated lipolysis.
5. An optimal fat cell preparation exhibits very low basal glucose transport activity, and has virtually no lipids on its surface.
6. The method is designed for a routine preparation of cells/membranes which will yield appro $7 \times 10^7$ cells from 10–12 rats. The method can be scaled-up to accomodate a maximum of 96 rats (4–6 persons are needed for dissecting and mincing of the tissue), and can also be scaled-down to prepare fat cells and membranes from 1–2 rats.
7. The conditions cannot be transferred to the preparation of fat cells from other species (e.g., mouse); all parameters have to be optimized accordingly.
8. Keep all conditions of the homogenization, particularly temperature, volume of cells per homogenization, and motor speed, highly reproducible. Cooling of cells below 15°C produces an increase in basal glucose transport. Also, cooling of cells increases the viscosity of the homogenate, and thereby tightens the grinder.
9. Grinders should be tested before use, and the same grinder should be used in each series of experiments. A *prima facie* criterion for the quality of the preparation is the yield of plasma membranes and the transporter abundance in basal membranes as compared with membranes from insulin-stimulated cells (*see* **Fig. 1**); grinders that are too tight give rise to low yields and increase basal GLUT4 content.
10. The pellet under the 38.5% sucrose cushion consists of mitochondria and nuclei, which can be further separated by centrifugation on a 50% sucrose cushion.
11. The purity of the membranes can be assessed with marker proteins, e.g., Ras (plasma membrane marker), GLUT4 (for low-density microsomes of basal preparations), cytochrome-C oxidase (for endoplasmic reticulum [ER]) and several others, as described previously (**5**).

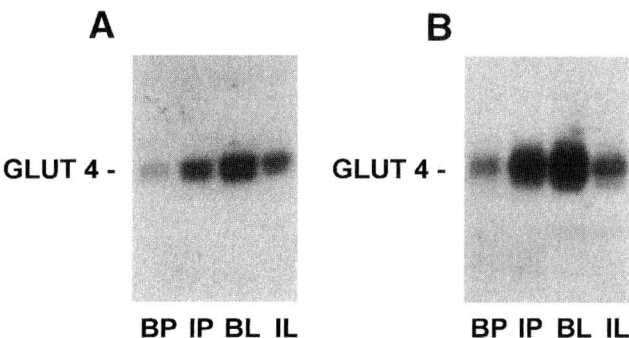

Fig. 1. Characterization of adipocyte membrane fractions with the effect of insulin on the subcellular distribution of the glucose transporter, GLUT4. **(A)** Membranes from rat adipocytes. **(B)** Membranes from differentiated 3T3-L1 cells. BP, basal plasma membranes; IP, plasma membranes from insulin-treated cells; BL, basal low-density microsomes; IL, low-density microsomes from insulin-treated cells.

12. The yield of membrane proteins per rat is approx 0.25 mg plasma membranes, 0.17 mg low-density microsomes, and 0.13 mg high-density microsomes.
13. The plasma membrane fraction consists of sealed vesicles that can be used for transport assays *(6)*.
14. Culture and differentiation of 3T3-L1 cells has been described in detail before in many other publications. A clone from American Type Culture Collection or from an experienced laboratory, should be used, and not more than six passages should be used without further subcloning. Another important parameter that determines the quality of the differentiation is the batch of fetal calf serum that must be tested.
15. Crosscontamination of the fraction appears to increase considerably with the vigor of the homogenization. Thus, the pestle must not be driven by a motor. In order to increase grinder clearance, the grinder vessel is equilibrated on ice; the pestle is kept at room temperature.
16. In contrast to plasma membranes from adipocytes, the membranes isolated from 3T3-L1 cells do not contain sealed vesicles. Therefore, a reconstitution method *(7)* has to be employed for assay of transport activity.
17. Pure plasma membranes can be obtained with the so-called "sheet assay" *(3)*. Culture dishes are sonicated in order to break up the cells, and to remove cytosol and intracellular organelles. Portions of the plasma membranes adhere to the culture dish, and can be used for immunofluorescence (e.g., with anti-GLUT4 antiserum), or can be isolated. The disadvantage of this method is that all intracellular organelles are lost, and that the yield of plasma membrane protein is low.

## References

1. Cushman, S. W. and Wardzala, L. J. (1980) Potential mechanism of insulin action on glucose transport in the isolated rat adipose cell. Apparent translocation of intracellular transport system to the plasma membrane. *J. Biol. Chem.* **255,** 4758–4762.

2. Weiland, M., Schürmann, A., Schmidt, W., and Joost, H.-G. (1990) Development of the insulin-sensitive glucose transport in murine 3T3-L1 adipocytes: Role of the two glucose transporter species and their subcellular localization. *Biochemical J.* **270,** 331–336.

3. Olson, A. L., Knight, J. B., and Pessin, J. E. (1997) Syntaxin 4, VAMP2, and/or VAMP3/cellubrevin are the functional target membrane and vesicle SNAP receptors for insulin-stimulated GLUT4 translocation in adipocytes. *Mol. Cell. Biol.* **17,** 2425–2435.

4. Weber, T. M., Joost, H.-G., Simpson, I. A., and Cushman, S. W. (1988) Methods for assessment of glucose transport activity and the number of glucose transporters in isolated rat adipose cells and membrane fractions, in *Receptor Biochemistry and Methodology,* vol. 12B: *Insulin Receptors. Biological Responses, and Comparison to the IGF-I Receptor.* (Kahn, C. R. and Harrison, L., eds.), A. R. Liss, New York, pp. 171–187.

5. Simpson, I. A., Yver, D. R., Hissin, P. J., Wardzala, L. J., Karnieli, E., Salans, L. B., and Cushman, S. W. (1983) Insulin-stimulated translocation of glucose transporters in the isolated rat adipose cell: Characterization of subcellular fractions. *Biochim. Biophys. Acta* **763,** 393–407.

6. Joost, H.-G., Weber, T. M., and Cushman, S. W. (1988) Qualitative and quantitative comparison of glucose transport activity and glucose transporter concentration in plasma membranes from basal and insulin-stimulated rat adipose cells. *Biochem. J.* **249,** 155–161.

7. Schürmann, A., Doege, H., Ohnimus, H., Monser, V., Buchs, A., and Joost, H.-G. (1997) Role of conserved arginine and glutamate residues on the cytosolic surface of glucose transporters (GLUT) for transporter function. *Biochemistry* **36,** 12,897–12,902.

# 6

## Quantification of Lipid-Related mRNAs by Reverse Transcription-Competitive Polymerase Chain Reaction in Human White Adipose Tissue Biopsies

### Hubert Vidal

## 1. Introduction

Investigation of the in vivo regulation of gene expression in human subcutaneous white adipose tissue (WAT) relies on the ability to estimate the changes in specific mRNA levels in biopsies taken before and after a designed intervention (i.e., diet, exercise, hyperinsulinemic clamp, and so on). Such study is limited by the size of the samples that could be taken, and by the high lipid content of adipose tissue (AT), which renders the isolation of total RNA difficult. However, it is possible to obtain highly pure RNA preparations from low amounts of AT, using commercially available kits that are based on the selective binding of RNA molecules on silica-gel supports. Nevertheless, the low yield in total RNA with fat tissue requires a highly sensitive method to quantify specific mRNA molecules, in order to estimate the expression levels of the genes of interest. An adequate and powerful method is the quantification of mRNAs by reverse transcription followed by competitive polymerase chain reaction (RT-cPCR), which relies on the addition of a known amount of an exogenous DNA molecule (called competitor) to the amplification mixture, after the reverse transcription step *(1–3)*. In this method, however, the efficiency of the reverse transcription reaction is not controlled. We have therefore determined experimental conditions that allow 100% efficiency of cDNA synthesis during the reverse transcription step *(4)*. This is possible when using a specific antisense primer (*see* **Note 1**) and a thermostable reverse transcriptase to perform the reaction at elevated temperature (*see* **Note 2**).

From: *Methods in Molecular Biology, vol. 155: Adipose Tissue Protocols*
Edited by: G. Ailhaud © Humana Press Inc., Totowa, NJ

During a competitive amplification, the ratio between the target and the competitor molecules remains constant during all of the reaction. Therefore, its quantification at the end of the reaction allows determination of the initial amount of target, since the initial amount of competitor is known *(4,5)*. In addition, because this ratio is maintained even after the exponential phase of the PCR *(5)*, the amplification can be continued for 30, 35, or more cycles. This allows one to obtain sufficient amounts of PCR products that can be easily analyzed, even on an agarose gel stained with ethidium bromide *(3,4)*.

The crucial issue of the RT-cPCR method is the design of the competitor DNA molecule. Because one of the determining parameters of PCR efficiency is the hybridization of the primers to the templates, the target and competitor must be amplified using the same set of sense and antisense primers. In addition, the target and the competitor should generate distinguishable PCR products, to allow the determination of their ratio at the end of the reaction. Different types of competitors have therefore been developed, generally by modifying the cDNA sequence of the target by either a small deletion or addition of an unrelated sequence, or by modification of a restriction site. In past years, we have set up and validated RT-cPCR assays for the quantification of several different mRNAs that are expressed in human AT (*see* **Note 3**).

## 2. Materials
### 2.1. Total RNA Extraction
1. Silica-based commercial kit for the preparation of total RNA from animal tissues (i.e., RNeasy® mini kit from Qiagen).
2. Mortar and pestle.
3. Liquid nitrogen.
4. Microcentrifuge.

### 2.2. RT-Competitive PCR
1. Thermal-cycler.
2. Thermostable reverse transcriptase (*Tth* DNA polymerase 5 U/µL; Promega), delivered together with 10X reverse transcriptase buffer, 25 m$M$ MnCl$_2$, 25 m$M$ MgCl$_2$, and 10X chelate buffer.
3. *Taq* DNA polymerase (5 U/µL; Life-Technologies).
4. 10 m$M$ deoxynucleoside triphosphates (dNTP; Life-Technologies).
5. Antisense and Cy-5-labeled (5') sense oligonucleotides (Eurogentec).
7. RNase-free water.
8. 10-, 100-, and 200-µL filter tips (ART).

### 2.3. Analysis of the PCR Products
1. ALFexpress™ DNA sequencer (Pharmacia).
2. ALFwin fragment analyzer software (Pharmacia).

3. Ready Mix gel ALF-grade.
4. 10X TBE buffer: 121 g/L Tris, 51 g/L boric acid, 3.75 g/L EDTA.
5. Loading buffer: 0.3 g blue dextran in 100 mL formamide.

## 3. Methods

### 3.1. Total RNA Extraction

#### 3.1.1. Preparation of Tissue Sample

The low stability of the mRNA molecules in crude tissue requires that the WAT biopsies are taken under the least-traumatic conditions, and are rapidly frozen in liquid nitrogen (LN). Needle (15-gage) aspiration under local anesthesia (1% lidocain without adrenaline) allows to take about 300–500 mg fat tissue at the level of umbilicus. The tissue is removed in less than 5 min, rapidly washed with 10 mL ice-cold saline buffer, set in an aluminium paper sheet, and immediately frozen. The tissue can be stored several months at –80°C.

When used, the tissue is first ground in a mortar that is kept in LN. For total RNA preparation, about 150 mg frozen powder is sampled in a 1.5-mL Eppendorf tube (*see* **Note 4**). The remaining pulverized tissue is stored at –80°C until further use.

#### 3.1.2. RNA Extraction with the RNeasy Minikit

Total RNA is prepared according to the instructions of the manufacturer (*see* **Note 5**), with the addition of an extra extraction step to increase the yield in RNA:

1. Homogenization of the powdered tissue in 600 μL of the lysis buffer, which is supplied with the kit.
2. Centrifuged at room temperature for 3 min (8000*g*).
3. Remove 500 μL of the infranatant (between the fat cake and the pellet) and store it in a clean tube at 4°C.
4. Submit the remaining tissue to a second round of extraction with 350 μLof lysis buffer, vortexed for 20 s, and centrifuged again (3 min, 8000*g*).
5. Remove 300 μL of the infranatant, and combine it with the first 500 μL.
6. Add 800 μL of 70% EtOH to the pooled solutions and load (no more than 700 μL at a time) the RNeasy mini spin column.
7. Centrifuge the column for 15 s (10,000*g*) at room temperature. The flow-through is discarded, and the remaining lysis solution is then applied on the same column.
8. When all the solution is loaded, the column is washed three times and the RNA is eluted in 40 μL (2 × 20 μL) of RNase-free water, according to the instructions of the manufacturer (*see* **Note 6**).

The whole procedure takes about 2 h to prepare RNA from six tissue samples simultaneously.

### 3.1.3. Quantification of the Total RNA Preparation

Total RNA is quantified spectrophotometrically using a mini-quartz cell (volume of 70 µL; 1-cm path length). The samples are generally diluted 20-fold (4 µL in 80 µL water) before quantification. Using the procedure described above, the absorbance ratio (260/280 nm) is between 1.9 and 2.0, and the yield in total RNA is $2.1 \pm 0.9$ µg/100 mg of AT ($n = 42$). The total RNA can be stored at $-80°C$ until use (*see* **Note 7**).

## 3.2. Reverse Transcription

Specific first-strand cDNA is synthetized from 0.1 or 0.2 µg total RNA with 2.5 U thermostable reverse transcriptase (*Tth* DNA polymerase, Promega) in 1X reverse transcriptase buffer (10 m$M$ Tris-HCl, pH 8.3, 90 m$M$ KCl), 1 m$M$ $MnCl_2$, 0.2 m$M$ dNTPs, and 15 pmol of the specific antisense primer, in a final volume of 20 µL. The medium is overlaid with mineral oil, and incubated for 3 min at 60°C, followed by 15 min at 70°C, in the thermocycler. The reaction is stopped by heating for 5 min at 99°C. After chilling on ice, 4 µL water are added to the reverse transcriptase medium from which 20 µL are sampled for cDNA quantification by competitive PCR.

## 3.3. Competitive PCR

The 20 µL reverse transcriptase medium are added to a PCR master mix of 1X chelate buffer (10 m$M$ Tris-HCl, pH 8.3, 100 m$M$ KCl, 0.75 m$M$ EGTA, 5% glycerol) containing 0.2 m$M$ dNTPs, 5 U*Taq* polymerase (Life-Technologies), 45 pmol of the specific sense primer, and 30 pmol of the antisense primer. The final volume is 100 µL. Then four aliquots of 20 µL each are transferred to 0.5-mL tubes, each containing 5 µL of a defined working solution of the specific DNA competitor. The tubes are closed and transferred to a preheated thermocycler (94°C). After 120 s at 94°C, the PCR mixtures are subjected to 30–40 cycles of PCR amplification with a cycle profile including denaturation for 40 s at 94°C, hybridization for 40 s at 58°C and elongation for 60 s at 72°C. The sense primer is 5'-labeled with a fluorescent probe (Cy-5), to allow the analysis of the PCR products with an automated laser fluorescence DNA sequencer (ALFexpress, Pharmacia).

## 3.4. Analysis of PCR Products

At the end of the PCR, the medium is diluted (20- to 40-fold) in the loading buffer and the PCR products are separated by electrophoresis, using a 4% denaturing polyacrylamide gel on the DNA sequencer (*see* **Note 8**). The areas under curve of the competitor and target picks are analyzed using the ALFwin fragment analyzer software. To determine the initial concentration of the target cDNA, the logarithm of the peak surface ratio (competitor/target) is plotted vs the logarithm of the amount of competitor added into the PCR medium. At the competition equivalence point, the initial concentration of the target corresponds to the initial concentration of competitor (*see* **Note 9**). It is important to perform different amplification reactions with amounts of competitor flanking the initial amount of target cDNA to determine this equivalence point accurately (*see* **Note 10**). Taking into account the different dilution factors and the amount of total RNA added in the reverse transcriptase reaction, the initial amount of target mRNA can be determined in a mol/μg of total RNA (*see* **Note 11**).

## 4. Notes

1. The use of random and/or oligodT primers to generate cDNAs generally results in lower reverse transcription efficiency. For example, we have found that the efficiency of cDNA synthesis of the glucose transporter GLUT4 mRNA in human AT RNA preparation was five times lower with oligodT primers than with a specific GLUT4 antisense primer.
2. The use of thermostable reverse transcriptase to perform the reverse transcription reaction at elevated temperature ($\geq 70°C$) overcomes the problems of RNA secondary structures, which could reduce the efficiency of cDNA synthesis. Moreover, it increases the specificity of primer hybridization. One of the more efficient enzymes is the *Tth* DNA polymerase that functions as a reverse transcriptase (RNA-dependent DNA polymerase) in the presence of manganese.
3. We have constructed competitors and validated the RT-cPCR assays for the mRNA that codes for the hormone-sensitive lipase, the lipoprotein lipase, the acylation-stimulating protein, the fatty acid transport protein-1, the peroxisome proliferator-activated receptor γ, and leptin (*6*).
4. Make a small hole with a needle in the cap of the Eppendorf tube before freezing it in LN to avoid a sudden opening during thawing.
5. Do not use more than 200 mg AT with the RNeasy mini kit otherwise, the column will be overloaded with contaminants (may be lipids) that dramatically reduce the yield in total RNA. For larger amounts, it is recommended to use the midi or maxi kits.
6. At the end of the purification, the eluate is systematically centrifuged (1 min, 10,000*g*), then the RNA suspension is transfered to a clean Eppendorf tube, to eliminate possible fines of silica.
7. We have observed that some preparations of total RNA from AT can be degraded when they are stored at $-80°C$ for several months. We therefore recommended using the RNA rapidly after preparation.
8. Other methods than electrophoresis on a DNA sequencer can be used to analyze the PCR products. For example, we (and others) have successfully used agarose gel stained with ethidium bromide (*3,4*).
9. Because the competitor DNA is a double-stranded molecule and the target cDNA is a single-stranded molecule, a factor of two must be included in the calculation.
10. We generally performed four competitive PCRs after a single reverse transcription reaction.
11. One attomole ($10^{-18}$ mol) corresponds to about 600,000 molecules (or copies) of RNA, taking the Avogadro constant.

## References

1. Ferré, F. (1992) Quantitative or semi-quantitative PCR: reality versus myth. *PCR Methods Appl.* **2,** 1–9.
2. Gilliland, G., Perrin, S., Blanchard, K., and Bunn, H. F. (1990) Analysis of cytokine mRNA and DNA: detection and quantitation by competitive polymerase chain reaction. *Proc. Natl. Acad. Sci. USA* **87,** 2725–2729.
3. Bouaboula, M., Legoux, P., Pésségué, B., Delpech, B., Dumont, X., Piechaczyk, M., Casellas, P., and Shire, D. (1992) Standardization of mRNA titration using a polymerase chain reaction method involving co-amplification with a multispecific internal control. *J. Biol. Chem.* **267,** 21,830–21,838.
4. Auboeuf, D. and Vidal, H. (1997) Use of the reverse transcription-competitive polymerase chain reaction to investigate the in vivo regulation of gene expression in small tissue samples. *Anal. Biochem.* **245,** 141–148.
5. Raeymaekers, L. (1995) Commentary on the practical applications of competitive PCR. *Genome Res.* **5,** 91–94.
6. Lefebvre, A. M., Laville, M., Vega, N., Riou, J. P., van Gaal, L., Auwerx, J., and Vidal, H. (1998) Depot-specific differences in adipose tissue gene expression in lean and obese subjects. *Diabetes* **47,** 98–103.

# 7

# Glycogen Synthase Activity in Adipose Tissue

## Methods for Freeze-Clamping and Assay

**Heidi K. Ortmeyer**

## 1. Introduction

One of the best markers of in vivo insulin-action at insulin sensitive tissues is an increase in glycogen synthase (GS) activity. GS is the rate-limiting enzyme of GS. The activity of glycogen synthase is increased by dephosphorylation, which converts the dependent form (dependent on glucose 6-phosphate [G6P]) to the independent form. GS activity can be increased by an increase in the activities of protein (serine/threonine) phosphatases, which dephosphorylate the enzyme, and/or by a decrease in the activities of protein (serine/threonine) kinases, which phosphorylate the enzyme.

In skeletal muscle of normal rhesus monkeys, in vivo insulin, during a euglycemic hyperinsulinemic clamp causes a significant increase in the independent activity of GS (measured in the presence of 0.1 m$M$ G6P), without a change in the total activity of the enzyme (measured in the presence of 10 m$M$ G6P) *(1)*. Therefore, insulin increases the fractional activity of GS (independent activity relative to total activity). In subcutaneous (sc) adipose tissue (AT) *(2)* and in liver *(3)* of normal rhesus monkeys, in vivo insulin during a euglycemic hyperinsulinemic clamp causes significant increases in independent, fractional, and total activity of GS.

The effect of in vivo insulin to increase GS activity has been shown to be positively correlated to whole-body, insulin-mediated glucose disposal rates during a euglycemic hyperinsulinemic clamp in skeletal muscle *(1)*, sc AT *(2)*, and in liver *(3)* of rhesus monkeys. Although the effect of insulin to increase GS activity is impaired in skeletal muscle *(1)* and in sc AT *(2)* of insulin-resistant monkeys, insulin action on liver and omental AT of insulin-resistant monkeys appears normal *(4)*.

From: *Methods in Molecular Biology, vol. 155: Adipose Tissue Protocols*
Edited by: G. Ailhaud © Humana Press Inc., Totowa, NJ

The method by which AT GS activity can be examined in frozen AT utilizing the *ex situ* freeze-clamp technique is provided.

## 2. Materials
### 2.1. Freeze-Clamps (For ex situ Freeze-Clamping) (Fig. 1)

1. Vise Grip Model 11SP (American Tool). The Vise Grip is 11 in. long (11), and is a locking C-clamp with swivel pads.
2. Attach 6-cm diameter by 8-mm deep aluminum block (drill-tapped holes) to swivel pad at end with "11SP" engraving.
3. Attach identical block to other swivel pad. Use roll pins to prevent swiveling.
4. 6-cm-diameter blocks will freeze-clamp approx 1 g of fat tissue.

### 2.2. Stock Solutions

1. 1 $M$ phenylmethylsulfonyl fluoride (PMSF) stock solution: dissolve 0.871 g PMSF (based on a mol wt of 174.2) in 5 mL of anhydrous isopropyl alcohol. Store at 4°C.
2. Homogenization buffer: 50 m$M$ tricine, 10 m$M$ EDTA, 100 m$M$ NaF. For 100 mL of buffer, dissolve 0.896 g tricine (based on a mol wt of 179.2), 0.416 g EDTA (based on a mol wt of 416.2), and 0.420 g NaF (based on a mol wt of 41.99) in 80 mL $H_2O$. Adjust pH to 7.5 at 4–7°C. Add 1 µL/mL 2-mercaptoethanol to buffer on day of assay. Bring volume to 100 mL. Store in 5-mL aliquots at –80°C.
3. GS reaction mixture: 200 m$M$ EDTA, 500 m$M$ Trizma base, 250 m$M$ potassium flouride (KF). For 50 mL buffer, dissolve 4.162 g EDTA, 3.028 g Trizma base (based on a mol wt of 121.1), and 1.177 g KF (based on a mol wt of 94.13) in 40 mL $H_2O$. Adjust pH to 7.8 at 30°C. Bring volume to 50 mL. Store at –80°C.
4. 100 m$M$ uridine 5'-diphosphoglucose (UDPG) stock solution: Dissolve 0.0657 g UDPG (based on a mol wt of 657.3) in 1 mL $H_2O$. Store at –80°C.
5. 100 m$M$ G6P stock solution: Dissolve 0.1411 g G6P (based on a mol wt of 282.1) in 5 mL $H_2O$. Store at –80°C.
6. 750 m$M$ KF stock solution: Dissolve 0.708 g KF (based on a mol wt of 94.13) in 10 mL $H_2O$. Store at –80°C.
7. 8% glycogen stock solution: Dissolve 4 g glycogen in 50 mL $H_2O$. Store at –80°C (*see* **Note 1**).
8. Uridine 5'-diphospho-[U-$^{14}$C] glucose: supplied as 50 µCi in 2.5 mL from DuPont NEN (Boston, MA, NEC-403, 250–360 mCi/mmol). Store at –80°C.
9. 66% EtOH solution: bring 1390 mL 95% EtOH to 2 L $H_2O$. Store 1 L at –20°C and 1 L at room temperature (RT).

### 2.3. Low/High G6P Reaction Mixture (5)

Use these mixtures to determine independent activity, total activity, and fractional activity of GS.

1. To make 12 mL final volume of the low G6P reaction mixture (final G6P concentration is 0.1 m$M$; independent activity of GS) and 12 mL final volume of the high G6P reaction mixture (final G6P concentration is 10 m$M$; total activity of GS), prepare the

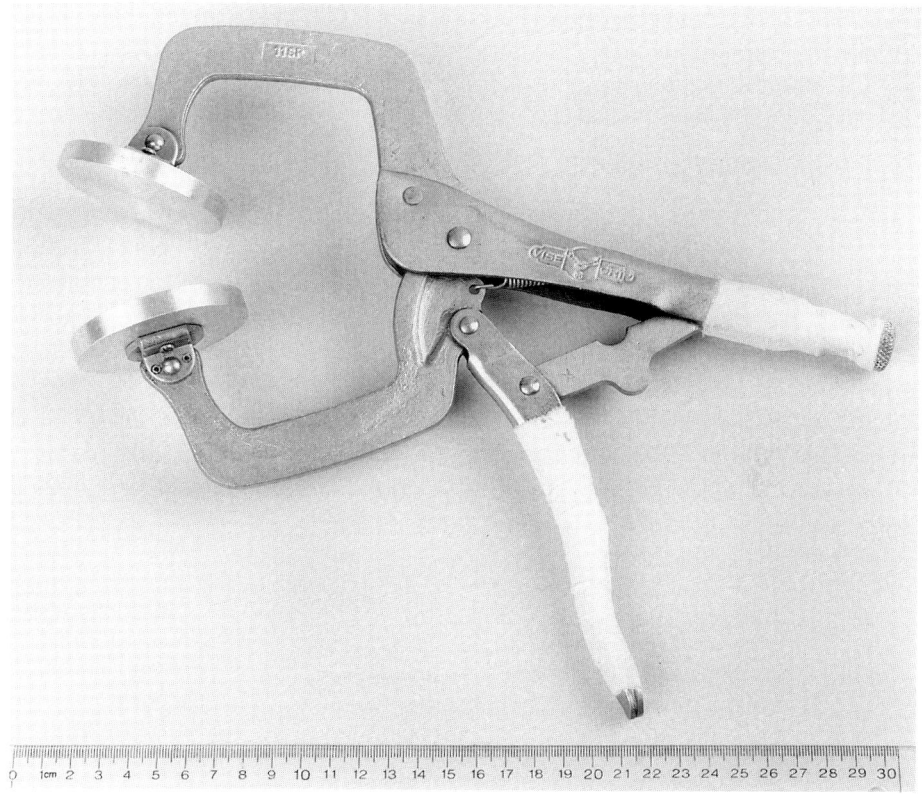

Fig. 1. Vise Grip Model 11SP (American Tool) with 6-cm diameter by 8-mm deep aluminum blocks.

following: to each tube (one labeled "low" and the other "high," both with "radioactive" labeled tape) add 1 mL 750 m$M$ KF, 1.2 mL reaction mixture, 1.5 mL 8% glycogen, 24 μL 100 m$M$ UDPG, 268 μL $^{14}$C UDPG, and 6.808 mL H$_2$O.

2. To the "low" tube add 12 μL 100 m$M$ G6P and 1.188 mL H$_2$O.

3. To the "high" tube add 1.2 mL 100 m$M$ G6P.

4. Check that 60 μL of each radioactive mixture contains approx 60,000 dpm, and that the dpm difference between the "low" and "high" mixture is ≤1%. If the radioactivity differs by more than 1%, add appropriate amount of $^{14}$C UDPG to the mixture with less radioactivity and the same volume of H$_2$O to the other mixture. Repeat until low and high mixtures are similar in dpm. Use the mean dpm of the two mixtures for calculations (*see* **Subheading 3.4.2.**).

5. Can store at 4°C for several weeks while running assay. For long-term storage freeze at –80°C.

6. Final concentrations: 87.5 m$M$ KF, 20 m$M$ EDTA, 50 m$M$ Trizma base, 0.2 m$M$ UDPG, and 1% glycogen.

7. Functions: KF (and NaF), phosphatase inhibitor; EDTA, kinase inhibitor; Trizma base (and tricine), buffer; 2-mercaptocthanol, reducing agent, PMSF, protease inhibitor; UDPG, substrate for GS; G6P, allosteric activator of GS; glycogen, stabilizer; [$^{14}$C]-UDPG, label.

## 2.4. Kinetic Studies of GS

In order to determine the apparent affinity of GS for G6P $K_a$ and the maximal activity ($V_{max}$) of AT GS, measure GS activity at various G6P concentrations. The G6P $K_a$ of GS is an estimate of the phosphorylation state of the enzyme (*6*).

Solutions are made exactly as in the previous section for the 0.1 m$M$ and the 10 m$M$ G6P. For the following G6P concentrations, substitute the appropriate amount of 100 m$M$ G6P and H$_2$O: 0.25 m$M$ G6P, 30 μL G6P + 1.17 mL H$_2$O; 0.50 m$M$ G6P, 60 μL G6P + 1.14 mL H$_2$O; 0.75 m$M$ G6P, 90 μL G6P + 1.11 mL H$_2$O; 1.0 m$M$ G6P, 120 μL G6P + 1.08 mL H$_2$O; 2.5 m$M$ G6P, 300 μL G6P + 900 μL H$_2$O; 5 m$M$ G6P, 600 μL G6P + 600 μL H$_2$O; 7.5 m$M$ G6P, 900 μL G6P + 300 μL H$_2$O.

# 3. Methods
## 3.1. Tissue Freeze-Clamping

1. Vise Grip is frozen in liquid nitrogen (*LN*$_2$) and kept in LN$_2$ until time of clamping. Only the aluminum blocks need to be frozen.
2. AT is placed (by surgeon) in the center of the nonswiveling aluminum block of the Vise Grip immediately after excision (≤ 3 s).
3. Vise Grip is clamped shut, and the aluminum blocks are immersed in LN$_2$ until tissue is frozen (boiling ceases).
4. Tissue is stored between –80°C and –196°C (cryostorage) until assay (*see* **Note 2**).

## 3.2. Tissue Homogenization

1. The following methods are based on using 1 g of AT. The method can be scaled-up or -down by calculating a 20% wt/vol solution (homogenization buffer) (*see* **Note 3**).
2. Add 2.5 μL 1 *M* PMSF to 5 mL ice-cold homogenization buffer immediately before use. Add buffer to cold 30 mL Wheaton Potter-Elvehjem tissue grinder (with Teflon pestle) in ice bucket. Place 1 g of frozen fat in tissue grinder.
3. Homogenize tissue, using an overhead stirrer at high-speed (setting of 10 on a Wheaton overhead stirrer) for 10 strokes in cold room. Keep tissue grinder in ice bucket.
4. Spin homogenate for 2 min at 4°C at 10,000*g*.
5. Carefully remove infranatant (solution below visible lipid layer) to a clean tube.
6. Keep infranatant on ice until assay (*see* **Notes 4** and **5**).

## 3.3. Glycogen Synthase Low/High Assay (see Notes 6 and 7)

1. Run assay in triplicate; label three glass 12 × 75-mm "L" tubes and three "H" tubes.
2. Run one blank for each G6P concentration.

3. Add 60 μL appropriate G6P reaction mixture to each tube.
4. Place tubes in 30°C water bath (with shaker) for 5 min.
5. Turn shaker on (gentle shake).
6. Add 30 μL sample (infranatant) to each tube in 15-s intervals. Add 30 μL H₂O (or nothing) to blank tubes.
7. Stop reaction with 2 mL ice-cold 66% EtOH after 20 min in 15-s intervals (*see* **Note 8**).
8. Remove tubes to room temperature.
9. Place Gelman glass fiber filters (Type A/E 25 mm) (one per tube) in Petri dish. Wet with 66% EtOH (RT).
10. Place filters on filtration instrument (e.g., Hoefer Scientific, 25-mm FH225V Ten Place Filter Holder ) which is connected to a vacuum pump (e.g., Cole-Parmer, Air Cadet Vacuum/Pressure Station).
11. Place 1-lb wt on each filter paper. Turn on vacuum.
12. Pour entire sample over filter paper. Rinse tube twice with 5 mL 66% EtOH (RT) and pour over filter paper. Wash each filter paper with 5 mL EtOH four more times (*see* **Note 8**).
13. Apply vacuum for an additional 2 min.
14. Place dried paper in 20-mL scintillation vial with 10 mL liquid scintillation cocktail (LSC) (e.g., Ready Safe, Beckman), shake vials, and count using a β-counter.
15. *See* **ref. 7** for the filtration assay.

## *3.4. Calculations (Table 1)*

1. In order to determine GS independent (activity in the presence of 0.1 m$M$ G6P) and total activity (activity in the presence of 10 m$M$ G6P), the protein content in the infranatant must be determined. Protein content can be determined by the Bradford method *(8)*. A 100-fold dilution of the infranatant (10 μL infranatant diluted to 1 mL H₂O) is usually sufficient for the value to fall within the standard curve (1–25 μg/mL BSA) when using 5 mL homogenization buffer per gram adipose tissue.
2. To calculate G6P reaction mixture dpm/nmole UDPG: dpm in 60 μL of G6P reaction mixture (**Subheading 2.3.5.**) divided by 12 (nmol UDPG in 60 μL of G6P reaction mixture). The number should be close to 5000.
3. To calculate sample dpm/min/mL: (median dpm of sample [or mean dpm of sample] minus blank dpm) divided by (reaction time in minutes × sample vol. in mL). Calculate the activity in the presence of 0.1 m$M$ and 10 m$M$ G6P.
4. To calculate sample enzyme activity (nmol/min/mL): dpm/min/ml divided by dpm/nmol UDPG. Calculate the activity in presence of 0.1 m$M$ and 10 m$M$ G6P.
5. To calculate sample enzyme specific activity (nmol/min·mg protein): enzyme activity divided by protein content, in mg/mL. Calculate for activity in presence of 0.1 m$M$ and 10 m$M$ G6P.
6. Fractional activity: activity in the presence of 0.1 m$M$ G6P divided by the activity in the presence of 10 m$M$ G6P. Multiply value by 100 to report as a percentage.
7. To determine the G6P Ka of GS and the $V_{max}$ of GS, plot G6P concentration vs G6P concentration divided by GS activity *(9,10)*. **Figure 2** shows the graph and calculations for monkey O-8 frozen sc AT.

**Table 1**
**Sample Calculations for Determining Independent, Total, and Fractional Activity of GS[a]**

| Tube | dpm | −Blank (avg) | dpm/min/mL | Enzyme activity (nmol/min/mL) | Enzyme-specific activity (nmol/min·protein) |
|------|-----|------|------|------|------|
| L blank | 59 | | | | |
| H blank | 65 | | | | |
| L sample | 3049 | | | | |
| L sample | 3076 | 3014 | 5023 | 0.91 | 0.69 |
| L sample | 3206 | | | | |
| H sample | 19,292 | | | | |
| H sample | 19,836 | 19,774 | 32,957 | 5.95 | 4.51 |
| H sample | 20,379 | | | | |
| L 60 μL | 66,551 | 66,490 | | | |
| H 60 μL | 66,571 | 66,509 | | | |

[a] dpm in 60 μL = 66,500 (average of L and H); difference between L and H G6P mixtures = 0.03% (want ≤1%); dpm/nmol UDPG = 66,500/12 = 5542; protein concentration = 1.32 mg/mL; reaction time = 20 min; sample amount = 0.03 mL; independent activity = 0.69 nmol/min·mg protein; total activity = 4.51 nmol/min·mg protein; fractional activity = 0.153 × 100 = 15.3%.

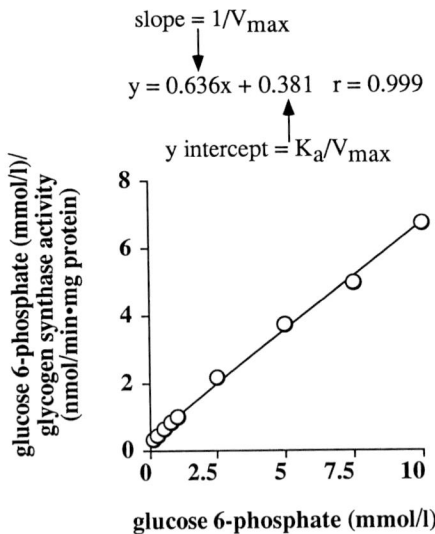

Fig. 2. G6P concentration(s) vs (s)/GS specific activity [v] in basal sc AT from Monkey 08 freeze-clamped *ex situ*. Using this equation, GS maximal specific activity is calculated to be 1.572 nmol/min·mg protein and the G6P *Ka* of GS is calculated to be 0.599 m*M*.

## 4. Notes

1. If the purity of the glycogen is uncertain, it can be purified using the following procedure: Filter 5 g glycogen diluted in 25 mL $H_2O$ through a column packed with a mixed bed resin (e.g., M 8032, Sigma). Wash the column with 25 mL $H_2O$. Precipitate the glycogen by adding ice-cold 95% EtOH to give a final concentration of 66% (111 ml 95% EtOH to 50 mL glycogen solution will give 66% final EtOH concentration, and a final volume of 161 mL). Add 14 mg lithium bromide. Pour this solution over a large Whatman filter paper. Wash with 66% EtOH, 95% EtOH, and then acetone. Let the glycogen dry until there is no odor of acetone.

2. GS fractional activity determined in frozen AT is similar to activity measured in fresh (nonfrozen) AT. GS fractional activity that was determined in fresh sc AT obtained from monkey O-8 in 1992 was 19%. A portion of the sample was freeze-clamped *ex situ* and stored at $-196°C$. The GS fractional activity after 7 yr of cryostorage was 19%.

3. When it is more convenient to determine GS activity in fresh tissue, the following changes in protocol are suggested. Place freshly obtained AT in RT Kreb's-Ringer's phosphate buffer solution. Remove blood and obvious connective tissue. Place remaining sample in fresh buffer and spin for 5 min at 1200*g* (RT). Remove sample (weigh) and proceed as in **Subheading 3.2.1.** Use a reaction time of 10 min, instead of 20 min.

4. The infranatant can be frozen up to at least 2 wk at $-80°C$, without loss of activity.

5. The infranatant can be filtered through a Sephadex G-25 column, to rule out any interference that salts or other low-mol-wt molecules might have on GS activity. The Micro Bio-Spin chromatography column (Micro Bio-Spin 6 column, Tris, Bio-Rad) is especially suited for removal of low-mo-wt substances.

6. As little as 100 mg AT can be used to measure independent, total, and fractional activity of GS (and total protein content).

7. The GS assay that has been presented is optimized for rhesus monkey (*Macaca mulatta*) sc and omental AT. For other species, the reaction should first be run at various times and the time plotted against percent substrate consumption of UDPG (dpm in sample divided by dpm in 60 µL G6P reaction mixture). An appropriate reaction time can then be selected to assure that the percent dpm falls on the linear portion of the percent dpm vs reaction time curve. The percent dpm vs reaction time curve for frozen sc AT (rhesus monkey) is shown in **Fig. 3**.

8. To dispense the EtOH, it is helpful to attach dispensers with tubing to the bottles holding the EtOH, set at 2 mL (for the cold EtOH) and 5 mL (for the RT EtOH).

## Acknowledgments

Many thanks to Dr. Charles Schwartz, Dr. Joe Larner, Dr. Barbara Hansen, Teerin Meckmongkol, and Walter Knapik.

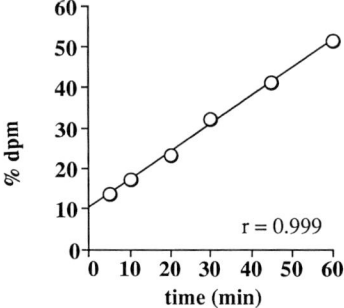

Fig. 3. Percent substrate consumption of UDPG (in the high G6P reaction mixture) in AT infranatant prepared from frozen tissue, as measured by disintegrations per minute (dpm).

## References

1. Ortmeyer, H. K., Bodkin, N. L., and Hansen, B. C. (1993) Insulin-mediated glycogen synthase activity in muscle of spontaneously insulin-resistant and diabetic rhesus monkeys. *Am. J. Physiol.* **265,** R552–R558.
2. Ortmeyer, H. K., Bodkin, N. L., and Hansen, B. C. (1993) Adipose tissue glycogen synthase activation by in vivo insulin in spontaneously insulin-resistant and Type 2 (non-insulin-dependent) diabetic rhesus monkeys. *Diabetologia* **36,** 200–206.
3. Ortmeyer, H. K., Bodkin, N. L., and Hansen, B. C. (1997) Insulin regulates liver glycogen synthase and glycogen phosphorylase activity reciprocally in rhesus monkeys. *Am. J. Physiol.* **272,** E133–E138.
4. Ortmeyer, H. K. and Bodkin, N. L. (1998) Lack of defect in insulin action on hepatic glycogen synthase and phosphorylase in insulin-resistant monkeys. *Am. J. Physiol.* **274,** G1005–G1010.
5. Guinovart, J. J., Salavert, A., Massague, J., Ciudad, C. J., Salsas, E., and Itarte, E. (1979) Glycogen synthase: a new activity ratio assay expressing a high sensitivity to the phosphorylation state. *FEBS Lett.* **106,** 284–288.
6. Roach, P. and Larner, J. (1976) Rabbit skeletal muscle glycogen synthase II. Enzyme phosphorylation state and effector concentrations as interacting control parameters. *J. Biol. Chem.* **251,** 1920–1925.
7. Oron, Y. and Larner, J. (1979) A modified rapid filtration assay of glycogen synthase. *Anal. Biochem.* **94,** 409–410.
8. Bradford, M. M. (1976) Rapid and sensitive method for the quantitation of micro-gram quantities of protein utilizing the principle of protein-dye binding. *Anal. Biochem.* **72,** 248–254.
9. Henderson, P. (1993) Statistical analysis of enzyme kinetic data, in *Enzyme Assays: A Practical Approach* (Eisenthal, R. and Danson, M., eds.), IRL Press, Oxford, pp. 277–316.
10. Ortmeyer, H. K., Huang, L., Larner, J., and Hansen, B. C. (1998) Insulin unexpectedly increases the glucose 6-phosphate Ka of skeletal muscle glycogen synthase in calorie-restricted monkeys. *J. Basic Clin. Physiol. Pharmacol.* **9,** 309–323.

# 8

## Assays of Lipolytic Enzymes

**Cecilia Holm, Gunilla Olivecrona, and Malin Ottosson**

### 1. Introduction

Free fatty acids (FAs) derived from adipose tissue (AT) are, quantitatively, the most important fuel in mammals, and provide more than half of the caloric needs during periods when dietary substrates are lacking. Hormone-sensitive lipase (HSL) catalyzes the rate-limiting step in the lipolysis of stored triglycerides (TGs) in AT, i.e., the hydrolysis of triacylglycerol to diacylglycerol (for a review on HSL, *see* **ref.** *1*). It also catalyzes the subsequent hydrolysis of diglycerides (DGs) and monoglycerides (MGs). A second lipase, monoglyceride lipase (MGL), is required to obtain complete hydrolysis of MGs. Selective immunoprecipitation of this lipase, from an AT preparation containing both HSL and MGL, leads to marked reduction in the glycerol release and accumulation of MGs *(2,3)*. FAs from the action of HSL and MGL are either released into the circulation or, alternatively, re-esterified directly. In humans, on a typical Western diet, *de novo* lipogenesis is rarely observed, and storage of TGs is mostly accomplished through re-esterification of preformed FAs derived from the blood. Lipoprotein lipase (LPL), on the endothelial cell, catalyzes the hydrolysis of TGs from chylomicrons and very low density lipoproteins (VLDL, for reviews on LPL, *see* **refs.** *4–7*). The released fatty acids are either transported into AT for re-esterification and storage or, alternatively, returned into the circulation. MGs, particularly, 2-monoacylglycerols from the action of LPL, are believed to be hydrolyzed by MGL after their diffusion into the adipocytes. HSL and LPL are thus the key enzymes in the mobilization and deposition of TGs in AT; MGL aids both of these enzymes in completing the hydrolysis of MGs to FAs and glycerol (*see* **Fig. 1**).

HSL is not a very specific enzyme. It hydrolyzes TGs, cholesteryl esters (CE), DGs, MGs, retinyl esters, and steroid esters (lipoidal derivatives), as

From: *Methods in Molecular Biology, vol. 155: Adipose Tissue Protocols*
Edited by: G. Ailhaud © Humana Press Inc., Totowa, NJ

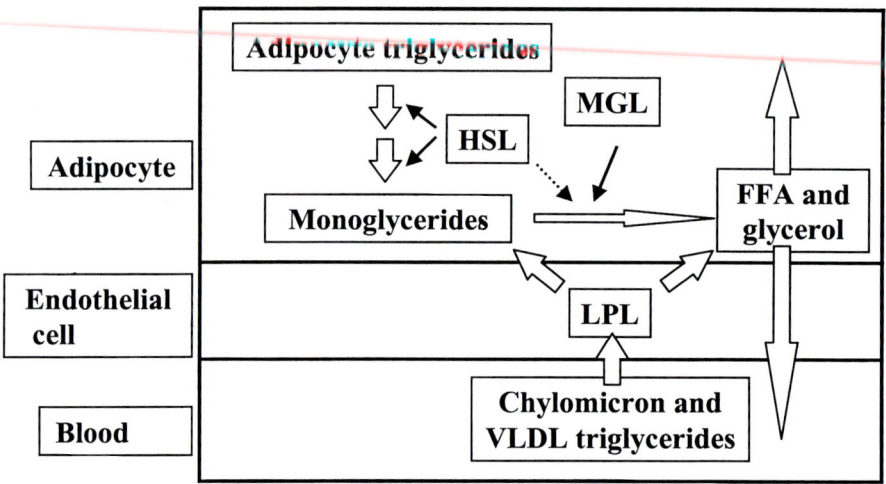

Fig. 1. Role of AT lipases in the flow of lipidic energy substrates. HSL, hormone sensitive lipase; MGL, monoglyceride lipase; LPL, lipoprotein lipase; FFA, free FAs; VLDL, very low density lipoprotein.

well as water-soluble esters, such as *p*-nitrophenyl esters *(8–11)*. Noteworthy, however, is that HSL lacks phospholipase activity. The relative maximal activities for triolein (TO), 1,2-diolein (DO), 1-mono-olein (MO), cholesterol oleate (CO), and *p*-nitrophenylbutyrate (pNPB) is 1:10:1:1.5:20 *(8,9)*. The high CE hydrolase activity of HSL is believed to be of importance in macrophages and steroid-producing tissues.

Regarding positional specificity, HSL has a preference, although not absolute, for the 1(3)-ester bonds *(12)*. The activity of HSL is under hormonal and neural control through a mechanism involving phosphorylation by protein kinase A (PKA). In response to stimulation of adipocytes by catecholamines, this kinase phosphorylates HSL at three serine residues, leading to increased enzyme activity and translocation of HSL from the cytosol to the lipid droplet *(13,14)*. Insulin, which is the most important antilipolytic hormone, counteracts phosphorylation and activation of HSL, mostly via a reduction in cAMP levels, accomplished through activation of phosphodiesterase 3B *(15)*. Upon phosphorylation of HSL in vitro by PKA and monitoring of changes in enzyme activities using emulsified substrates of long-chain acylglycerols or water-soluble ester substrates, only the activity against TGs and CEs are increased; the activity against the other substrates remains unaltered.

In contrast to HSL, MGL is more specific, with monoacylglycerol as its only known lipid substrate *(3)*. MGL has no positional specificity *(3)*. Furthermore, the activity of MGL appears not to be subject to acute regulation.

LPL is synthetized by adipocytes as a fairly early marker for differentiation. In contrast to HSL and MGL, LPL is a secretory protein. Its functional site is at the vascular side of the endothelial cell, from where it can reach its substrate in the TG-rich plasma lipoproteins. LPL is not phosphorylated, but is regulated by other, less defined posttranslational mechanisms that affects the formation of active dimers. Transcriptional regulation also occurs, chiefly for long-term adaptations *(5,16)*. LPL hydrolyzes a number of ester substrates. In contrast to HSL, it does not hydrolyze CE. LPL is a relatively good phospholipase (phospholipase $A_1$). LPL does not hydrolyze 2-MGs, but is dependent on isomerization of MGs to the 1(3)-ester form before complete hydrolysis occurs. Alternatively, 2-MGs enter the cell, where they are hydrolyzed by MGL (*see above*). The relative maximal rates for hydrolysis of TO, DO, MO, dioleoylphosphatidylcholine, retinyl oleate, and *p*-nitrophenyl esters are roughly 100:200:100:10:1:1. For hydrolysis of TGs with long-chain FAs, LPL is dependent on a protein provided with the substrate lipoproteins (apolipoprotein CII [apoCII]). Activity against DGs, MGs, and phospholipids is also stimulated by apoCII, but activity against more hydrophilic substrates (*p*-nitrophenyl esters, tributyrin) is not.

This chapter describes methods for measurement of HSL, MGL, and LPL in AT using synthetic lipid substrates and antibodies (Abs). It includes methods for preparation of AT extracts for these measurements, as well as a method to quantitate LPL mRNA. The principle for the different activity assay systems is the same, namely, the release of [3]H- (or [14]C)-labeled oleic acid (OA) from emulsions or micellar preparations of the respective [3]H-labeled lipid substrate *(17–21)*. The [[3]H]OA is separated from the remaining substrate as potassium (K) [[3]H]oleate into the upper phase of a liquid–liquid partition system, and determined by liquid scintillation *(22)*. Bovine serum albumin (BSA) is used as FA acceptor. The routine substrate used for measurement of HSL activity is a monoether analogue of dioleoylglycerol, 1(3)-mono-oleoyl-2-*O*-mono-oleylglycerol (MOME), which has several advantages relative to other substrates *(23)*. Upon hydrolysis of this substrate, no secondary substrate is formed for MGL. Under the conditions used for the assay (pH 7.0 in the absence of apoCII) essentially no LPL activity is measured. These features optimizes specificity of the assay. Furthermore, because of the substrate specificity of HSL, the use of a diacylglycerol analog increases the sensitivity of the assay. It should be noted, however, that the only substrates that can be used to monitor the increase in activity that occurs upon phosphorylation of HSL by PKA are TGs and CEs with long chain FAs. The substrate used for measurement of MGL activity is a micellar preparation of 1-mono-oloeylglycerol.

LPL activity is routinely measured using emulsified trioleoylglycerol. The emulgator can be phospholipids, gum arabic, or Triton X-100 *(20,21)*. The phospholipid-stabilized emulsions have the disadvantage that phospholipids are also a substrate for the enzyme. TGs are, however, strongly preferred by LPL even though they make up only a few percent of the surface layer of the emulsion *(24)*. Emulsions stabilized by gum arabic or Triton X-100 are usually more sensitive to inhibition by other surface-active proteins in the tissue extracts, and by the detergents that are used for efficient extraction of LPL.

## 2. Materials

### 2.1. Preparation of AT Samples for the Measurement of HSL and MGL Activity

1. Homogenization buffer: 0.25 $M$ sucrose, 1 m$M$ EDTA, pH 7.0, 1 m$M$ dithioerythritol (DTT), 20 µg/mL leupeptin, 2 µg/mL antipain, and 1 µg/mL pepstatin (*see* **Note 1**). This buffer is either made fresh or stored in aliquots at –20°C.

### 2.2. Assay of HSL Activity Using Diacylglycerol Analog MOME or DO as Substrate

1. 1(3)-mono-[$^3$H]oleoyl-2-$O$-mono-oleylglycerol ([$^3$H]-MOME) and MOME are synthesized as described *(23)* (*see* **Note 2**). A commercially available alternative to MOME, providing the same sensitivity, but lower specificity, is DO (oleoyl-1-$^{14}$C), which is available from American Radiolabel (St. Louis, MO; cat. no. ARC 767, 1–5 mCi/mmol) and unlabeled 1,2-DO is available from Sigma (St. Louis, MO) and Nu Chek Prep. (Elysiane, MN). The radioactive substance is dissolved at 5 mg/mL in toluene, and the nonradioactive substance in hexane or heptane at 50 mg/mL. The substances are stable for at least 6 mo when stored at –20°C (*see* **Note 3**).
2. Phosphatidylcholine (PC) from egg yolk and phosphatidylinositol (PI) from soybean can be obtained from Sigma (cat. nos. P-3556 and P-5954, respectively). A 20 mg/mL solution with weight ratios of PC:PI of 3:1 is prepared in chloroform, and stored at –20°C (stable for at least 12 mo).
3. 0.1 $M$ potassium phosphate (K$_2$HPO$_4$), pH 7.0. Store at –20°C, or store at 4°C after sterile filtration, and make fresh weekly.
4. Sonicator with titanium tip (Branson Sonifier 250, or equivalent) (*see* **Note 4**).
5. 20% defatted BSA (cat. no. 775 827 from Boehringer Mannheim or A0281 from Sigma) in 0.1 $M$ K$_2$HPO$_4$, pH 7.0. Store at –20°C.
6. 20 m$M$ K$_2$HPO$_4$, 1 m$M$ EDTA, 1 m$M$ DTT, and 0.02% defatted BSA, pH 7.0. Store at –20°C without the BSA. Add the BSA after thawing from a 20% stock.
7. Extraction mixture: methanol:chloroform:heptane (10:9:7). Store in a light-protected (brown) glass bottle in a ventilated cupboard (*see* **Note 5**).
8. 0.1 $M$ potassium carbonate (K$_2$CO$_3$), 0.1 $M$ boric acid, pH 10.5. Store at room temperature (RT).
9. Liquid scintillation solution (Ready-Safe from Beckman, Fullerton, CA, or equivalent).

### 2.3. Assay of HSL Activity using TO as Substrate

1. TO (1-oleoyl-9,10-$^3$H) (Amersham, Buckinghamshire, UK; cat. no. TRA 191, 5–20 Ci/mmol) is dissolved at 5 mg/mL in toluene, and TO (Sigma, cat. no. T-1740) is dissolved at 50 mg/mL in hexane or heptane. The lipids are stable for at least 6 mo at –20°C (*see* **Note 3**).
2. **Items 2–9** under **Subheading 2.2.**

### 2.4. Assay of HSL Activity using CO as Substrate

1. Cholesteryl [1- $^{14}$C]oleate (Amersham, cat. no. CFA 256, 50–62 mCi/mmol) is dissolved at 5 mg/mL in toluene and cholesteryl oleate (Sigma, cat. no. C-9253) is dissolved in heptane or hexane at 5 mg/mL. Stable for at least 6 mo at –20°C (*see* **Note 3**).
2. **Items 2–9** under **Subheading 2.2.**

### 2.5. Assay of MGL using MO as Substrate

1. 1-mono-oleoyl[$^3$H]glycerol or 1-mono[$^3$H]oleoylglycerol is synthesized as described *(25)* and dissolved at 20 mg/mL in toluene. 1-mono-oleoylglycerol, synthesized as described *(25)* or commercially available from for instance Sigma (M 7765) dissolved at 20 mg/mL in heptane or hexane. The substances are stable for at least 6 mo when stored at –20°C (*see* **Note 3**).
2. 0.2 *M* Tris-HCl, pH 8.0 (4°C), 0.8% (w/v) Triton X-100.
3. Sonicator with titanium tip (Branson Sonifier 250 or equivalent).
4. 20 m*M* Tris-HCl, pH 7.0 (22°C), 0.2% (w/v) Triton X-100, 1 m*M* EDTA, 1 m*M* DTT.
5. Extraction mixture/buffer (for substrate with label in glycerol): chloroform:methanol:heptane (1.25:1.41:1) and 2% (w/v) NaCl. Extraction mixture/buffer (for substrate with label in FA): methanol:chloroform:heptane (10:9:7) and 0.1 *M* $K_2CO_3$, 0.1 *M* boric acid, pH 10.5.
6. Liquid scintillation solution (Ready-Safe, from Beckman, or equivalent).

### 2.6. Measurement of HSL and MGL Protein Using Western Blot Analysis

1. Blocking solution: Tris-buffered saline (TBS) (20 m*M* Tris-HCl, 137 m*M* NaCl, pH 7.6) with 5% defatted milk powder.
2. Washing solution: TBS with 2.5% defatted milk powder and 0.25% Tween.
3. Affinity-purified rabbit anti-HSL and anti-MGL Abs (1 mg/mL).
4. Donkey horseradish peroxidase-conjugated antirabbit immunoglobulin G (IgG) (Amersham).
5. (ECL) detection system and Hyperfilm ECL (Amersham).

### 2.7. Preparation of AT Samples for Measurement of LPL Activity

1. Homogenization buffer: 25 m*M* $NH_3$, 1% Triton X-100, 0.1% sodium dodecyl sulfate (SDS), 1 mg BSA/mL, 5 U heparin/mL (from solution used for anticoagu-

lant treatment in the clinic, 5000 U/mL corresponding to 33.3 mg/mL, Löwens, Malmö, Sweden), and 1 pill of protease inhibitors (Complete Mini, Boehringer, no. 1836153) per 50 mL buffer, or the following concentrations of individual inhibitors: leupeptin (10 µg/mL), pepstatin (1 µg/mL, both from the Peptide Institute, Osaka, Japan), and Trasylol (25 U/mL, from Sigma). Store at –20°C (*see* **Note 6**).

2. A mechanical homogenization devise, such as a Polytron (PT-MR 3000, Kinematica, Switzerland) with a rotating probe (Aggregat PT-DA 3012/2, 11 mm diameter).

## 2.8. Assay of LPL Activity Using Phospholipid-Stabilized Emulsion of Soybean TGs

1. Intralipid 10%, from Fresenius-KABI (former Pharmacia, Uppsala, Sweden), a stable emulsion used clinically for parenteral nutrition. Contains soybean TGs (100 g/L), egg yolk phospholipids (12 g/L), and 22 g glycerol/L. In our hands, this lipid mixture gives the most reproducible data. It is less sensitive to inhibition by proteins and detergents than other synthetic emulsions. If Intralipid 10% is not available, any similar type of commercial emulsion of long-chain TGs can probably be used. Another alternative is to emulsify pure lipids from Sigma (*see* **Subheadings 2.2.**, **item 2** and **2.3.**, **item 1**).
2. [$^3$H]-TO (*see* **Subheading 2.3.**, **item 1**).
3. Sonicator (*see* **Subheading 2.2.**, **item 4**, MSE Soniprep 150 with the small 23 KHz probe, no. 38121-1154, 10 mm diameter).
4. Assay medium: 0.3 $M$ Tris-HCl, 0.2 $M$ NaCl, 0.2 mg heparin/mL, and 120 mg BSA/mL (fraction V, Sigma, no. 3401). Adjust pH to 8.5 with HCl, and store frozen (–20°C) in convenient aliquots (*see* **Note 7**).
5. Serum as source of apoCII. We use serum from rats fasted over night. Blood is taken into empty glass tubes. After coagulation at room temperature for 1 h, the tubes are stored overnight at +4°C for clot retraction. Serum is recovered after centrifugation (1000$g$ for 10 min), followed by heat inactivation for 30 min at 56°C. To inactivate any possible remaining lipase activity, the serum is treated overnight at 10°C with 100 µg phenylmethylsulfonyl fluoride (PMSF)/mL. After dialysis against 10 m$M$ Tris-HCl, pH 7.4, containing 0.15 $M$ NaCl, 2 m$M$ EDTA, and 0.1 m$M$ NaN$_3$, and brief centrifugation to remove precipitated material, the serum is stored at –20°C. Human serum can be used as an alternative (*see* **Note 8**).
6. Water bath (gently shaking) set at 25°C.
7. Ventilated bench or hood.
8. Organic solvents for extraction: *see* **Subheading 2.2.**, **item 7**.
9. 0.1 $M$ K$_2$CO$_3$, pH 10.5 (or *see* **Subheading 2.2.**, **item 8**).
10. Liquid scintillation solution (Opti Phase-Hi Safe 3, Wallac Sverige).

## 2.9. Assay of LPL Protein in Tissue Samples

1. Microtiter plates (96-well, MaxiSorp, Nunc).
2. Buffer for coating of: 10 m$M$ phosphate, pH 7.4, 0.15 $M$ NaCl.

3. Buffer for washing of plates: PBS-Tween, EC Diagnostics AB, Uppsala, Sweden, 1 pill/L (10 m*M* phosphate, pH 7.4, 0.15 *M* NaCl, 0.05 % (v/v) Tween-20, 0.003 *M* KCl).
4. Buffer for incubation with antigen: same as under **Subheading 2.9., item 2**, but also containing 15 % (v/v) glycerol and 1 mg heparin/mL.
5. Buffer for dilution of standard: same as under **Subheading 2.9., item 3**, but also containing 70 mg BSA/mL (Sigma fraction V).
6. Buffer for incubation with detection antibodies: same as under **Subheading 2.9., item 1**, but with 0.1% (v/v) Tween-20 and 4% (w/v) BSA.
7. Peroxidase-labeled goat antimouse IgG (A2554, Sigma).
8. 0.05 *M* citrate-phosphate buffer, pH 5.0.
9. Orthophenylenediamine (OPD) tablets (2 mg/tablet, Dako, S-2045).
10. $H_2O_2$, 30%.

### 2.10. Measurement of LPL mRNA in Human AT

#### 2.10.1. Linearization of the Probe

1. The probe: A *Nae*I-*Nco*I fragment of 292 base pairs, derived from the 5′-region of the cDNA encoding the human LPL gene, ligated into the *Hinc*II site of the pGEM4Z vector (*16,26*); store at –80°C.
2. Restriction enzyme *Hin*dIII and corresponding buffer (Gibco), store at –20°C.
3. Phenol (Tris-saturated), store at +4°C.
4. Chisam (chloroform:isoamyl alcohol, 24:1), store at RT.
5. EtOH (95%), store at –20°C.
6. NaOAc (3 *M*), store at RT.
7. RNase free $H_2O$, store at RT.

#### 2.10.2. [³²P]-Labeling Antisense LPL RNA Probe for RPA (In Vitro Transcription)

1. Linearized LPL template, store at –20°C.
2. 5X transcription buffer, 100 m*M* DTT, 10 m*M* ATP, 10 m*M* GTP, 10 m*M* UTP, 10 m*M* CTP, 20 U RNasin, 20 U DNase, 15–20 U/mL SP6 RNA-polymerase (all from Promega), store at –20°C.
3. [³²P]-CTP (Amersham), store at –20°C.
4. Nick Column (Pharmacia), store at RT.
5. *See* **Note 9**.

#### 2.10.3. [³⁵S]-Labeling Antisense 18S RNA Probe for Ribonuclease Protection Assay (In Vitro Transcription)

1. The linearized pTRI internal control template, pTRI RNA 18S, 80 bp (Ambion), store at –20°C.
2. 5X transcription buffer, 100 m*M* DTT, 10 m*M* ATP, 10 m*M* GTP, 10 m*M* UTP, 10 m*M* CTP, 20 U RNasin, 20 U DNase, 15–20 U/mL T7 RNA-polymerase (all from Promega), store at –20°C.

3. [$^{35}$S]-UTP (Amersham), store at –20°C.
4. NH$_4$OAc (5 M), store at –20°C.
5. EtOH (95%), store at –20°C.
6. Nick Column (Pharmacia), store at RT.
7. See **Note 9**.

### 2.10.4. Hybridization of Probe and Sample RNA

1. Sample total RNA, store at –80°C.
2. [$^{32}$P]-labeled antisense LPL RNA probe, store at –80°C until just prior to use.
3. [$^{35}$S]-labeled antisense 18S RNA probe: For storage, see above.
4. The RPAII kit (Ambion), store at –20°C.
5. EtOH (95%), store at –20°C.

### 2.10.5. RNase Digestion of Hybridized Probe and Sample RNAs

1. All reagents provided in the RPAII kit (Ambion).

### 2.10.6. Separation and Detection of Protected Fragments

1. Precast 6% Tris boric acid/EDTA (TBE)-Urea gel (Novex Electrophoresis).
2. TBE running buffer (1X).
3. Gel loading buffer II included in the RPAII kit (Ambion).

## 3. Methods

### 3.1. Preparation of AT Samples for Measurement of HSL and MGL

For measurement of HSL and MGL activity, AT samples are homogenized in 2–3 vol homogenization buffer, and centrifuged at 110,000g, at 4°C for 45 min, to obtain fat-free infranatants. From hormonally quiescent primary adipocytes and AT, approx 70–80% of the HSL activity is recovered in the infranatant, 15–20% in the floating fat cake, and 5–10% in the pellet fraction *(8)*. However, it is possible to have a different distribution from this, because of prior stimulation of the adipocytes/AT with lipolytic agents, since it has been shown that HSL translocates from the cytosol to the lipid droplet upon lipolytic stimulation of adipocytes *(14)*. In the adipocyte cell line, 3T3-L1, a very small fraction of HSL is found in the floating fat cake (>5%); a considerable fraction is found in the 110,000g pellet (40–60%). An early report has shown a redistribution of HSL to the pellet fraction, upon lipolytic stimulation of 3T3-L1 cells *(27)*. More recent immunohistochemical studies have demonstrated that HSL translocates to the surface of the multiple small lipid droplets found in 3T3-L1 cells, upon lipolytic stimulation *(28,29)*. In order to estimate the HSL activity of the fat cake and pellet fractions, these can be resuspended in homogenization buffer and assayed (*see* **Note 10**).

### 3.2. Assay of HSL Activity Using Diacylglycerol Analog MOME or DO as Substrate

Below is a description of the assay for HSL activity, using MOME as substrate, followed by descriptions of other lipid substrate assays for HSL. Since these other assays are very similar to the MOME assay, only differences in the assay procedures are described. All steps are performed at RT, using room-tempered solutions, unless otherwise indicated. The assays are linear, with increasing amounts of enzyme up to at least 10% substrate hydrolysis, when purified HSL samples are used. However, for unknown reasons, the linearity is very poor when crude tissue samples are assayed. We therefore recommend a careful titration of each system, before comparisons regarding to the HSL activity can be made between different AT samples (*see* **Note 11**).

The substrate concentration of the MOME substrate is 5 m$M$ and the molar ratio between MOME and phospholipid is approx 25:1.

1. In a 4-mL glass vial ($14 \times 45$ mm), mix MOME (12.1 mg, including $16 \times 10^6$ cpm of [$^3$H-MOME]) and phospholipids (30 μL, 20 mg/mL PC:PI, 3:1), and evaporate the solvents under a gentle stream of $N_2$.
2. Transfer the vial with the solvent-free lipids to a desiccator and leave under vacuum for at least 15 min, to ensure the absence of residual solvent.
3. Add 2 mL of 0.1 $M$ $K_2HPO_4$, pH 7.0, and sonicate at a setting for 1–2 min for two cycles, each of 1 min duration, with a 1-min interval between the two cycles. The sonicator tip should be approx 1 cm below the surface during the sonication (*see* **Note 12**).
4. Add an additional 1.6 mL 0.1 $M$ $K_2HPO_4$, pH 7.0, and continue the sonication on ice for $4 \times 30$ s, with a 30-s interval between the cycles.
5. After completed sonication, add 0.4 mL of 20% BSA in 0.1 $M$ $K_2HPO_4$, pH 7.0. Use immediately, or store on ice until used (*see* **Note 13**).
6. For each assay, mix in $13 \times 100$-mm glass tubes, 100 μL HSL sample, diluted in 20 m$M$ $K_2HPO_4$, 1 m$M$ EDTA, 1 m$M$ DTT, and 0.02% defatted BSA, pH 7.0, with 100 μL substrate. Incubate for 10–30 min at 37°C without shaking.
7. Add 3.25 mL extraction mixture, and vortex briefly (1–2 s) to terminate the reaction. Add 1.05 mL of 0.1 $M$ $K_2CO_3$, 0.1 $M$ boric acid, pH 10.5, and vortex briefly (1–2 s).
8. Centrifuge at 800$g$ for 20 min, and use 1 mL of the upper phase for scintillation counting, together with 10 mL water-miscible liquid scintillant (*see* **Note 14**).
9. One unit of enzyme activity is defined as 1 μmol of FAs released per min at 37°C. Enzyme activity can be calculated knowing the partition coefficient of K oleate to the upper phase (1.9 at 22°C; i.e., 71.5% of the OA is recovered in the upper phase) and the total volume of the upper phase (2.45 mL) (*see* **Note 15**).

### 3.3. Assay of HSL Activity, Using TO as Substrate

5.9 mg TO, including $50 \times 10^6$ cpm [$^3$H]TO, and 0.6 mg phospholipid mix (30 μL, 20 mg/mL PC:PI, 3:1) are used for each 4-mL substrate. This gives a final substrate concentration of 1.67 m$M$ (5 m$M$ regarding hydrolyzable FAs) and a molar ratio between TO and phospholipid of approx 9:1. The substrate is prepared as described under **Subheading 3.2.**, except that 1 mL (instead of 1.6 mL) is added at **step 4**, and 1 mL of 20% BSA in 0.1 $M$ K$_2$HPO$_4$, pH 7.0, is added at **step 6** (*see* **Note 16**).

### 3.4. Assay of HSL Activity Using CO as Substrate

1.17 mg CO, including $20 \times 10^6$ cpm [$^3$H]CO (or [$^{14}$C]CO), and 1.42 mg phospholipid mix (71 μL, 20 mg/mL PC:PI, 3:1) are used for each 4-mL substrate. The final substrate concentration is 0.45 m$M$ and the molar ratio between CO and phospholipid is 1:1. The substrate is prepared as described under **Subheading 3.2.**, with the same modifications as described under **Subheading 3.3.** Furthermore, both the first and second sonication step is performed at 37°C, using prewarmed solutions.

### 3.5. Assay of MGL using MO as Substrate

Below is a description of how to make a 10-m$M$ micellar MO substrate, which is $^3$H-labeled in either the glycerol part or in the FA. Please note that the extraction system is different, depending on which part is labeled. The assay is linear with increasing amounts of enzyme up to at least 10% substrate hydrolysis (*see* **Note 17**).

1. Aliquot 14.2 mg MO, including $8 \times 10^6$ cpm of [$^3$H]MO, in a 4-mL glass vial ($14 \times 45$ mm), and evaporate the solvents under a gentle stream of N$_2$.
2. Transfer the vial with solvent-free lipids to a desiccator, and leave under vacuum for at least 15 min, to ensure the absence of residual solvent.
3. Add 2 mL of 0.2 $M$ Tris-HCl, pH 8.0 (4°C), 0.8% (w/v) Nonipol TD 12 and sonicate for 2 min at setting 2, with the sonicator tip 1 cm below the surface. Use the substrate within 3–4 h of preparation (*see* **Note 18**).
4. For each assay, mix in $13 \times 100$ mm glass tubes 100 μL MGL sample, diluted in 20 m$M$ Tris-HCl, pH 7.0 (22°C), 0.2% (w/v) Nonipol TD 12, 1 m$M$ EDTA, 1 m$M$ DTT, with 100 μL substrate. Vortex, and incubate at 22°C for 10 min.
5. Add 3.25 mL of the appropriate extraction mixture (*see* **Subheading 2.5., item 5**), and vortex briefly (1–2 s) to terminate the reaction. Add 1.05 mL of the appropriate extraction buffer (*see* **Subheading 2.5., item 5**), and vortex briefly (1–2 s).
6. Centrifuge at 800$g$ for 20 min, and use 1 mL of the upper phase for scintillation counting, together with 10 mL water-miscible liquid scintillant (*see* **Note 14**).
7. One unit of enzyme activity of enzyme activity is defined as 1 μmol of FAs released per min at 22°C. Enzyme activity can be calculated knowing the partition coefficient of glycerol and OA, respectively, to the upper phase. Glycerol partitions to more than 99% to the upper phase in the extraction system described; 71.5% of the K oleate is recovered in the upper phase in its extraction system (partition coefficient 1.9 at 22°C).

### 3.6. Measurement of HSL and MGL Protein Using Western Blot Analysis

Among several different Western blot protocols tested, the protocol below, in our hands, yields the best results when analyzing AT samples for HSL and MGL. All steps are performed at RT. **Steps 1–6** are performed on an agitation table.

1. Block the nitrocellulose membrane for 2 h in blocking solution.
2. Wash the membrane for 15 min, followed by 2 × 5 min, in washing solution.
3. Incubate with affinity-purified anti-HSL or anti-MGL Abs, diluted to approx 0.2 µg/mL in washing solution for 2 h.
4 Wash for 15 min, followed by 4 × 5 min, in large volumes of wash solution.
5. Incubate with horseradish peroxidase-conjugated antirabbit IgG, diluted 2000 times, in wash solution.
6. Wash as in **step 4**.
7. Develop the membrane using the ECL system according to the instructions from the manufacturer, and expose to film (*see* **Note 19**).

### 3.7. Preparation of AT Samples for Measurement of LPL

In contrast to HSL and MGL, LPL is present both within cells and outside cells, e.g., in the connective tissue and on cell surfaces, such as on endothelial cells in capillaries and larger vessels. Endothelial LPL constitutes the functional pool of the enzyme, and should be the most relevant variable to measure. Technically, this is, however, not easy. Two methods have been used: First, incubation of tissue biopsies in heparin-containing medium, followed by measurement of heparin-releasable LPL in the medium *(16)* (this method is dependent on efficient and reproducible penetration of heparin into the tissue); second, measurement of total LPL activity in tissue homogenates, as well as in collagenase-isolated adipocytes from the same tissue. The difference (total LPL–cell LPL) gives an estimate of the extracellular LPL in the tissue *(30)*. This may be overestimated if some cells are damaged or lost during the procedure. Therefore, in most published studies on LPL, the total activity in an extract of homogenized tissue is measured *(31)*. Pieces of fresh AT are immediately put in 9 vol ice-cold homogenization buffer and homogenized until the tissue is fully disintegrated (approx 3–5 s). Then, the samples are briefly centrifuged (10 min at 10,000*g* at 4°C), and the clear subphase, below the fat cake, is carefully sampled (*see* **Note 20**). The extraction is almost complete. Re-extraction of the fat cake and pellet gives less than 10% additional LPL (*see* **Note 21**).

### 3.8. Assay of LPL Activity, Using Radiolabeled Intralipid as Substrate

Emulsions of TGs are the most commonly used substrates for quantification of LPL in tissue samples. MOME, or DGs, could theoretically also be used for measurement of LPL activity. The contribution of HSL to the activity must then be controlled, e.g., by the use of immunoinhibition of either of the two lipases. As for HSL (see **Subheading 3.2.**) the linearity with amount of tissue extract is poor, even with the Intralipid-based emulsion. We use a sample volume of only 1% of the total assay system. For samples with low activity, long incubation times are preferred, since linearity with time (during incubation at 25°C) is good for at least 2–3 h (see **Note 22**).

1. Evaporate [$^3$H]-labeled TO (about $0.3 \times 10^9$ cpm) under a gentle stream of $N_2$ on the walls of a round-bottomed 30-mm diameter glass vessel suitable for sonication.
2. Add 3 mL Intralipid (use a sterile syringe, avoid contamination of the flask).
3. Chill the vessel in ice water, and sonicate for 5 min in a 50% pulse mode. Standardize the conditions (see **Note 12**).
4. Inspect the emulsion. It does not change appearance on sonication but it becomes less stable, and can be stored at 4°C for 1 wk with reproducible results. The TG concentration of the emulsion is around 115 m$M$ and it should contain about 70% of the labeled TO added. For calculation of specific radioactivity, samples of the emulsion have to be counted for radioactivity.
5. Mix appropriate volumes of assay medium (see **Subheading 2.8., item 4**), heat-inactivated serum (**Subheading 2.8., item 5**) and labeled emulsion at RT. Use this mixture for analyses during 1 d only. The optimal amount of serum should be tested out for each batch of serum used. A typical composition is assay medium:serum:[$^3$H]-labeled Intralipid 10:1:1. Pipet 120 µL of the mix into tubes for assay (disposable glass, $13 \times 100$ mm, Kimbell). This will give a final substrate concentration of 5 mg TGs/mL. The optimal amount of emulsion for obtaining maximal activity should be tested out experimentally (usually 1–5 mg/mL).
6. Make up the total volume of the assay to 200 µL with sample, corresponding buffer or water. This emulsion can tolerate at least 20 µL homogenization buffer, but for tissue homogenates the sample volume has to be kept low to stay in the linear range of the assay (see **Note 11**). We routinely use 2 µL of the extracts or make 5- or 10-fold dilutions (in homogenization buffer) and take 10 or 20 µL to facilitate the use of automatic pipets.
7. Incubate tubes with all assay components for at least 5 min in water bath at 25°C, before the samples are added. The incubation is then continued on gentle shaking for the desired time (see **Note 22**). For longer times (more than 1 h) the tubes should be covered to limit evaporation.
8. Run blank incubations, with homogenization buffer only, and preferably also reference samples (see **Note 23**). Make duplicate or triplicate assays for each sample.
9. Stop the reaction after the desired time with organic solvents and extract the released FAs (see **Subheading 3.2., items 7–9**; see **Note 24**).
10. Relate outcome of assay to extraction efficiency for the FAs, the specific radioactivity of the substrate, incubation time, and sample volume, and express the results as described in **Subheading 3.2., item 9**. The results can be expressed per g tissue by multiplying by a factor of 10 (e.g., mU/g AT, see **Note 25**).

### 3.9. Measurement of LPL Protein Mass by Sandwich ELISA

Methods for measurement of LPL protein by immunoprecipitation and Western blots were recently described in this series by Doolittle and Ben Zeev *(32)*. ELISA assays for LPL have been developed in several laboratories, and one such assay was recently described by Vilella and Joven *(33)*. Only slight differences are made for application of the method to AT extracts. The most crucial requirement for the assay is a good and specific Ab. We have used polyclonal Abs raised in chickens against LPL from bovine milk for assay of LPL in AT from humans, mink, rat, and mouse. Different batches of Abs have slightly different specificity. For that reason, the best Ab for each application has to be tested out, e.g., first by Western blotting. Because of the high degree of conservation of LPL between animal species, few epitopes are recognized, and the antisera obtained are rather weak. The procedure for raising Abs in chickens and isolation of IgG, as well as affinity purification, was previously described in detail *(32)*. It was pointed out that LPL should be denatured before immunization to expose more epitopes. Our experience is that, for detection of LPL in tissue extracts by ELISA, injection of native LPL results in Abs that are more convenient, because they detect both native and denatured forms of the enzyme equally well. For specific detection of the bound antigen on the coated microtiter plates, the use of a peroxidase-labeled mAb (5D2), raised against bovine LPL, has been very helpful *(33)*. For analyses of mouse LPL, a peroxidase-labeled, affinity purified, polyclonal chicken IgG, raised against bovine LPL, must be used instead of the mAb (the same as for coating or preferably a different batch, which may detect additional epitopes). Samples to be measured for LPL mass can be stored long-term in the freezer, since inactivation of LPL should not change the immunoreactivity. Tissue extracts can first be thawed and analyzed for LPL activity, then analyzed for LPL protein mass at some later stage.

1. Coat microtiter plates with affinity-purified polyclonal Ab (5 µg/mL, 100 µL/well). Cover the plate, and incubate 4 h at 37°C.
2. Wash 3× with PBS-Tween. Keep plates on crushed ice.
3. Make dilutions of standard (0–40 ng/mL, at least 10 dilutions) and samples, depending on their LPL activity (usually five logarithmic dilutions, *see* **Note 26**) in microcentrifuge tubes on ice (total volume, e.g., 200 µL). One-fifth of the sample (40 µL) should contain tissue homogenate or corresponding buffer, standard ,or buffer from **Subheading 2.9., item 4**. The remaining volume (160 µL) should be the buffer described under **Subheading 2.9., item 3**.
4. Add 100 µL/well of standards (triplicates) and of samples (duplicates).
5. Cover the plate, and incubate overnight at 4°C.
6. Wash 4× with PBS-Tween.
7. Incubate with mAb (diluted 1:10,000, 100 µL/well) for 3 h at RT.
8. Wash 4× with PBS-Tween.

9.  Detect with peroxidase-conjugated goat antimouse IgG (diluted 1:8000, 100 μL/well), 3 h at RT in the dark.
10. Wash 4× with PBS-Tween.
11. Develop by addition of substrate (1 OPD pill/5 mL citrate-phosphate buffer, add 2 μL 30% $H_2O_2$). Incubate at RT in the dark.
12. Stop the reaction by addition of 50 μL 3 $M$ $H_2SO_4$ when sufficient yellow color has developed, and read absorbance at 490 nm. Calculate a regression curve from the standard, and express samples in ng LPL/mL of undiluted sample (*see* **Notes 25** and **27**).

## 3.10. Measurement of LPL Messenger RNA

Northern blotting, quantitative reverse transcriptase-polymerase chain reaction (RT-PCR, described for LPL mRNA in **ref. *34***) and ribonuclease protection assay (RPA) are methods suitable for measurement of LPL mRNA levels. This subheading describes the measurement of human AT LPL mRNA, using RPA. 18S ribosomal RNA is used as an internal standard (*see* **Note 28**). RPA is a sensitive technique for the detection and quantification of RNA species, usually mRNA, in a sample mixture of total RNA. For the RPA, a labeled RNA probe is synthesized, which is complementary to a specific part of the sample RNA to be analyzed (for this particular RPA, probes complementary to LPL mRNA and 18S RNA are synthesized). The probe fragment is inserted into one of the common transcription vectors containing a suitable promoter, such as the SP6 or T7 promoter, and the corresponding SP6 or T7 RNA polymerase is used to generate an RNA transcript of high specific activity. The labeled probe is then mixed with the sample RNA and incubated under conditions favoring hybridization of complementary transcripts. After hybridization, the mixture is treated with ribonucleases degrading single-stranded, unhybridized probe. Labeled probe that hybridized to complementary RNA in the sample mixture will be protected from ribonuclease digestion, and can be separated on a polyacrylamide gel, and visualized by autoradiography. The intensity of the protected fragment is proportional to the amount of complementary RNA in the sample mixture, if the probe is present in molar excess over the target fragment during the hybridization reaction.

### 3.10.1. Linearization of the Probe

1.  Linearize the pGEM4Z vector, containing the LPL insert (292 bp), with *Hin*dIII. Add to a microcentrifuge tube: x μL DNA (10 μg), 2 μL buffer, y μL $H_2O$ (to give a final volume of 20 μL), and 1 μL enzyme (*Hin*dIII).
2.  Incubate at 37°C overnight.
3.  Add 10 μL phenol and 10 μL chisam: mix well.
4.  Centrifuge at 10,000*g* for 5 min.
5.  Move the supernatant to a tube with 20 μL chisam: mix.
6.  Centrifuge at 10,000*g* for 1 min.

7. Move the supernatant to a tube with 50 μL EtOH (95%) and 2 μL NaOAc (3 *M*). Invert the tube several times, and precipitate at –20°C for 15 min.
8. Centrifuge at 10,000*g* for 15 min at +4°C.
9. Remove the supernatant, and wipe the walls inside the tube.
10. Resuspend the pellet in 20 μL H$_2$O, and leave the tube for 5 min at RT.
11. Analyze the fragment on a 1% agarose gel.

### 3.10.2. [$^{32}$P]-Labeling Antisense LPL RNA Probe for RPA (In Vitro Transcription)

1. Add to a microcentrifuge tube (*see* **Note 29**): 4 μL 5X transcription buffer; 2 μL 100 m*M* DTT; 4 μL ATP-, GTP-, and UTP-mixture (2.5 m*M* each); 2.4 μL 100 μ*M* CTP; 0.5 μL RNasin (20 U); *x* μL template (0.2–1 μg); and *y* μL H$_2$O to give a total volume of 20 μL, when [$^{32}$P]-CTP and SP6 RNA-polymerase are added.
2. Mix and centifuge quickly.
3. Add 2.5 μL of [$^{32}$P]-CTP (Amersham, 800 Ci/mmol, 20 mCi/mL) and 1 μL SP6 RNA-polymerase.
4. Incubate at 37°C for 60 min.
5. Add 1 μL DNase (10 U).
6. Incubate at 37°C for 15 min.
7. Inactivate DNase, 65°C for 5 min, then put the tube on ice.
8. Purify the probe using the Nick Column (Pharmacia). Count 1 μL of the second fraction in a scintillation counter (gives 100,000–300,000 cpm).

### 3.10.3. [$^{35}$S]-Labeling Antisense 18S RNA Probe for RPA (In Vitro Transcription)

1. Add to a microcentrifuge tube (*see* **Note 29**): 4 μL 5X transcription buffer; 2 μL 100 m*M* DTT; 6 μL ATP-, GTP-, CTP-, and UTP-mixture (2.5 m*M* each); 0.5 μL RNasin (20 U); *x* μL template (2 μg); and y μL H$_2$O to give a total volume of 20 μL, when [$^{35}$S]-UTP and T7 RNA-polymerase are added.
2. Mix and centifuge quickly.
3. Add 2 μL [$^{35}$S]-UTP (Amersham, 1000 Ci/mmol, 10 mCi/mL) diluted 1:5 in H$_2$O and 2 μLT7 RNA-polymerase.
4. Incubate at 37°C for 3 h.
5. Add 1 μL DNase (10 U).
6. Incubate at 37°C for 15 min.
7. Inactivate DNase, 65°C for 5 min, then put the tube on ice.
8. Purify the probe using the Nick Column (Pharmacia).
9. To the second fraction (400 μL), add 40 μL NH$_4$OAc (5 *M*) and 800 μL of ethanol (95%), precipitate at –20°C for 1 h.
10. Centrifuge at 10,000*g* for 15 min at +4°C.
11. Remove the supernatant, and wipe the walls inside the tube.
12. Resuspend the pellet in 20–30 μL H$_2$O.
13. Count 1 μL of the second fraction in a scintillation counter (gives 150,000–200,000 cpm).

### 3.10.4. Hybridization of Probe and Sample RNA (see **Note 30**)

1. Mix 2–10 μg AT total RNA with about 100,000 cpm of LPL probe and about 200,000 cpm of 18S probe in 1.5-mL microcentrifuge tubes. Include two control tubes with the same amount of labeled LPL probe as above and 10 μg yeast RNA.
2. Co-precipitate the probes and RNA in 0.5 $M$ NH$_4$OAc and 2.5 vol of EtOH (95%), mix, and place the tubes in –20°C freezer for 15 min.
3. Centrifuge at 10,000$g$ for 15 min at +4°C.
4. Remove the EtOH supernatant.
5. Dissolve the pellet in 20 μL of hybridization buffer, mix well for about 5–10 s, then centrifuge a few seconds to collect the liquid at the bottom of the tube.
6. Denature the RNA by incubating the tubes at 90°C for 5 min, mix, and centrifuge briefly.
7. Allow the probes and complementary RNA in the sample RNA to hybridize in an incubator at 42°C overnight.

### 3.10.5. RNase Digestion of Hybridized Probe and Sample RNAs (see **Note 30**)

1. Dilute the RNase A/RNase T1 mix in RNase digestion buffer 1:70.
2. Add 200 μL diluted RNase mixture to each tube containing sample RNA, and to one of the yeast RNA control tubes (negative control). To the other yeast RNA control tube add 200 μL RNase digestion buffer without RNase (positive control) (*see* **Note 31**).
3. Vortex, and centrifuge the tubes briefly, incubate at 37°C for 30 min to digest unprotected single-stranded RNA.
4. Add 300 μL RNase inactivation/precipitation solution to each tube. Vortex, and centrifuge the tubes briefly.
5. Put the tubes in a –20°C freezer for at least 15 min.

### 3.10.6. Separation and Detection of Protected Fragments

1. Prepare a denaturing polyacrylamide gel. We use a precast 6% TBE-urea gel (Novex Electrophoresis) and TBE running buffer (1X).
2. Centrifuge the tubes at 10,000$g$ for 15 min at +4°C to pellet the precipitated products.
3. Remove all supernatant, and dry the pellet in RT.
4. Dissolve the pellet in 6 μL gel loading buffer II included in the RPAII kit (Ambion), vortex, and centrifuge the tubes at 10,000$g$ for 1 min.
5. Heat the tubes for 3–4 min at 90°C, to denature the RNA.
6. Load the samples (only 1–2 μL positive control), run at 130 V for about 1 h, or until the Bromphenol blue (BFB) dye band is near the bottom of the gel.
7. Transfer the gel to a Whatman 3MM filter paper, and cover with plastic wrap.
8. Dry the gel in a gel-drying apparatus.
9. Expose the gel to a phosphoimager screen for 2 h or overnight.
10. Analyse the signals of the protected fragments using densitometry. Calculate the ratio LPL mRNA:18S RNA for each sample.

## 4. Notes

1. The composition of the homogenization buffer has been optimized to obtain the major part of HSL and MGL in the fat-free infranatant fraction. Increasing the ionic strength of the buffer leads to increased partition of HSL and MGL to the fat-cake fraction. Detergents should not be included in this buffer, although detergent solubilization is a necessary subsequent step for the purification of HSL and MGL from tissue and cell homogenates. The cocktail of protease inhibitors has been optimized for crude adipose tissue samples. Diisopropyl fluorophosphate, PMSF, and similar substances should not be used as protease inhibitors, because they are potent inhibitors of HSL and MGL. Both HSL and MGL are dependent on the presence of reducing substances, such as DTT, to retain full enzymatic activity. Because of this, 1 m$M$ DTT is routinely included in all buffers used for HSL and MGL. DTT can be substituted by 0.1 m$M$ β-mercaptoethanol.

2. The synthesis of the diacylglycerol analog, MOME, may not be feasible in every laboratory. As stated, 1,2-DO is a commercially available alternative to MOME. It presents problems with specificity when assaying for HSL in crude tissue homogenates, since, upon hydrolysis of DO, MO will be formed, which will be hydrolyzed efficiently by MGL.

3. The blanks in all the HSL assays should ideally be below 1% of total substrate hydrolysis. When the blanks exceed this value, it is recommended that the lipid substances are purified free from FAs (formed through spontaneous hydrolysis) using silicic acid chromatography. Regarding MGL assays, the blanks in the assay, based on 1-mono-oleoyl[$^3$H]glycerol, are, and should normally be, below 1% of total substrate hydrolysis; the blanks in the assay, based on 1-mono[$^3$H]oleoyl glycerol, generally are higher, because of the partition of some nonhydrolyzed MO to the upper phase. More accurate determinations of released FAs require, e.g., thin-layer chromatography on silica acid.

4. The quality of the sonicator and sonicator tip is vital for the quality of the lipid substrates. The titanium tip used in our laboratory has a diameter of 3 mm, and is exchanged when cavities appear or when performance is unsatisfactory. The sonicator tip should be calibrated against sonic pressure, since a constant sonifier energy output is essential to assure reproducibility in the preparation of substrates. The setting of 1–2 for the Branson Sonifier 250 corresponds to 2.5–3 g sonic pressure.

5. It is important that the extraction mixture is prepared with high precision from solvents of high purity (*pro analysi*), since any deviations in its composition will affect the partition coefficient of OA in the system. When assaying samples of very low activity, OA (3 g/L) should be included in the extraction mixture as a carrier.

6. LPL is stabilized in this buffer both by heparin and by the detergents. The mixed micelles of Triton X-100 and SDS reduces the denaturing effect of SDS, and are excellent for stabilization. Another aspect is that this buffer is compatible with the activity assay and the ELISA.

7. BSA can cause problems. It is not necessary to use FA-free BSA. Some more pure preparations are strongly inhibitory. Check different batches. It is only

necessary to have sufficient binding capacity for the released FAs. Calculate on five binding sites per molecule BSA. If the binding capacity is exceeded, the lipolysis rate levels off with time, because of product inhibition.

8. Check local routines for handling of human blood in the laboratory. Select healthy donor.

9. There are several commercial in vitro transcription kits that can be used, instead of separate materials and reagents.

10. Before a study of HSL and MGL is initiated, it is recommended to establish their partition to the different fractions in the given system. Also, because of the well-documented translocation of HSL to the lipid droplet (fat cake fraction), upon lipolytic stimulation of adipocytes, it is important to consider HSL in all fractions, or, alternatively, establish that the parameter of study does not affect the distribution of HSL (and MGL) between the different fractions. The high lipid content of the fat-cake fraction makes it difficult to reliably measure the HSL and MGL activity of this fraction. By diluting the fraction to at least the same volume as the infranatant fraction, a rough estimation can be obtained. A more reliable way of checking the distribution is to perform quantitative Western blot analysis of aliquots of the different fractions. For these analyses, the fat-cake fractions should be solubilized in a SDS sample buffer containing 20% SDS *(8)*.

11. When assaying (partially) purified preparations of HSL, the assays described are linear, up to approx 10% of total substrate hydrolysis. However, for crude AT preparations, as well as all other crude tissue and cell, the HSL assays are linear in a much more narrow range. The reason for this is unknown. The poor linearity presents a problem when trying to compare HSL activities in crude samples. It is, therefore, very important to establish the range of linearity for a given system at the initiation of a study, and furthermore, to similar conditions, when comparing the activity between samples of, e.g., μL homogenate in the assay and % substrate hydrolysis.

12. Practice and experience are required to create a perfect lipid emulsion, and the sonication step is therefore likely to be the bottleneck in performing assays based on lipid emulsion substrates. For all the assays described, the sonicator tip should be placed 1 cm below the surface during the sonication, although it could be helpful to initially sonicate for a few seconds with the tip just at the surface, to release the dried lipids from the bottle of the glass vial. Bubbles should never be created during the sonication. If flakes of lipid remain after the completed sonication procedure, these may disappear by extending the sonication for a few 30-s periods. If not, the substrate should be discarded. Repeated problems with the formation of flakes is a sign of problems with the sonicator tip (or sonicator), which should then be calibrated or, if cavitated, exchanged. Four mL of substrate is enough for 36 assays (usually 18 samples in triplicates), 3 blanks, and determination of the substrate specific radioactivity ($3 \times 25$ μL). If more substrate is needed, we recommend making several 4-mL aliquots, which are combined after completed sonication, since, in our experience, it is difficult to get good emulsions when attempting to sonicate large volumes.

13. Sonicated lipid substrates should be prepared fresh and used within 3–4 h of preparation (2 h for the CO substrate).

14. To maximize the efficiency of the assay, and to minimize the exposure of the worker to radioactive solvents, it is very helpful to have an automatic diluter placed on a ventilated bench or hood for the transfer of 1 mL of the upper phase, together with 10 mL scintillation solution to a scintillation vial. Regardless of whether an automatic diluter is used, or the upper phase is transferred manually, it is important that the wall of the assay tube is not touched, since it is covered with a thin film of substrate.

15. In laboratories where the assay procedure has been made more efficient through the use of dispensers for the addition of extraction mix and buffer, automatic diluter (*see* **Note 13**), and so on, approx 100 samples can be processed in 2.5–3 h (excluding the time for the liquid scintillation counting).

16. The TO assay described under **Subheading 3.3.**, represents the routine assay for measurement of TG lipase activity, performed around the optimal pH for HSL (i.e., 7.0). However, at least for rat HSL, the use of a 0.5 m$M$ substrate (instead of 1.67 m$M$) at pH 8.3 (instead of 7.0) optimizes the difference in activity measured for unphosphorylated and PKA-phosphorylated (activated) HSL although the absolute values are only about 35–40% of those measured at pH 7.0. At pH 8.3 the substrate is prepared in a Tris-buffer, instead of a phosphate buffer (10 m$M$ Tris-HCl, 5 m$M$ NaCl, 0.5 m$M$ EDTA, pH 8.3), and the HSL samples are diluted in 10 m$M$ Tris-HCl, 5 m$M$ NaCl, 0.5 m$M$ EDTA, 1 m$M$ DTT, 0.02% defatted BSA.

17. The problems with linearity when assaying crude tissue homogenates that have been observed regarding HSL (*see* **Note 11**), has not been observed regarding MGL, in which linearity is observed up to at least 10% substrate hydrolysis, regardless of the purity of the samples.

18. Upon completed sonication, the MO substrates should be clear. If not, the substrates should be sonicated for a few more 30-s intervals. If this does not help, the substrate should be discarded. Repeated problems in making a good substrate is probably a sign of poor quality of the sonicator tip, which then should calibrated or, alternatively, exchanged.

19. For quantitative purposes, it is very important to establish the linearity range in the ECL system. Increased linearity range can be obtained by using a radioactive detection system instead, i.e., incubation with [125]I-labeled protein A (Amersham) at $0.1 \times 10^6$ cpm/mL at **step 5** in the protocol described under **Subheading 3.6.**, followed by washes (**step 6**) and phosphoimage analysis.

20. In most cases, these extracts can be stored frozen (–70°C) for later analyses, but the stability of the LPL activity with time should be checked in each case. Samples of rat AT retain more than 90% activity after 1 mo.

21. Buffers without detergents, such as the one used for extraction of HSL and MGL, or similar buffers without glycerol, extract only 25–50% of the total LPL, and the instability of the LPL activity in such extracts is usually a problem *(30)*. In older literature, acetone-diethyl ether extraction of AT was the dominating method. The dried protein powders were then extracted in regular buffers without detergents. Compared to the method recommended here, recovery of LPL activity from acetone-ether powders is only 10–30%, and could be even lower, depending on the technique for recovery of the protein.

22. LPL is thermolabile. Although the incubation mixture contains heparin and BSA for stabilization, there is considerable loss of activity at 37°C, not the least initially, before the enzyme has established contact with the lipid emulsion. LPL is much more stable at 25°C, at which the assay is usually linear with time for hours, unless the substrate is significantly consumed or the FA-binding capacity of BSA is exceeded. Theoretically, the maximal catalytic activity should be only half at 25°C, compared to 37°C.

23. Standard samples that are perfectly stable on storage are difficult to find. Ideally they should be from the same source as the samples assayed. LPL is, however, very stable in postheparin plasma (plasma from blood taken 10 min after injection of 100 U heparin/kg body wt). Plasma samples can be stored frozen in aliquots for 1–2 yr with almost full retention of activity. Another commonly used standard is bovine skim milk frozen at –70°C. The main purpose of the standard is to control the interassay variability.

24. If the blank values are too high in relation to the sample values (for samples with very low LPL activity, e.g., human AT), the more time-consuming extraction procedure, based on Dole's method, should be used (*see* protocol 8, **ref. *20***).

25. The weight of AT depots (at least in laboratory animals) changes dramatically with nutritional state, because of deposition/mobilization of fat. Therefore, LPL activity as well as other variables, should preferably be related to some stable parameter, e.g., tissue content of DNA *(31)*.

26. For standard, LPL from the same animal source should be used. The regression curves for standard and samples may otherwise not be parallel. Bovine LPL can be easily purified from milk, but, for other sources, this may not be easy. One possibility is to make a semipurification of active LPL by heparin–Sepharose chromatography of postheparin plasma or extracts of tissue homogenates. Then use a specific LPL activity of 400 U/mg to calculate the LPL content of the standard. This is of course only an approximation.

27. If dilutions fall outside the linear range, repeat the assay with more appropriate dilutions. Tissues may contain large amounts of catalytically inactive LPL. Therefore, the estimate of LPL protein from activity measurements may sometimes be underestimated.

28. Control templates like β-actin, glyceraldehydphosphate dehydrogenase (GAPDH), 28S RNA, and 18S RNA are used to produce probes that serve as internal controls in RPAs and Northern blots. They will hybridize to RNAs that are expressed at relatively constant levels. They are ideal for use in multiple-probe assays for simultaneous detection of an internal control RNA and the RNA(s) being studied.

29. Wear gloves. Use RNase-free materials and reagents. Purified RNA samples and RNA probes prepared by in vitro transcription are always susceptible to RNase digestion.

30. We have used the RPAII kit (Ambion), and followed the standard procedure described in the instruction manual. However, we do not use the GlycoBlue solution included in the kit.

31. Negative control: Probe + yeast RNA is digested with RNase, leaving little or no background signal. Positive control: Probe + yeast RNA without any RNase shows a single band representing full length probe.

## References

1. Langin, D., Holm, C., and Lafontan, M. (1996) Adipocyte hormone-sensitive lipase: a major regulator of lipid metabolism. *Proc. Nutr. Soc.* **55,** 93–109.
2. Fredrikson, G., Tornqvist, H., and Belfrage, P. (1986) Hormone-sensitive lipase and monoacylglycerol lipase are both required for complete degradation of adipocyte triacylglycerol. *Biochim. Biophys. Acta* **876,** 288–293.
3. Tornqvist, H. and Belfrage, P. (1976) Purification and some properties of a monoacylglycerol-hydrolyzing enzyme of rat adipose tissue. *J. Biol. Chem.* **251,** 813–819.
4. Bensadoun, A. (1991) Lipoprotein lipase. *Annu. Rev. Nutr.* **11,** 217–237.
5. Enerbäck, S. and Gimble, J. M. (1993) Lipoprotein lipase gene expression: physiological regulators at the transcriptional and post-transcriptional level. *Biochim. Biophys. Acta.* **1169,** 107–125.
6. Olivecrona, G. and Olivecrona, T. (1995) Triglyceride lipases and atherosclerosis. *Curr. Opin. Lipidol.* **6,** 291–305.
7. Zechner, R. (1997) The tissue-specific expression of lipoprotein lipase: implications for energy and lipoprotein metabolism. *Curr. Opin. Lipidol.* **8,** 77–88.
8. Fredrikson, G., Strålfors, P., Nilsson, N.Ö., and Belfrage, P. (1981) Hormone-sensitive lipase of rat adipose tissue. Purification and some properties. *J. Biol. Chem.* **256,** 6311–6320.
9. Østerlund, T., Danielsson, B., Degerman, E., Contreras, J. A., Edgren, G., Davis, R. C., Schotz, M. C., and Holm, C. (1996) Domain-structure analysis of recombinant rat hormone-sensitive lipase. *Biochem J.* **319,** 411–420.
10. Lee, F. T., Adams, J. B., Garton, A. J., and Yeaman, S. J. (1988) Hormone-sensitive lipase is involved in the hydrolysis of lipoidal derivatives of estrogens and other steroid hormones. *Biochim. Biophys. Acta* **963,** 258–264.
11. Wei, S., Lai, K., Patel, S., Piantedosi, R., Shen, H., Colantuoni, V., Kraemer, F. B., and Blaner, W. S. (1997) Retinyl ester hydrolysis and retinol efflux from BFC-1b adipocytes. *J. Biol. Chem.* **272,** 14,159–14,165.
12. Fredrikson, G. and Belfrage, P. (1983) Positional specificity of hormone-sensitive lipase from rat adipose tissue. *J. Biol. Chem.* **258,** 14,253–14,256.
13. Anthonsen, M., Rönnstrand, L., Wernstedt, C., Degerman, E., and Holm, C. (1997) Identification of novel phosphorylation sites in hormone-sensitive lipase that are phosphorylated in response to isoproterenol and govern activation properties *in vitro*. *J. Biol. Chem.* **273,** 215–221.
14. Egan, J. J., Greenberg, A. S., Chang, M.-K, Wek, S. A., Moos, Jr., M. C., and Londos, C. (1992) Mechanism of hormone-stimulated lipolysis in adipocytes: Translocation of hormone-sensitive lipase to the lipid storage droplet. *Proc. Natl. Acad. Sci. USA* **89,** 8537–8541.

15. Manganiello, V. C., Taira, M., Degerman, E., and Belfrage, P. (1995) Type III cGMP-inhibited cyclic nucleotide phosphodiesterases (PDE3 gene family). *Cell. Signal.* **7,** 445–455.

16. Ottosson M., Vikman-Adolfsson K., Enerbäck S., Olivecrona G., and Björntorp P. (1994) The effects of cortisol on the regulation of lipoprotein lipase activity in human adipose tissue. *J. Clin. Endocrinol. Metab.* **79,** 820–825.

17. Fredrikson, G., Strålfors, P., Nilsson, N.Ö., and Belfrage, P. (1981) Hormone-sensitive lipase from adipose tissue of rat. *Meth. Enzymol.* **71,** 637–646.

18. Holm, C., Contreras, J. A., Verger, R., and Schotz, M. C. (1997) Large-scale purification and kinetic properties of recombinant hormone-sensitive lipase from baculovirus/insect cell systems. *Meth. Enzymol.* **284,** 272–284.

19. Holm, C. and Østerlund, T. (1998) Hormone-sensitive lipase and neutral cholesteryl ester lipase, in *Methods in Molecular Biology: Lipase and Phospholipase Protocols* (Reue, K. and Doolittle, M., eds.), Humana, Totowa, NJ, pp. 109–121.

20. Bengtsson-Olivecrona, G. and Olivecrona, T. (1992) Assay of lipoprotein lipase and hepatic lipase, in *Lipoprotein Analysis. A Practical Approach* (Converse, C. and Skinner, E. R., eds.), IRL, Oxford, pp. 169–185.

21. Olivecrona, T. and Olivecrona, G. (1997) Determination and clinical significance of lipoprotein lipase and hepatic lipase, in *Handbook of Lipoprotein Testing* (Rifai, N., Warnick, G. R., and Dominiczak, M. H., eds.), AACC, Washington, DC, pp. 373–391.

22. Belfrage, P. and Vaughan, M. (1969) Simple liquid-liquid partition system for isolation of labeled oleic acid from mixtures with glycerides. *J. Lipid Res.* **10,** 341–344.

23. Tornqvist, H., Björgell, P., Krabisch, L., and Belfrage, P. (1978) Mono-acylmonoalkylglycerol as a substrate for diacylglycerol hydrolase activity in adipose tissue. *J. Lipid Res.* **19,** 654–656.

24. Deckelbaum, R. J., Ramakrishnan, S., Eisenberg, S., Olivecrona, T., and Bengtsson-Olivecrona, G. (1992) Triglyceride and phospholipid hydrolysis in human plasma lipoproteins: Role of lipoprotein lipase and hepatic lipase. *Biochemistry* **31,** 8544–8551.

25. Tornqvist, H., Krabisch, L., and Belfrage, P. (1974) Simple assay for mono-acylglycerol hydrolase activity of rat adipose tissue. *J. Lipid Res.* **15,** 291–294.

26. Ottosson, M., Vikman-Adolfsson, K., Enerbäck, S., Elander, A., Björntorp, P., and Edén, S. (1995) Growth hormone inhibits lipoprotein lipase activity in human adipose tissue. *J. Clin. Endocrinol. Metab.* **80,** 936–941.

27. Hirsch, A. H. and Rosen, O. M. (1984) Lipolytic stimulation modulates the subcellular distribution of hormone-sensitive lipase in 3T3-L1 cells. *J. Lipid Res.* **25,** 665–677.

28. Londos, C., Gruia-Gray, J., Brasaemle, D. L., Rondinone, C. M., Takeda, T., Dwyer, N. K., Barber, T., Kimmel, A. R., and Blanchette-Mackie, E. J. (1996) *Int. J. Obesity Relat. Metab. Disord.* **20, Suppl. 3,** S97–S101.

29. Clifford, G. M., McCormick, D. K., Vernon, R. G., and Yeaman, S. J. (1997) Translocation of hormone-sensitive lipase and perilipin in response to lipolytic enzymes. *Biochem. Soc. Trans.* **25,** S672.

30. Semb, H. and Olivecrona, T. (1986) Nutritional regulation of lipoprotein lipase in guinea pig tissues. *Biochim. Biophys. Acta.* **876,** 249–255.
31. Bergö, M., Olivecrona, G., and Olivecrona, T. (1997) Regulation of adipose tissue lipoprotein lipase in young and old rats. *Int. J. Obes.* **21,** 980–986.
32. Doolittle, M. H. and Ben-Zeev, O. (1999) Immunodetection of lipoprotein lipase:Antibody production, immunoprecipitation, and Western blotting techniques, in *Methods in Molecular Biology, Vol. 109, Lipase and Phospholipase Protocols* (Doolittle, M. H. and Reue, K., eds.), Humana , Totowa, NJ, pp. 215–237.
33. Vilella, E. and Joven, J. (1999) In vitro measurement of lipoprotein and hepatic lipases, in *Methods in Molecular Biology, Vol. 110, Lipoprotein Protocols* (Ordovas, J. M., ed.), Humana, Totowa, NJ, pp. 243–251.
34. Ranganathan, G. and Kern, P. A. (1999) Techniques for the measurement of lipoprotein lipase messenger RNA, in *Mehthods in Molecular Biology, Vol. 109, Lipase and Phospholipase Protocols* (Doolittle, M. H. and Reue, K., eds.), Humana, Totowa, NJ, pp. 329–335.

# 9

## Assays of Lipogenic Enzymes

### Raymond Bazin and Pascal Ferré

### 1. Introduction

The deposition of fat in adipose tissue (AT) is an important aspect of energy metabolism. AT fat originates either from ingested fatty acids (FAs) or from *de novo* synthesis. The synthesis of FAs is active in several tissues (liver, mammary gland, lung, intestine, brown and white ATs); however, in rodents, white AT is the major site of lipogenesis, and contributes to more than 50% of total FA biosynthesis. Lipogenesis from glucose is regulated by the nutritional state: It is decreased to a minimum in carbohydrate deficiency, and is considerably enhanced during carbohydrate availability.

Several different regulatory processes are involved in the control of FA synthesis, including substrate and coenzyme availability, regulation of key enzymatic steps by allosteric and covalent modification mechanisms, and regulation of the quantity of key enzymes.

This chapter focuses first on acetyl coenzyme A (CoA) carboxylase (ACC; EC 6.4.1.2) and fatty acid synthase (FAS; EC 2.3.1.85), which are directly involved in FA synthesis from two-carbon units. ACC catalyzes the irreversible formation of malonyl-CoA from acetyl-CoA. It contains biotin as its prosthetic group, covalently bound in amide linkage to the $\varepsilon$-amino group of a lysine residue on one of the three subunits of the enzyme molecule. In eukaryotes, ACC contains the biotin carboxylase and the transcarboxylase activities, as well as the biotin carboxyl carrier on a single polypeptide chain of 265 kDa. ACC polymerizes in forms of molecular masses in the range of 4000–8000 kDa; This polymerization is promoted by citrate *(1)*. In AT, formation of malonyl-CoA has been found to be the rate-limiting reaction in the synthesis of FAs from acetyl-CoA *(2)*. The energy needed for this reaction is provided by adenosine triphosphate (ATP).

From: *Methods in Molecular Biology, vol. 155: Adipose Tissue Protocols*
Edited by: G. Ailhaud © Humana Press Inc., Totowa, NJ

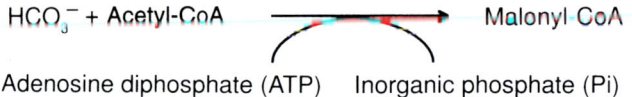

$$HCO_3^- + Acetyl\text{-}CoA \longrightarrow Malonyl\text{-}CoA$$

Adenosine diphosphate (ATP)    Inorganic phosphate (Pi)

This reaction transfers the carboxyl group derived from bicarbonate to acetyl-CoA, to yield malonyl-CoA. The method measures the incorporation of [14]C-labeled into malonyl CoA, as described by Martin and Vagelos *(2)*, but with Tris-acetate buffer *(3)*. Citrate is added in the medium to promote maximal activation (polymerization) of ACC *(1)*. Since this method could also measure the activity of other carboxylases, such as pyruvate carboxylase, prior to the assay, pyruvate in the homogenate supernatant is eliminated by filtration on columns containing Sephadex G-25 (*see* **Note 1**).

FAS is a multifunctional protein (Mr 240,000) that catalyzes the synthesis of long-chain FAs from acetyl-CoA and malonyl-CoA. Its activity is correlated with the rate of FA synthesis. This protein is active as a dimer (Mr 480,000). Reduced nicotinamide adenine dinucleotide phosphate (NADPH) is the electron carrier needed to reduce double bonds and carbonyl groups of several intermediates produced during synthesis of FA by the FAS complex. Allosteric effectors of FAS are not known, but the concentration of the enzyme is highly sensitive to nutritional and hormonal regulation *(4)*.

$$Acetyl\text{-}CoA + 7\ Malonyl\text{-}CoA \longrightarrow Palmitate + 7\ CO_2 + 8\ CoA + 6\ H_2O$$

14 NADPH, H$^+$    14 NADP$^+$

The FA chain is built up from condensation of acetyl-CoA and malonyl-CoA, followed by two steps of reduction, which require NADPH as electron donor. In the assay, the oxidation of NADPH at 340 nm is measured for 10 min at 37°C *(5)*.

Two other enzymes are closely related to the lipogenic pathway: citrate lyase, also called citrate cleavage enzyme (CCE; EC 4.1.3.8); and malic enzyme (ME; EC 1.1.1.40). Nearly all the acetyl-CoA used in FA synthesis is formed in the mitochondria, from pyruvate oxidation and from the catabolism of the carbon skeletons of amino acids. Because the mitochondrial inner membrane is impermeable to acetyl-CoA, an indirect shuttle transfers acetyl group equivalents across the inner membrane. Intramitochondrial acetyl-CoA first reacts with oxaloacetate to form citrate. Citrate then passes into the cytosol, in which cleavage by CCE regenerates acetyl-CoA and oxaloacetate. This reaction is driven by the investment of energy from ATP. Oxaloacetate is reduced by cytosolic malate dehydrogenase (MDH) to malate, then part of the malate produced in the cytosol is used to generate cytosolic NADPH, through the activity of ME. In adipocytes, cytosolic NADPH is mostly generated by ME; however, NADPH is also supplied by the reactions of the pentose phosphate pathway.

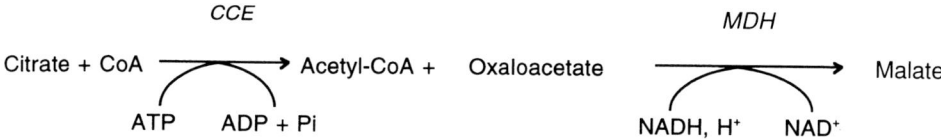

Citrate is cleaved by CCE to give oxaloacetate, which in turn is converted into malate by MDH. Thus, the assay is based on the measurement of NADH oxidation at 340 nm (**6**).

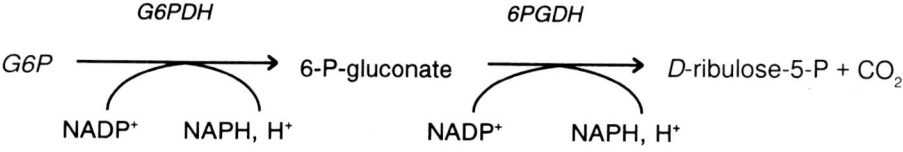

In the presence of $NADP^+$ as co-factor, the decarboxylation of malate produces pyruvate. The assay is performed according to Ochoa (**7**), but with 100 m$M$ malate. Formation of NADPH is measured at 340 nm.

Glucose-6-phosphate dehydrogenase (G6PDH; EC 1.1.1.49) activity determines the entry of G6P into the pentose phosphate pathway, and $NADP^+$ is the electron acceptor of this reaction. In another step of this pathway, 6-phosphogluconate (6PG) undergoes dehydrogenation and decarboxylation by 6-phosphogluconate dehydrogenase (6PGDH; EC 1.1.1.44), a reaction that generates a second molecule of NADPH.

| | G6PDH | | 6PGDH | |
|---|---|---|---|---|
| *G6P* | → | 6-P-gluconate | → | *D*-ribulose-5-P + $CO_2$ |
| | $NADP^+$  NAPH, H⁺ | | $NADP^+$  NAPH, H⁺ | |

The enzymatic dehydrogenation of G6P forms 6-P-glucono-δ lactone, an intramolecular ester, which is hydrolyzed to the free acid 6-P-gluconate by a lactonase. In the next step, 6-P-gluconate undergoes dehydrogenation and decarboxylation by 6PGDH to form the ketopentose, D-ribulose-5-P, a reaction that generates a second molecule of NADPH.

Determination of these two activities is based on the measurement of NADPH at 340 nm (8). In a first reaction, both G6PDH and 6PGDH activities are measured in the presence of G6P and 6PG (see **Note 2**). The activity of 6PGDH alone is determined in the presence of 6PG, and the G6PDH one is obtained by difference between the sum of dehydrogenase activities and 6PGDH activity.

Describe here are methods developed to measure these lipogenic enzymes, either directly involved in FA synthesis (FAS and ACC) or closely related to this pathway (CCE, ME, G6PDH, and 6PGDH). These enzymes are found in the cytosol. This location segregates synthetic processes from degradative reactions, many of which are taking place in the mitochondrial matrix.

## 2. Materials

1. Solution 1: 60 m$M$ Tris-acetate/8 m$M$ MgCl$_2$ buffer containing 2 m$M$ dithiotreitol (DTT), 20 m$M$ potassium (K) citrate, and 10 mg/mL FA-free bovine serum albumin at pH 7.4.
2. Solution 2: 15 m$M$ ATP in solution 1 at pH 7.4.
3. Solution 3: 2.5 m$M$ acetyl-CoA in solution 1 at pH 7.4.
4. Solution 4: In a flask containing 5.5 mmol NaHCO$_3$ in 15 mL dH$_2$O, and closed with a rubber cap, add, through the cap (with a needle and a syringe), 5 mL 100 m$M$ $^{14}$C-NaHCO$_3$ (1 μCi/μmol). The solution then contains 300 mmol and 25 μCi $^{14}$C-NaHCO$_3$/mL.
5. Solution 5: 100 m$M$ potassium phosphate buffer, at pH 6.5, containing 200 μ$M$ NADPH and 100 μ$M$ acetyl-CoA.
6. Solution 6: 100 m$M$ potassium phosphate buffer, at pH 6.5, containing 600 μ$M$ malonyl-CoA.
7. Solution 7: 200 m$M$ Tris-HCl buffer containing 20 m$M$ MgCl$_2$, at pH 8.7, containing 20 m$M$ K citrate, 1 m$M$ DTT, 1 U/mL MDH, 0.5 m$M$ CoA, and 0.2 m$M$ NADH.
8. Solution 8: 200 m$M$ Tris-HCl buffer containing 20 m$M$ MgCl$_2$, at pH 8.7, containing 60 m$M$ ATP.
9. Solution 9: 200 m$M$ Tris-HCl buffer containing 20 m$M$ MgCl$_2$ buffer, at pH 7.4, containing 2 m$M$ NADP
10. Solution 10: 200 m$M$ Tris-HCl buffer containing 20 m$M$ MgCl$_2$, at pH 7.4, containing 100 m$M$ malate.
11. Solution 11: 200 m$M$ Tris-HCl buffer containing 20 m$M$ MgCl$_2$, at pH 7.4, containing 2 m$M$ NADP.
12. Solution 12: 200 m$M$ Tris-HCl buffer containing 20 m$M$ MgCl$_2$, at pH 7.4, containing 1 m$M$ G6P and 1 m$M$ 6PG.
13. Solution 13: 200 m$M$ Tris-HCl buffer containing 20 m$M$ MgCl$_2$, at pH 7.4, containing 1 m$M$ 6PG.
14. Solution 14: Homogenization buffer: 0.25 $M$ sucrose containing 1 m$M$ DTT, 1 m$M$ EDTA, and a mixture of several proteases inhibitors (Complete, Boehringer Mannheim) at pH 7.4.
15. Solution 15: 60 m$M$ Tris-acetate buffer containing 5 m$M$ MgCl$_2$ at pH 7.4.
16. Solution 16: 200 m$M$ Tris-HCl buffer containing 20 m$M$ MgCl$_2$ at pH 8.7.

17. Solution 17: 200 m$M$ Tris-HCl buffer containing 20 m$M$ MgCl$_2$ at pH 7.4.
18. Potter homogenizers (all-glass).
19. Double-beam spectrophotometer equipped with a recorder.

## 3. Methods
### 3.1. Preparation of Homogenates

1. Homogenize 50–100 mg AT or $2.10^6$ isolated adipocytes in 1 mL of ice-cold homogenization sucrose buffer (solution 14).
2. Centrifuge homogenates at 105,000$g$ at 0°C for 60 min.
3. Carefully remove the clear supernatant (below the fat cake), and either use immediately or freeze at –20°C for further determination of lipogenic enzyme activities.

### 3.2. Assay Procedures
#### 3.2.1. Acetyl-CoA Carboxylase

1. In glass scintillation vials, distribute 0.5 mL solution 2.
2. Add 0.1–0.3 mL homogenate supernatant after filtration on Sephadex G 25 (*see* **Note 1**).
3. Seal the flasks with rubber caps.
4. Deliver, with a syringe, 0.1 mL solution 4.
5. Incubate for 20 min at 37°C.
6. Start the reaction with 0.2 mL solution 3, and incubate 2–4 min at 37°C.
7. Stop the reaction by addition of ice-cold 1 $M$ HCl.
8. Remove last traces of $^{14}CO_2$ by gassing the vials with $CO_2$ for 45 min, or by evaporation to dryness at 80°C and the subsequent addition of 1 mL water.
9. Add 10 mL scintillation fluid appropriate for aqueous solutions. Correct all radioactivity counts for blank run in the absence of acetyl-CoA (solution 3). Verify that the reaction is linear with respect to time and to sample concentration.

#### 3.2.2. Fatty Acid Synthase

1. In semimicrocuvets (1 mL) for spectrophotometer with 1 cm light path, mix 700 μL solution 5 and 10–100 μL homogenate supernatant. Wait for temperature equilibrium at 37°C (5 min).
2. Start the reaction with 100 μL solution 6, and run for 10 min at 37°C against blank (100 μL water instead of 100 μL malonyl-CoA solution).
3. Calculations: Determine the amount of NADPH oxidized (molar extinction coefficient of NADPH at 340 nm: $\varepsilon = 6.22 \times 10^3$ $M$/cm; *see* **Note 3**). Results are expressed as the amount of enzyme needed to catalyze the oxidation of 1 n$M$ NADPH.

#### 3.2.3. Citrate Cleavage Enzyme

1. In semimicrocuvets (1 mL) for spectrophotometer mix 700 μL solution 7, and 100 μL homogenate supernatant. Wait for temperature equilibrium at 37°C (5 min).
2. Start the reaction with 100 μL solution 8, and run for 10 min at 37°C, against blank (100 μL water, instead of 100 μL ATP solution).

### 3.2.4. Malic Enzyme

1. In semimicrocuvets (1 mL) for spectrophotometer mix 700 µL solution 9 and 10–20 µL homogenate supernatant. Wait for temperature equilibrium at 37°C (5 min).
2. Start the reaction with 100 µL solution 10, and run for 10 min at 37°C, against blank (100 µL water, instead of 100 µL malate solution).

### 3.2.5. G6P- and 6PGDHs

#### 3.2.5.1. SUM OF G6PDH AND 6PGDH

1. In semimicrocuvets (1 mL) for spectrophotometer, mix 700 µL solution 11 and 50–100 µL homogenate supernatant. Wait for temperature equilibrium at 37°C (5 min).
2. Start the reaction with 100 µL solution 12, and run for 10 min at 37°C, against blank (100 µL water, instead of the solution of G6P and 6PG).

#### 3.2.5.2. 6PGDH

1. In semimicrocuvets (1 mL) for spectrophotometer, mix 700 µL solution 11 and 50–100 µL homogenate supernatant. Wait for temperature equilibrium at 37°C (5 min).
2. Start the reaction with 100 µL solution 13, and run for 10 min at 37°C, against blank (100 µL water, instead of 100 µL 6PG solution).

## 4. Notes

1. To eliminate pyruvate from the medium, homogenate supernatant is treated by gel filtration on Sephadex G25 Medium. $10 \times 100$ mm propylene columns are packed with Sephadex G 25, and rinsed with water. 1 mLhomogenate supernatant is applied on the top of the column, and elution is made with sucrose homogenization buffer. Take care to minimize dilution of homogenate (maximum, twofold).
2. Lactonase activity is brought about by the homogenate supernatant.
3. The Bouguer-Lambert-Beer law states that the extinction, $E$, is proportional both to the light path, $d$, and to concentration, $c$, of the absorbing substance ($E = \varepsilon \times d \times c$). The proportionality constant $\varepsilon$ is the extinction of the substance in question at a concentration of unity with a light path of 1 cm. The dimension found for $\varepsilon$ is $cm^2/mol$ (e.g., for NADPH or NADH, $\varepsilon_{340nm} = 6.22 \times 10^6$ $cm^2/mol$). It is more convenient to express the concentration in mol/L. With the unit mol/L ($M$), for the concentration, $c$, the molar extinction coefficient, has the dimension $M/cm$. The extinction coefficient based on 1 $mol/cm^3$ is 1000 times the molar extinction coefficient based on 1 mol/L; for NAD(P)H, $\varepsilon_{340nm} = 6.22 \times 10^3$ $M/cm$.

## References

1. Beaty, N. B. and Lane, M. D. (1983) The polymerization of acetyl-CoA carboxylase. *J. Biol. Chem.* **258**, 13,051–13,055.
2. Martin, D. B. and Vagelos, P. R. (1962) The mechanism of tricarboxylic acid cycle regulation. *J. Biol. Chem.* **237**, 1787–1792.

3. Allred, J. B. and Roehrig, K. L. (1980) Inhibition of rat liver acetylCoA carboxylase by chloride. *J.Lipid Res.* **21,** 488–491.
4. Volpe, J. J. and Vagelos, P. R. (1976) Mechanisms and regulation of biosynthesis of saturated fatty acids. *Physiol. Rev.* **56,** 339–417.
5. Halestrap, A. P. and Denton, R. M. (1973) Insulin and the regulation of adipose tissue acetyl coenzyme A carboxylase. *Biochem. J.* **132,** 509–513.
6. Cottam, G. L. and Srere, P. A. (1969) The sulfhydryl groups of citrate cleavage enzyme. *Arch. Biochem. Biophys.* **130,** 304–311.
7. Ochoa, S. (1955) Malic enzyme. *Method Enzymol.* **1,** 739-753.
8. Glock, C. E. and McLean, P. (1953) Further studies on the properties and assay of glucose-6-phosphate dehydrogenase and 6-phosphogluconate dehydrogenase of rat liver. *Biochem. J.* **55,** 400–408.

# 10

# Assays of Adrenergic Receptors

*Including Lipolysis and Binding Measurements*

## Christian Carpéné

## 1. Introduction

   Adipocytes express the three major types of adrenoceptors (ARs), $\alpha_1$-, $\alpha_2$-, and $\beta$-ARs, each of them being further divided into three subtypes. With the exception of $\alpha_{1A}$-, $\alpha_{1B}$-, and $\alpha_{1D}$-ARs, which have not been described to directly influence cAMP production in fat cells, all the other AR subtypes are known to interact with the regulation of lipolytic activity through stimulatory ($\beta_1$-, $\beta_2$-, and $\beta_3$-ARs) or inhibitory ($\alpha_{2A/D}$-, $\alpha_{2C}$-, and putatively $\alpha_{2B}$-ARs) G-protein-coupled mechanisms (reviewed in **ref. 1**). This explains why, besides their direct quantification by radioligand-binding analyses, the assays on ARs described in this chapter will also include the measurement of one of the biological effects they regulate: lipolysis.

   Vertebrates present large interspecies variations in the adrenergic receptivity of adipocytes (receptor number and lipolytic responsiveness). In birds, catecholamines (CATs) play a minor role in lipid mobilization; in rodents, fat cells exhibit high $\beta_3$-adrenergic responsiveness (except in guinea-pigs) and, in humans, subcutaneous adipocytes express large amounts of $\alpha_2$-ARs, (particularly in the femoral region of obese women) *(2)*. The philogeny of these species-specific differences is not elucidated. However, changes in the adrenergic responsiveness of adipose tissue (AT) are mostly studied in a limited number of species (obese rodents or humans) and are related to environmental and hormonal conditions, or to pharmacological treatments. To characterize such changes, it is thus necessary to determine the potency and efficacy of CATs (or selective AR-agonists) on biological response(s), and to quantify the ligand–receptor interactions in AT, according to the principles of quantitative pharmacology. Therefore, lipolyis measurements are often used to define the maximal

From: *Methods in Molecular Biology, vol. 155: Adipose Tissue Protocols*
Edited by: G. Ailhaud © Humana Press Inc., Totowa, NJ

effect (or intrinsic activity, when one subset of AR is concerned) and the $EC_{50}$ (molar concentration of an agonist that produces 50% of the maximal effect) of the studied drugs; binding experiments allow to calculation of the number of receptors ($B_{max}$) and the dissociation equilibrium constant ($KD$) for the chosen ligand(s). Therefore, receptor occupancy/response can be deduced for ARs, and putative changes in adipocyte adrenergic receptivity evidenced.

## 1.1. Available Adrenergic Selective Agonists and Antagonists

**Table 1** summarizes the AR subtypes already described in adipocytes, several agents, and radioligands. This list is not exhaustive, and is limited, to help in choosing the compounds to be used among the numerous adrenergic agonists and antagonists available. Furthermore, adrenaline, noradrenaline, and some poorly selective agonists (e.g., isoprenaline, mixed β) or antagonists (propranolol, mixed β; phentolamine, mixed α) can also be useful for obtaining information on the overall adrenergic responsiveness. For radioligands, the higher the affinity the better, because a lower concentration of radioligand can be used in binding assay, and results in a lower level of nonspecific binding (NSB). Typically, the affinities of $^3$H-radioligands are in the n$M$ range; the maximal effect of agonists on lipolysis is obtained between 1 and 10 μ$M$. Finally, there is no absolutely selective compound for any given AR subtype, and each subtype should not be characterized by only one selective agent, but by the relative order of potency of a complete set of various drugs.

## 1.2. Determination of Glycerol Released by Isolated Adipocytes

Only lipolysis of isolated adipocytes will be considered here. The triglyceride breakdown occurring during in vitro lipolysis generates both free fatty acids, di- and mono-glycerides, and glycerol. Since adipocytes lack glycerokinase (GK) activity, the measurement of the extracellular glycerol released by the fat cells in the incubation medium is sufficient for the determination of lipolytic activity, although it does not provide enough information about re-esterification processes. Generally, the incubation period for lipolysis assays is 60–90 min, since glycerol accumulation increases linearly with time within this period.

Enzymatic determination of glycerol is based on its phosphorylation by GK (EC 2.7.1.30).

Glycerol + adenosine triphosphate → Glycerol 3-phosphate + adenosine diphosphate

and its subsequent transformation by glycerol-3-phosphate dehydrogenase (G3PDH, EC 1.1.1.8.),

Glycerol 3-phosphate + NAD → Dihydroxyacetone phosphate + NADH

where NAD = nicotinamide adenine dinucleotide and NADH = reduced NAD.

**Table 1**
**Recommended Adrenergic Agents for Assays of ARs**[a]

| Receptor subtype | Agonist | Antagonist | Radioligand |
|---|---|---|---|
| $\beta_1$-AR | Dobutamine | Betaxolol | [3H]-CGP 12177 |
|  | TO 509 | CGP 20712A |  |
| $\beta_2$-AR | Salbutamol | ICI 118551 | [$^3$H]-ICI 118551 |
|  | Procaterol |  | [$^3$H]-CGP 12177 |
| $\beta_3$-AR | CL 316243 | SR 59230A | [$^{125}$I]Iodocyanopindolol |
|  | BRL 37314 | Burpranolol | [$^3$H]-CGP 12177 |
| $\alpha_{2A}/\alpha_{2D}$ | UK 14304 | Methoxy-idazoxan | [$^3$H]-RX 821002 |
| $\alpha_{2C}$ | Clonidine | MK 912 | [$^3$H]-MK912 |

[a] CGP 12177 is considered as a $\beta_1$-/$\beta_2$-antagonist, and as an agonist for $\beta_3$- and putative $\beta_4$-ARs. Dexmedetomidine can stimulate $\alpha_2$B-ARs, the presence of which is disputable in adipocytes. Oxymetazoline and prazosin can activate and block the $\alpha_1$-ARs, respectfully. RX 821002 is also called methoxy-idazoxan.

The reaction is rendered irreversible by the presence of hydrazine ($NH_2$-$NH_2$), which reacts with the ketone function of dihydroxyacetone phosphate. Thus, spectrophotometric determination at 340 nm of the NADH formed is proportional to the amount of glycerol present in the assay, as first described by Wieland *(3)* (*see* **Note 1**).

## 1.3. AR Binding Assays

Crude membrane preparations can be readily prepared from freshly isolated adipocytes, and are suitable for binding assays using separation of free and bound radioligand by vacuum filtration through glass-fiber filters. The most important step in binding assay is the determination of specific binding to the AR of interest. In effect, specific binding is calculated as the difference between total binding (TB) and NSB. TB is binding to the receptors studied plus any other binding. NSB is determined in the presence of an excess of unlabeled drug (at least 1000× higher than the concentration of the radioligand) to occupy fully the receptors, and includes the binding to any other receptor sites in the preparation, the adsorption to filters themselves, or the dissolution in membrane lipids. Characteristically, NSB increases linearly with the concentration of radioligand while specific binding is saturable, upon reaching $B_{max}$. The most frequently used linear transformation of saturation data is termed "Scatchard analysis," but deserves to be more properly called "Rosenthal plot" since the "bound/free vs bound" plot useful for receptor binding studies was first used by Rosenthal *(4)*. This representation allows calculation of the number of receptors ($B_{max}$) and *the affinity constant ($K_D$)*.

For binding assays aimed at determining the number of ARs present in fat cell membranes, or even in competition experiments with unlabeled compounds, the following conditions have to be settled in preliminary experiments:

1. Membrane preparations must be sufficiently homogenous for the detection of reliable amounts of labeled receptors. Obviously, increasing protein concentration will increase receptor number and the ratio of specific to NSB, but the ligand bound to the receptors does not exceed 10% of the added radioligand, and the chosen membrane concentration must be in the range in which the amount of specific binding is linearly related to protein concentration.
2. Incubation time needs to be sufficient to reach equilibrium or at least steady state. Another crucial component of the binding assay is the separation of bound from free radioligand: It must be sufficiently fast, prevent significant dissociation of the receptor–ligand complex, and needs to be reproducible. This is generally achieved using the vacuum filtration technique.

## 2. Materials
### 2.1. Preparation of Buffers and Enzymatic Reagents for Functional Adipocytes

1. Krebs-Ringer medium buffered with bicarbonate plus HEPES, and containing albumin (KRBHA buffer) must be freshly prepared from stock solutions of salts and buffers (stored at 6°C) to give: 120 m$M$ NaCl, 4.8 m$M$ KCl, 2.5 m$M$ CaCl$_2$, 1.2 m$M$ KH$_2$PO$_4$, 1.2 m$M$ MgSO$_4$, 15 m$M$ NaHCO$_3$, 10 m$M$ HEPES. This solution must be gassed at least 5 min with carbogen (95% O$_2$, 5% CO$_2$). Thereafter, carefully dissolve bovine serum albumin, fraction V (3.5 g/100 mL) and glucose (108 mg/100 mL); place at 37°C; and adjust at pH 7.4 with NaOH before use.
2. HGM buffer: Prepare, under fume hood, 52 g hydrazine hydrate (approx 50 mL), 15 g glycine, 2 mL 1 $M$ MgCl$_2$ in a total volume of 1 L; pH 9.0–10.0. This is stable at 6°C for up to 12 mo (*see* **Note 2**).
3. Dole and Meinertz extraction solvent: 50 mL heptane, 200 mL isopropanol, 5 mL 1 $N$ H$_2$SO$_4$. Store refrigerated in glass bottle only (*see* **Note 2**).
4. GK (85 U/mL) and G3PDH (1700 U/mL) solutions (Boehringer Mannheim, Germany).
5. Glycerol standard at 10 m$M$. Store refrigerated.
6. One layer of a piece from polyamide-nylon voile pantyhose fixed with elastic rubberband on a truncated 25-mL syringe is a convenient, disposable, and inexpensive material to filter AT after collagenase digestion.
7. Polyethylene vials (5 mL), disposable polystyrene microcuvets, 50-mL polypropylene conical tubes (Falcon, Becton Dickinson, Franklin Lakes, NJ). Use plasticware vessels only in all the steps concerning functional fat cells.
8. UV-visible spectrophotometer and shaking water bath.

### 2.2. Preparation of Adrenergic Compounds at Suitable Concentrations

Most of the adrenergic compounds listed in **Table 1** can be commercially purchased from Sigma, RBI Biochemical (Natick, MA), Tocris Cookson

(Bristol, UK), or equilavent. They also can be requested from their originaly companies. Their dose-dependent effects are tested at final concentrations between $10^{-10}$ and $10^{-4}$ $M$. For lipolysis measurements, adrenergic compounds and drugs should be aliquoted and stored frozen as 100× more concentrated than the final maximal doses tested (typically $10^{-4}$ $M$), since they will be added as 10-μL portions in 1 mL cell suspension. Thus, prepare several ml of $10^{-2}$ $M$ of the adrenergic compounds of interest (in water or appropriate vehicle: EtOH, dimethyl sulfoxide [DMSO]), store as 100- to 500-μL aliquots or use immediately (especially for the CATs) for preparation of the different dilutions to be tested. For instance, mixing 100 μL $10^{-2}$ $M$ stock solution plus 900 μL water is a convenient way to prepare enough compound at $10^{-3}$ $M$ to be tested at final concentration of $10^{-5}$ $M$ (as 10-μL portions), or to be further diluted (100 μL plus 900 μL water) to obtain $10^{-6}$ $M$ final, and so on. For drugs, as for radioligands, dilutions can be done on ice, and must be prepared immediately before the incubations.

## 2.3. Binding of AR on Membrane Preparations

1. Lysing buffer: 2 m$M$ Tris-HCl, 1 m$M$ KHCO$_3$, 2.5 m$M$ MgCl$_2$. Store refrigerated as 10X. Before cell lysis, dilute 10-fold, then, to the appropriate volume for the planned experiment, add 76.1 mg/100 mL of EGTA and 100 μL/100 mL thawed aliquots of antiproteases (1000X), in order to reach the final concentrations: 0.1 m$M$ 4-(2-aminoethyl)benzenesulfonyl fluoride (AEBSF), 0.1 m$M$ benzamidine. Adjust pH to 7.5. Because cooling conditions used to protect from proteolysis induce the formation of a coalescent fat cake that could trap most of the membranes and deeply alter the fractionation yield, it is necessary in the case of adipocytes, to conduct the first steps of membrane peparation at room temperature in the presence of antiproteases in the lysing medium.
2. Binding buffer: 50 m$M$ Tris-HCl, 0.5 m$M$ MgCl$_2$, pH 7.5. Store refrigerated as 10X.
3. Washing buffer: 10 m$M$ Tris-HCl, 0.5 m$M$ MgCl$_2$, pH 7.5. Store refrigerated as 10X.
4. Vacuum filtration device: Manifold (Millipore, Bedford, MA), Cell harvester (Skatron, Tranby, Norway), or equivalent.
5. Glass microfiber filters GF/C, in strips or circles (Whatman, Maidstone, UK).
6. Kit for protein determination, commercially purchased (Bio-Rad, Hercules, CA, DC protein assay, or equivalent).
7. Scintillation cocktail for aqueous solutions.
8. β-counter or γ-counter.
9. Refrigerated centrifuge capable of 40,000$g$.

## 2.4. Preparation of Radioligands for Membrane Receptor Assays

Radioligands are available from Amersham or New England Nuclear or from any other purchaser. Most of them are furnished in EtOH–acid aqueous solution, so they can be stored cool but not frozen. Because binding assay tubes will contain 50 μL buffer, 50 μL competitor, 50 μL membranes, and 50 μL radioligand, it is necessary to prepare dilutions of radioligand 4× more concentrated than the expected final concentrations. For instance, [³H]-RX 821002, an $\alpha_2$-AR ligand generally purchased at 56 Ci/mmol (1 mCi/mL) and at 17,800 n$M$, is needed at a maximal final concentration of 10 n$M$. Then diluting 4.5 µL stock solution in 2000 µL binding buffer gives [³H]-RX 821002 solution at 40 n$M$, and approx 255,000 dpm/50 µL. This volume is sufficient to obtain at least three tubes at 10 n$M$ (for TB, NSB, and determination of total radioactivity [TOT] present in 50 µL), the remainder being used to prepare lower concentrations. Typically, a saturation study is conducted with 12 concentrations ranging from 0.02 to 20 n$M$.

## 3. Methods

### 3.1. Adipocyte Isolation

1. AT (from intra-abdominal or subcutaneous white fat depots of rodents) is dissected, roughly minced with scissors, and digested in KRBHA buffer containing 1.5 mg/mL collagenase (approx 10 mg collagenase/g fresh AT) for 30–45 min at 37°C in a shaking water bath (100 cycles/min). For human AT, collagenase concentration is 1 mg/mL, and digestion lasts 25 min (*see* **Note 3**).
2. After digestion, filter the AT through the nylon mesh system fixed on a truncated syringe. Collect the filtrate in 50-mL polypropylene conical tubes, and leave at 37°C without shaking. Discard the medium under the floating fat cells (containing the stromal-vascular fraction) by aspiration, and replace with 10–15 mL KRBHA without collagenase. Wash so 3×.
3. The floating packed cells are adjusted to a suitable dilution with KRBHA, for immediate distribution into the assay vials (approx 30 mg cell lipid/mL are obtained when 1 vol cells is added to 9 vol medium).

### 3.2. Incubation of Adipocyte Suspensions for Lipolysis Assays (see Note 4)

1. One milliliter fat cell suspension (under constant agitation) is added in the plastic vials containing 10 µL drugs to be tested or 10 µL water (basal lipolysis) or vehicle. This starts the incubation period, which will be conducted at 37°C under gentle shaking. Immediately stop lipolysis in several tubes, in order to determine the amount of glycerol present in the medium before incubation (zero-time).
2. After 60 or 90 min, stop incubation by cooling the assay vials in pilled ice. After 10–15 min of cooling, carefully pipet 200 µL medium below the coalescent fat cake formed by cooled floating fat cells, and transfer into microcuvets for glycerol determination.
3. Prepare, in parallel to the assays, zero-time, and standards containing 0.00, 0.05, 0.10, and 0.15 µmol glycerol in 200 µL (from a 1:10 dilution of the stock glycerol standard).

4. Dissolve in 100 mL HMG buffer the following reagents: 75 mg ATP; 37.5 mg NAD; 300 μL G3PDH; 100 μL GK. Then add 1 mL of this enzymatic reagent to all the microcuvets (*see* **Note 2**).
5. Incubate 40 min at room temperature standards, zero-time, and assays. Determine the amount of glycerol released by measuring optical density at 340 nm.

## 3.3. Quantification of Amount of Fat Cells Present During Incubation

1. Determining the mass of cellular lipids in representative samples is one convenient and inexpensive way to quantify the amount of adipocytes present during incubation. For that, 1 mL cell suspension is randomly added, during its distribution, into 10-mL glass tubes. Add 2 mL Dole and Meinertz extraction solvent to each tube. Stop up, and shake vigorously. Add 2 mL heptane, and shake.
2. Let the formation of two phases take place before transferring 2 mL of the upper organic phase into a glass vial, the exact weight of which has been previously noted.
3. Evaporate to dryness under fume hood or in other adequate equipment, and gravimetrically determine the weight of remaining lipids after total evaporation of heptane.
4. This weight corresponds to 2 mL taken up from 2.4 mL (heptane plus heptane in extraction solvent); multiplying it by 1.2 gives the mean mass of cellular lipid incubated with the tested drugs. Results of lipolysis can be expressed as μmol of glycerol released/100 mg cellular lipid/60 min.

## 3.4. Calculations for Expression of Lipolytic Activity

1. Other methods for the quantification of fat cells present in the incubation can be alternatively chosen (*see* **Note 5**), especially when studying models with different degrees of adiposity.
2. Whatever the expression of results, according to mass of cell lipids or to number of adipocytes, the effects of adrenergic agents can be transformed to percent of maximal response or to increase over basal values. In this case, it is advised to define basal lipolysis and maximal lipolysis in response to the mixed β-agonist isoprenaline ($10^{-5}$ $M$), or preferably to nonadrenergic agents such as adrenocorticotropic hormone (ACTH) $10^{-6}$ $M$ (for rat adipocytes), forskolin $10^{-4}$ $M$ (for human or rat adipocytes), or dibutyryl cAMP (*see* **Notes 6** and **7**).
3. For any given drug, calculate the concentration giving half-maximal response ($pD_2 = -\log[EC_{50}]$) by linearization of the sigmoidal dose–response curve by the Hill plot: $\log (E\%/[100 - E\%])$ vs $\log (M)$ agent; E% being the percent of maximal effect obtained at each tested dose, basal = 0, and maximum = 100). This can be computed by any commercial program of scientific graphics.

## 3.5. Preparation and Storage of Crude Membranes

1. Vigorously shake, at room temperature, stopped tubes containing no more than 5 mL freshly isolated fat cells plus 30 mL lysing buffer, then collect the mixture in centrifuge tubes (40 mL) and spin at 1000*g* for 3 min in order to discard blood cells, stromal-vascular elements, or partially digested interstitial fragments contained in the pellet.

2.  Apply to the same centrifuge tubes, after equilibration with lysing buffer, a centrifugation of 40,000*g* at 15–20°C, for 15 min. Then, centrifuge tubes typically contain, from bottom to top, the pellet of crude membranes (plasma and intracellular membranes plus nuclear material), the cytosol diluted in infranatant medium, an interface consisting of phospholipids/water micelles or revesiculed cell fractions, and a supernatant constituted by an oil layer which should be discarded by aspiration through a vacuum system.

3.  At this step, the tubes can be placed on ice. Cleaning the tube walls with filter paper is often necessary before aspirating the pellets with a syringe and long needle. Then resuspend the pellets in fresh lysing buffer, either for being stored in cryotubes at –80°C (generally as 1–4 mL portions, up to 4 mo) or for being immediately washed in the following centrifugation. These pellets not only contain membrane proteins, but also some of the considerable amount of albumin present in KRBHA buffer, thus rendering protein determination at this step very unprecise.

4.  Centrifuge at 40,000*g* for 20 min, in order to exchange lysing medium with binding buffer (approx 30 mL). This last centrifugation can be conducted at 4–6°C, since the fatty supernatant has been discarded.

5.  Resuspend the pellet in an adequate volume of binding buffer for the planified experiment, either by vortexing the centrifuge tube or by 5–6 syringe aspirations through a needle (0.4 × 20 mm). Potterization of the suspended crude membranes is recommended with a Thomas (AA size) Teflon pestle, 6–10 strokes. Optimal protein concentration for binding studies ranges between 40 and 60 µg/50 µL (*see* **Notes 8** and **9**).

### 3.6. Incubation of Membrane Suspension with Radioligands and Competitors (see Note 10)

1.  For saturation studies, prepare assay tubes for TB (100 µL binding buffer), NSB (50 µL binding buffer plus 50 µL of an excess of cold adrenergic agent) (*see* **Note 11**). Then add 50 µL of the various radioligand concentrations in all the tubes (TB and NSB).

2.  For competition studies, use a single radioligand concentration (typically twofold the $K_D$ value), and distribute 50 µL in replicates of TB and NSB and in a series of tubes containing 50 µL of increasing concentrations of competitors.

3.  Distribute 50 µL of freshly prepared membrane suspension in the binding assay tubes. Incubate at 25°C until steady state, defined by preliminary kinetic studies, is reached (typically between 30 and 60 min).

4.  Reserve duplicate samples for protein determination (add 50 µL membrane suspension only) and for TOT determination (add 50 µL of each radioligand concentration tested, directly into scintillation vials).

### 3.7. Vacuum Filtration on Glass Fiber Filters

1.  Filtration protocols are slightly different, according to the vacuum system used, but all of them must be conducted at low temperature, and as rapidly as possible. Prior to filtration, it is necessary to cool the washing buffer to approx 6°C, presoak

the glass fibre filters (*see* **Note 12**), place them in the device, check that vacuum is operational, and note the order in which the assays will be filtered and the filters transferred into scintillation minivials.

2. When a Millipore Manifold is used, the assays (200 µL in plastic tubes of 5 mL) are diluted with 4 mL ice-cold washing buffer, and immediately vacuum-filtered. The filters are washed twice with 10-mL portions of washing buffer, placed into minivials containing 4 mL liquid scintillation cocktail, and counted in a scintillation spectrometer (or directly counted in a γ-counter, in the case of [$^{125}$I]-labeled ligands).

3. When a cell harvester device is used, adhere to the instructions, and check that the tubular suction system is fully efficient.

4. Count the TOT added in the vials for each concentration of ligand.

5. Determine the protein content in 50 µL membrane suspension.

## 3.8. Calculations for Binding Parameters

1. The linear transformation of saturation data requires, for each radioligand concentration, the calculation of the bound (TB – NSB) and the free (TOT – bound) amounts of the ligand (*see* **Note 13**).

2. Calculate from the Rosenthal plot (bound/free vs bound), the dissociation equilibrium constant ($K_D = -1$/slope, expressed in n$M$), which is an estimate of the affinity of the radioligand for the studied receptors and the $B_{max}$ (intersection with horizontal axis, expressed in fmol/mg protein), which corresponds to the density of AR/mg of membrane protein.

3. Therefore, definitive expression of saturation data can be obtained only after determination of the protein content in assays (*see* **Subheadings 3.5.–3.7.**).

4. For competition studies, express inhibition data as percent of specific bound. IC$_{50}$ can be calculated from sigmoidal displacement curves as detailed for lipolysis in **Subheading 3.4.**

5. A convenient way to verify the nanomolar concentration of a given radioligand under TOT, TB, or NSB forms is to divide the dpm in a given fraction by $444 \times$ SA (SA is the specific activity of the radioligand, expressed in Ci/mmol). This shortcut is valid only in the abovementionned conditions: 50 µL radioligand in a final incubation volume of 200 µL (*see* **Note 14**).

# 4. Notes

1. Other enzymatic determinations of glycerol can be used, especially in the case of lower incubation volumes and/or higher dilutions of adipocytes. The method of Bradley and Kaslow (**5**), using only GK, is more sensitive, but needs many more safety precautions, since it requires $^{32}$P-ATP. The bioluminescent assay described by Kather et al., using GK, G3PDH, and luciferase, is highly sensitive, but also more expensive, and needs a performant bioluminometer (**6**).

2. Safe handling of hydrazine hydrate and heptane is readily achieved by manipulating with gloves under a fume hood.

3. Different batches of collagenase or preparations of bovine serum albumin have to be tested on the same sample of AT in order to establish their influence on the

lipolytic activity. Variations in cell breakage and cell responsiveness can occur from batch to batch, depending on the furnisher, method of fractionation, degree of purity, or degree of delipidation (for albumin). The best batches are those that allow the greatest difference between basal and maximally stimulated lipolysis. After selection, they must be purchased in large quantities, in order to use the same batch throughout a given protocol. Alternatively, glucose transport activity is a good index for selecting collagenase batches, only few of them allow the isolation of adipocytes highly sensitive to insulin.

4. Brown adipocytes and cultured preadipocytes can also be used for lipolysis, membrane preparation, and binding studies. Incubation time can be limited to 45 min for brown adipocytes (100,000 cells/mL), and care should be taken to collect KRBHA medium without adipocytes for the subsequent glycerol determination. For cultured adipocytes, overnight serum deprivation is recommended before assays.

5. Other methods for quantification of the fat cells present in the cell suspension are based on microscopic counting of a representative aliquot, or on image analysis and cell-sizing in combination with methods aimed at determining the mass (*see* **Subheading 3.3.**) or the volume (lipocrit) occupied by the cells. Whatever the method, the most convenient manner to determine intrinsic activity and potency ($EC_{50}$) of the adrenergic agents is to express the lipolytic (or antilipolytic) responses as percent of maximal response to a reference drug (*see* **Subheading 3.4.**).

6. To better evidence the antilipolytic action of $\alpha_2$-adrenergic agonists, it is judicious to stimulate basal lipolysis. This can be achieved by the addition of nonadrenergic lipolytic agents, such as adenosine deaminase (4 IU/mL) or isobutylmethylxanthine ($10^{-5}$ $M$–$10^{-3}$ $M$).

7. When spontaneous lipolytic activity is a problem (elevated basal values, fluctuations in maximal responses), the addition of adenosine ($10^{-7}$ $M$) during adipocyte preparation can lower basal values and improve reproducibility, as reported by Londos et al. *(7)*. In this case, it is advisable to add adenosine deaminase (4 IU/mL) during incubation, in the presence of the tested adrenergic agents.The nonmetabolizable adenosine analog, (–)phenylisopropyladenosine, is a strong antilipolytic agent, and can be useful to reduce spontaneous glycerol release.

8. Protein yield during membrane preparation can be checked by measuring the protein content of the adipocyte suspension (carefully washed from the albumin present in the KRBHA buffer), and that of the membrane pellet. This recovery varies according to the anatomical location of the fat depot and the degree of adiposity. Typically, in rats weighing 200 g, approx 5% of the fat cell proteins are recovered in the crude membrane fraction.

9. When the level of endogenous CATs is suspected to be elevated (pharmacological treatments, stress conditions, and so on), it is of interest to remove the remaining CATs tightly bound to the receptors, even after the membrane preparation procedure. For this, crude membrane pellets must be extensively washed at least once in 30 mL Tris 50 m$M$, EDTA 0.5 m$M$, pH 7.5, repelleted, and resuspended in the selected binding buffer, before incubation with the radioligand. In lipolysis measurements, the remaining CATs bound on the fat cells can affect basal values, and determination of real basal and maximal levels of lipolysis must be conducted, as in **Note 7**.

10. Binding experiments can be conducted on intact fat cells. In this case, it is necessary to replace Tris-Mg as binding buffer by Krebs-Ringer medium or Hank's balanced salt solution containing 20 m*M* HEPES and 0.5 g/100 mL albumin. Although more physiological than assays on membrane preparation, binding on intact fat cells are complicated by ligand uptake and receptor internalization. It is possible to limit these phenomena with 0.03 mg/mL chloroquine.

11. Improving the definition of NSB must be obtained by conducting inhibition experiments with a variety of cold ligands. At reasonably high concentrations, they should all inhibit TB to the same extent and, at best, reach a plateau. When less than 30% of TB is nonspecific with various competitors, binding conditions are considered good enough (less than 10% is excellent). Whatever the AR subtype, 100–500 µ*M* adrenaline (or noradrenaline) itself must totally displace specific binding. If not, the presence of nonadrenergic binding sites, with high affinity for several adrenergic drugs, must be suspected. A well-known example of such sites in fat cells are the imidazoline binding sites *(8)*.

12. Reduction of NSB can be obtained by presoaking the glass fiber filters in the solution of adrenergic compound used for the determination of NSB or with 0.1% aqueous polyethylenimine. This can reduce the capacity of the ligand to bind to the filters themselves. Alternatively, the radioligand must be the last component added to the binding assays. Extensive washing during vacuum filtration is not a good solution, since it may also displace the ligand bound to the receptors.

13. Biologically active receptors vs total receptor population can be assessed by the comparison of agonist and antagonist binding. Whatever the subtype of AR, agonists readily label only a portion of the receptor population (the high-affinity-state receptors); antagonist ligands label all the available ARs. Nonhydrolysable guanosine triphosphate analogs can be used to determine the low-affinity state of the ARs for the agonists, essentially in competition experiments.

14. Other experimental conditions and binding calculations allowing the determination of β-AR are detailed in **ref.** *9*.

## References

1. Lafontan, M. and Berlan, M. (1993) Fat cell adrenergic receptors and the control of white and brown fat cell function. *J. Lipid Res.* **34,** 1057–1091.
2. Mauriège, P., Galitzky, J., Berlan, M., and Lafontan, M. (1987) Heterogenous distribution of beta- and alpha2-adrenoceptor binding sites in human fat cells from various fat deposits : functional consequences. *Eur. J. Clin. Invest.* **17,** 156–165.
3. Wieland, O. (1957) Eine enzymatiche methode zur bestimmung von glycerin. *Biochem. Z.* **239,** 313–319.
4. Rosenthal, H. E. (1967) Graphical method for the determination and presentation of binding parameters in a complex system. *Anal. Biochem.* **20,** 525–532.
5. Bradley, D. C. and Kaslow, H. R. (1989) Radiometric assays for glycerol, glucose, and glycogen. *Anal. Biochem.* **180,** 11–16.
6. Kather, H., Schroder, F., and Simon, B. (1982) Microdetermination of glycerol using bacterial NADH-linked luciferase. *Clin. Chim. Acta* **120,** 295–300.

7. Honnor, R. C., Dhillon, G. S., and London, C. (1985) cAMP dependent protein kinase and lipolysis in rat adipocytes. I. Cell preparation, manipulation and predictability in behavior. *J. Biol. Chem.* **260,** 15,122–15,129.
8. Carpéné, C., Marti, L., Hudson, A., and Lafontan, M. (1995) Nonadrenergic imidazoline binding sites and amine oxydases activities in fat cells. *Annals NY Acad. Sci.* **763,** 380–397.
9. Dunigan, C. D., Curran, P. K., and Fishman, P. M. (2000) Detection of β-adrenergic receptors by radiogland binding, in *Methods in Molecular Biology*, vol. 126 : Adrenergic Receptor Protocols (Machida, C. A., ed), Humana, Totowa, NJ, pp. 329–343.

# 11

## Transfection of Adipocytes and Preparation of Nuclear Extracts

### Isabelle Dugail

### 1. Introduction

The uniqueness of the adipose cell type must be considered before choosing a procedure for gene transfer into mature adipocytes. Because adipocytes isolated from adipose tissue (AT) are filled with so many lipids that cells cannot be plated on a culture support, because of their low density, and because such terminally differentiated cells do not divide, the procedure of choice for introducing DNA into fat cells is electroporation. The detailed protocol described below has been published in references (1,2). One must keep in mind however, that primary adipocytes isolated from the animal have a limited life-span in culture (not exceeding 1 wk), so that only transient transfection can be envisaged in such a system. When stable transfectants are needed, other cellular models must be used, such as established preadipose cell lines. Among the most popular, 3T3-L1, 3T3-F442A, and Ob17 can also be transfected by electroporation (3). A protocol adapted to these situations is also given. These techniques are simple to perform, but require special equipment for electric shock delivery. Finally, it is important to realize that only a small proportion of the cell population will incorporate foreign DNA, using these transfection procedures. High-efficiency gene transfer in mature adipocytes, using recombinant adenoviral systems is also possible (3,4), but will not be described here.

The preparation of nuclear extracts from isolated adipocytes from AT or from cultured adipose cell lines, is also described in this chapter. The procedure is derived from that initially described for hepatocytes by Dignam et al. (5). It is very short and easy to perform, but needs great amounts of starting materials (e.g., isolated mature adipocytes). However, when applied to adipose

From: *Methods in Molecular Biology, vol. 155: Adipose Tissue Protocols*
Edited by: G. Ailhaud © Humana Press Inc., Totowa, NJ

cell lines, nuclear protein recovery is achieved with good efficiency. The two protocols detailed below use mature adipocytes isolated from AT as the starting material, the adipocyte isolation procedure, not provided in this chapter, is derived from the original paper of Rodbell *(6)*.

## 2. Materials
### 2.1. Electroporation

1. A commercially available electroporator system and disposable electroporation cuvets (4-mm gap electrode).
2. Plasmid DNA: closed circular plasmid DNA. (prepared by the alkaline lysis method, followed by $CsCl_2$ gradient purification), dissolved in 10 m$M$ Tris-HCl, 1 m$M$ EDTA, pH 8.0, at a concentration greater than 1 μg/μL.
3. Sterile Dulbecco's modified Eagle's medium (DMEM) (supplied by Life Technology).
4. A thermostated magnetic stirrer.
5. Sterile DMEM supplemented with 10% fetal calf serum (FCS) and antibiotics.
6. Sterile 2-mL Eppendorf tubes.
7. Blunted needles.
8. 1-mL syringes.
9. Phosphate buffered saline (PBS).
10. Lysis buffer: 0.25 $M$ Tris-HCl, pH 8.0, 5 m$M$ dithiothreitol (DTT).

### 2.2. Nuclear Extracts

1. Hypotonic buffer: 10 m$M$ HEPES, pH 7.9, 10 m$M$ KCl, 1.5 m$M$ $MgCl_2$, 1 m$M$ DTT. Prepare fresh from stock solution at each preparation, add DTT (from a 0.5 $M$ stock) and protease inhibitors extemporaneously.
2. Buffer C for swelling: 10 m$M$ HEPES, 0.42 $M$ NaCl, 25% glycerol v/v, 1.5 m$M$ $MgCl_2$, 0.5 m$M$ EDTA. Prepare fresh from stock solution at each preparation, add DTT and protease inhibitors extemporaneously. NaCl is added from a stock 5 $M$ solution.
3. Protease inhibitors cocktail : Stock solutions are prepared as 1000X solutions and kept frozen at –20°C: 20 m$M$ leupeptin, 2 m$M$ pepstatin, 2 m$M$ aprotinin, 0.5 $M$ phenylmethyl sulfonyl fluoride (PMSF). Leupeptin and aprotinin are diluted in water, PMSF in isopropanol, and pepstatin in DMSO (*see* **Note 1**).
4. 20-mL syringe equipped with a long needle.
5. Small magnetic stirrers (3-mm long).
6. 0.2% Trypan blue in 0.15 $M$ NaCl.

## 3. Methods
### 3.1. Transient Transfection of Mature Adipocytes by Electroporation (Fig. 1)

1. Prepare isolated fat cells by collagenase digestion of AT.
2. Resuspend them in DMEM, by aspirating the infranatant and washing twice in warmed DMEM.

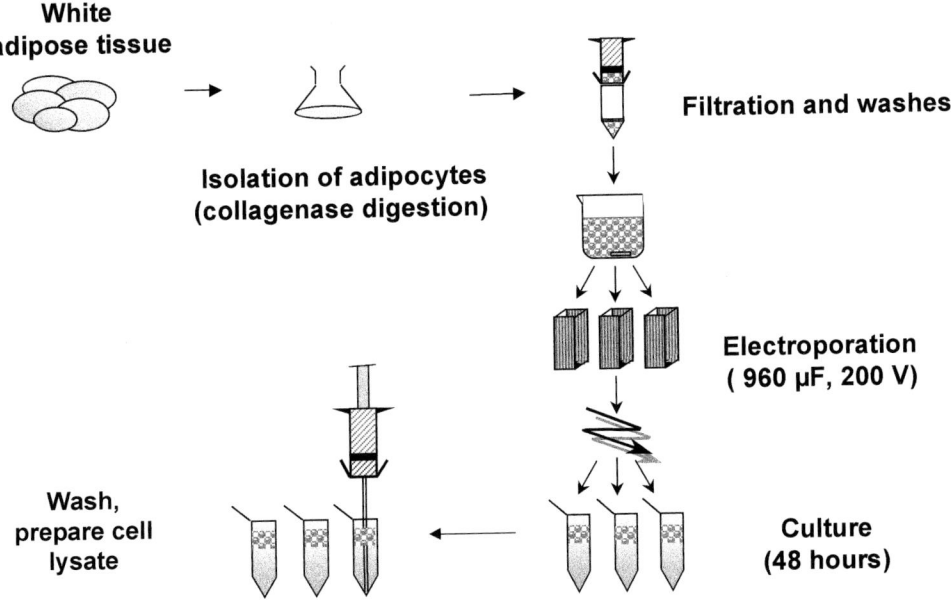

Fig. 1. Summary of the procedure for electroporation of mature adipocytes.

3. Adjust the total volume with DMEM so that the adipocryte (ratio between fat cell volume and DMEM) is about 1:2 to 1:5. Note that, for large fat cells, which give big fat cakes at the surface of the tube, a low adipocryte should be chosen (1:2) to increase the number of cells/mL (*see* **Notes 2** and **3**).

4. Pour the adipocyte suspension in a plastic beaker, and keep under agitation with a magnetic stirrer. A thermostated magnetic stirrer, in which the cell suspension could be kept at 37°C, should be used.

5. Prepare sterile electroporation cuvets and distribute plasmid DNAs, so that the total volume does not exceed 20 μL. When working with reporter genes, it is recommended to include a second plasmid in all cuvets, for normalization purposes. Each experimental point should be in triplicate, using three separated cuvets.

6. While keeping the fat cell suspension under agitation, distribute 200 μL cell suspension in each cuvet.

7. Proceed for the electric shock with the following settings: voltage: 200 V, capacitance: 960 μF (*see* **Fig. 2**). Each cuvet should be gently agitated just before the electric shock, so that the adipocytes do not stay at the surface during electroporation (*see* **Note 4**).

8. Distribute 1.5 mL warmed DMEM supplemented with 10% of FCS and antibiotics in 2-mL sterile Eppendorf tube (FCS is expensive, and can be substituted with 1% calf serum or 3% bovine serum albumin).

9. Using a plastic tip, transfer the electroporated cells from the cuvet to the Eppendorf tube containing culture medium.

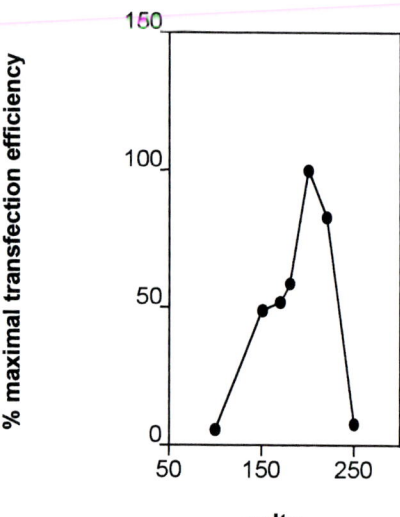

Fig. 2. Effect of voltage on the efficiency of transfection of mature rat adipocytes. Adipocytes from inguinal AT of 1-mo-old rats were isolated and transfected as described in **Subheading 3.**, with a constant amount of a plasmid encoding the chloramphenicol acetyl transferase (CAT) reporter gene under the control of the Rous sarcoma virus (RSV) promoter. Different voltages were used for electroporation, and the capacitance was kept constant: 960 μF. Specific CAT activities were measured, and results are expressed as percent of maximal activity. These results are from **ref. *1***.

10. Incubate transfected fat cells for 24–48 h in humudified 5% $CO_2$ 95% air atmosphere. Do not close the Eppendorf tube, so that the culture medium can equilibrate with $CO_2$. The fat cells should form a white 1-mm large fat cake at the surface of the culture medium.

## 3.2. Transient Expression Assay

Described here is a procedure to recover electroporated adipocytes and prepare a total cell lysate. When working on promoter studies using reporter genes, reporter gene activities should be assayed directly in those lysates.

1. Remove the adipocytes from the incubator, and place them in a 37°C water bath.
2. Remove the culture medium by aspiration. This should be done very carefully, so that the floating adipocytes are kept at the surface, which can be achieved by inserting a blunt needle in the Eppendorf, then letting the adipocytes stand for a few minutes and aspirating the infranatrant by inserting a 1-mL syringe on the needle.
3. Rinse twice with warm PBS, using the same procedure.
4. Eliminate the last wash, so that a minimal volume of infranatrant is kept.
5. Add 100 μL cold lysis buffer.
6. Break the cells by three rounds of freezing–thawing.

7. Centrifuge the Eppendorf tube in a cold microcentrifuge (10,000*g*, 15 min), to obtain a clear whole-cell lysate, which stands between the pellet and the fat cake.
8. Use the lysate to measure protein content and reporter gene activities.

### 3.3. Preparation of Adipocyte Nuclear Extracts

1. Prepare isolated adipocytes by collagenase digestion of AT. Use sufficient amount of starting material, so that at least 5 mL fat cell cake should be obtained.
2. Without disturbing the fat cells floating at the surface, get rid of as much buffer as possible. This can be easily done with a 20-mL syringe equipped with a long needle aspirating at the bottom of the tube.
3. Add cold hypotonic buffer. The volume should be at least fivefold that of the fat cell cake.
4. Break the cells by passing through a 21-gauge needle for 5×.
5. Pour into a cold conical polypropylene tube, and centrifuge for 10 min at 800*g* (*see* **Note 5**).
6. Discard the supernatant, taking care of removing all the fat from the tube.
7. Carefully dry the pellet. No hypotonic buffer should remain.
8. Evaluate the approximate volume of the pellet.
9. Add an equal volume of buffer C (usually 100–200 µL).
10. Resuspend the nuclei with a pipet tip.
11. Transfer to a 1.5-mL Eppendorf tube containing a small magnetic stirrer ( 3-mm long).
12. Place the tubes on ice, and allow the nuclear proteins to swell out of the nuclei, under gentle magnetic agitation for 1 h.
13. Centrifuge for 15 min at 4°C at full speed in a refrigerated microcentrifuge to pellet the nuclei.
14. The clear supernatant, consisting of the crude nuclear extract, is aliquoted and frozen at –80°C until use (*see* **Notes 6–8**).

## 4. Notes

1. Inhibitors of proteases are included in all buffers. We routinely use a cocktail of protease inhibitors from Boehringer, which provides ready-to-dilute tablets. Phosphatase inhibitors can also be used, such as sodium vanadate (2 m*M* final concentration) and sodium fluoride (50 m*M* final concentration).
2. Because cell concentration of a solution of isolated adipocytes cannot be determined rapidly using standard techniques, the crucial point is to correctly determine the resuspension volume as a function of the size of the fat cell cake, so that a sufficient amount of fat cells could be used for electroporation. *A posteriori* determinations of fat cell concentration can be performed in preliminary experiments. These techniques are described in Chapters 5, 10, 12, 15, and 24. In our hands, we found that a concentration of approx $2–3 \times 10^6$ cells/mL was correct.
3. When working with adipose cell lines, instead of isolated adipocytes, cells should by trypsinized according to standard techniques, and resuspended in DMEM at a concentration of $10 \times 10^6$ cells/mL (as determined with a Malassez chamber). Then the above procedure should be followed from **steps 5–7**. After electroporation, the cells from each cuvet are plated in three separated 35-mm culture dishes, and

incubated in culture medium supplemented with FCS and antibiotics. The medium is not changed during the first 48 h. It is also possible to transfect 3T3 adipose cell lines by the calcium phosphate co-precipitation technique *(7)*.

4. Most electroporation devices give time constant values, which represent the time (usually in seconds) needed for the capacity to discharge between the electrodes. This parameter is useful, because it can help to estimate whether or not the adipoctyte concentration is correct. The more adipocyte in the cuvet, the more lipids, the higher time constants will be obtained, because lipids behave as isolants. In our hands, time constants ranging from 40 to 80 were found to give good efficiencies.

5. The procedure is achieved in the cold from **step 5**.

6. This procedure yields to nuclear extracts with a protein concentration from 3 to 10 mg/mL.

7. If the nuclear extracts are to be used in DNA binding studies (gel shift experiments or DNase1 footprinting), we recommend that the last centrifugation should be faster and longer (100,000*g* for 30 min), in order to pellet any DNA liberated by broken nuclei. Such DNA in nuclear extract preparation may behave as a nonspecific competitor in binding studies.

8. At **step 10** the nuclei can be observed under the microscope as a twofold dilution in 0.2% trypan blue. They should appear as blue, round-shaped material. The clear supernatant consisting of the crude nuclear extract is aliquoted and frozen at –80°C until use (*see* **Notes 6** and **8**).

## References

1. Rolland, V., Dugail, I., Le Liepvre, X., and Lavau, M. (1995) Evidence of increased glyceraldehyde-3-phosphate dehydrogenase (GAPDH) and fatty acid synthase (FAS) promoter activities in transiently transfected adipocytes from genetically obese rats. *J. Biol. Chem.* **270,** 1102–1106.

2. Rolland, V., Le Liepvre, X., Jump, D., Lavau, M., and Dugail, I. (1996) A GC-rich region containing Sp1 and Sp1-like binding sites is a crucial regulatory motif for Fatty Acid Synthase Gene promoter activity in adipocytes. *J. Biol. Chem.* **271,** 21,297–21,302.

3. Boizard, M., Le Liepvre, X., Lemarchand, P., Foufelle, F., Ferre, P., and Dugail, I. (1998) Obesity-related overexpression of Fatty-acid synthase gene in adipose tissue involves SREBP transcription factors. *J. Biol. Chem.* **273,** 29,164–29,171.

4. Sharma, P. M., Egawa, K., Gustafson, T. A., Martin J. L., and Olefsky, J. M. (1997) Adenovirus-mediated overexpression of IRS-1 interacting domains abolishes insulin stimulated mitogenesis without affecting glucose transport in 3T3L1 adipocytes. *Mol. Cell Biol.* **17,** 7386–7397.

5. Dignam, J. D., Lebovitz, R. M., and Roeder. R. G. (1983) Accurate transcription initiation by RNA polymerase II in a soluble extract from isolated mammalian nuclei. *Nucleic Acids Res.* **11,** 1475–1489.

6. Rodbell, M. (1964). Metabolism of isolated fat cells. I: Effects of hormones on glucose metabolism and lipolysis. *J. Biol. Chem.* **239,** 375–380.

7. Moustaid, N., Beyer, R. S., and Sul. H. S. (1994) Identification of an insulin response element in the fatty acid synthase gene promoter. *J. Biol. Chem.* **269,** 5629–5634.

# 12

## Assay of Membrane Transport of Long-Chain Fatty Acids by Adipocytes

**Nada A. Abumrad**

### 1. Introduction

As a result of their poor aqueous solubility, Fatty acids (FAs) are quantitatively bound to plasma albumin in the circulation and to cytosolic FA-binding proteins inside the cell. The presence of FA-binding proteins on both sides of the plasma membrane complicates interpretation of FA uptake studies, and does not allow the detailed characterization accomplished with other uptake systems. However, despite these limitations, significant information can be obtained. In the past decade and a half, evidence has accumulated to document the existence of two components of FA uptake, a protein-facilitated component, referred to as "transport," and a passive diffusion component *(1)*. When assaying FA transport, it is important to keep in mind that the contribution of the transport component to cellular FA uptake will depend on several factors, including the concentration of free or unbound FA, the cell type, and so on.

Before describing protocols for assays of FA uptake by adipose tissue (AT), it is helpful to briefly outline the determinants of the uptake process. FAs complexed to albumin are used in most uptake assays in vitro, to overcome complications introduced by poor FA solubility. In the absence of albumin or other FA-binding proteins, only very low micromolar concentrations of long-chain FA can be used in assays *(2)*. These concentrations can be depleted by cells almost instantaneously, and, as a result, the linear portion of the uptake time-course will be extremely short and impossible to measure. Additional complications are FA adsorption to assay tubes and pipet walls, which can represent a significant fraction of added FA, and FA aggregation.

From: *Methods in Molecular Biology, vol. 155: Adipose Tissue Protocols*
Edited by: G. Ailhaud © Humana Press Inc., Totowa, NJ

## 1.1. FA:Albumin Ratios and Concentration of Unbound FA

When FA are added, complexed with albumin, uptake follows the concentration of free or unbound FA (ubFA), rather than the total FA concentration *(3,4)*. ubFA is determined by the molar ratio of FA to albumin.

Although FA dissociation from albumin is very fast, it could become limiting if the albumin concentration or the molar ratio used are too low. This can be avoided by keeping the concentration of the FA–albumin complex high, and by maximizing the volume of transport buffer relative to that of cell suspension *(3,4)*.

Concentration of ubFA is computed based on FA–albumin binding constants. Initial estimates provided by the detailed studies of Spector et al. *(5)* have been revised downward, based on novel approaches. The use of a fluorescent fatty acid binding protein (ADIFAB), included at low concentration in a solution of protein-bound FA, allowed direct estimation of ubFA from the shift in fluorescence for ADIFAB upon acid binding *(6)*. Current estimates of ubFA levels at circulating FA:albumin ratios are believed to be in the range of 5–50 n*M* *(7)*. For uptake assays, it is convenient to calculate ubFA based on the binding constants provided by the ADIFAB studies.

## 1.2. FA Transport vs FA Metabolism

Nonmetabolizable FAs are not readily available in radioactive form, so separating the transport step from metabolic steps remains an experimental challenge. In the case of adipocytes, omission of glucose and lowering the temperature to 23°C significantly impair esterification of the FA, so that most of what is taken up at early time-points remains unmodified (*see* **Note 1**). Some FA analogs have been shown to have very slow rates of cellular metabolism, e.g., fluorescent FAs that are esterified at a reduced rate by adipocytes *(8)*. Another example is that of β-methyl-iodophenyl-penta-decanoic acid (BMIPP), a FA that carries a methyl group at the β position, making it a poor substrate for β oxidation. However, BMIPP is esterified into triglycerides and phospholipids at rates comparable to native FA *(9)*. All FA analogs have to be validated as good substitutes for native FA, and care should be taken to show that they are recognized by the FA transport system examined. This has been shown to be the case for some fluorescent FA, and for BMIPP *(8,9)*.

## 2. Materials
## 2.1. Stocks of FA:Albumin

The isotopic solution used for transport assays consists of Krebs-Ringer HEPES (KRH) buffer, pH 7.5, containing the desired concentrations of FA complexed to plasma albumin (*see* **Notes 2–4**). The FA–albumin complexes to be used in the transport assays can be prepared in several ways, depending on the concentration and type of FA to be used.

1. For relatively more soluble, unsaturated, or shorter-chain FA (also good for low concentrations of less-soluble FA, such as palmitate and stearate, up to 100 $\mu M$). To prepare oleate–bovine serum albumin (BSA) complexes, add oleate directly from a concentrated stock in ethanol (usually 100–200 m$M$) to warm buffer (up to 40°C) with albumin. The buffer is placed on a stirrer, and the FA is added dropwise, with the pipet placed in the liquid, to allow gradual equilibration of FA with albumin in the buffer. Final concentrations of oleate of 100–150 $\mu M$ can be obtained this way.

   Add radioactive oleate, as for unlabeled FA, to the stirring buffer. A specific activity of 25,000–50,000 cpm/nmol works well for adipocytes. Allow buffer to stir for 20–30 min, then remove from stirrer, and take several aliquots for scintillation counting. Formation of the desired FA–albumin complex can be ascertained from a match of the counts obtained with those expected, and from good duplication between counted samples (*see* **Notes 3** and **4**).

2. For less-soluble FA, and for high total FA concentrations, dissolve the required amount of FA (with 1:1 NaOH) in water at 40–50°C. Once a clear solution is obtained, add albumin from a concentrated buffered stock (10–20%). Allow the mixture to equilibrate for about 20 min, with gentle stirring. Then dilute 1:1 with 2X concentrated KRH buffer, pH 7.5, to obtain the correct final salt concentrations.

## 2.2. Stop Solution

In order to measure transport reliably, radioactivity must not be lost from the cell during uptake, cell separation, or washing. For this, the use of a stop solution is important. KRH buffer at ice temperature, containing 200 $\mu M$ phloretin works well as stop solution. Typically, about 6 mL stop solution are needed per sample.

To stirring KRH buffer, add phloretin from a concentrated stock in EtOH to a final concentration of 200 $\mu M$. The phloretin is added slowly, in order to avoid its precipitation. Place the solution on ice, and use for assays, once it is completely chilled to ice temperature (*see* **Note 5**).

## 2.3. Preparation of Cell Suspension

Adipocytes were isolated as described elsewhere *(10)*. Inclusion of a large amount of albumin (2–4%) during the isolation procedure, minimal mechanical agitation of the cells, and keeping all centrifugations at low speed (<300$g$) and very short (30 s) are recommended, to prevent lysis, especially when working with human adipocytes. It is also useful to add adenosine to a final concentration of 200 n$M$, in order to inhibit basal lipolysis. This will help in keeping the adipocytes live longer, and, by inhibiting release of FA from the cells, will minimize changes in the FA:albumin ratios during the transport assay.

1. Wash the isolated cells twice with buffer containing 1% albumin, then twice with buffer containing a low concentration of FA-free albumin (0.1–0.2%). Reduction of the albumin in the cell suspension diminishes the change of the FA:albumin

ratio of the isotopic solution used in the assay. Remove the infranatant after the last wash, and add an amount of buffer equivalent to about half that of the packed cells, then determine cell density by what is referred to below as "lipocrit."

2. Determination of cell density, or lipocrit: it is best to keep cell density (v/v) as close as possible to 30%, to obtain transport rates that are comfortably above nonspecific values. Mix the cell suspension with the aid of an automatic pipet (1–5 mL capacity, depending on the suspension volume), with the plastic tip enlarged to avoid cell breakage. In our experience, this is the best way to obtain reproducible sampling, if care is taken to remove the sample immediately after mixing the cells and before they begin to float. Rapidly pipet a known volume (7–8 μL) of mixed cell suspension into one end of a microcapillary, which is then sealed using wax or paraffin. Centrifuge the capillary for 1 min in a microcentrifuge at ~1800*g*. Then divide the volume of packed floated cells by the total volume (liquid plus cells) to estimate cell density. This is easily done, using a magnifying glass, and placing the capillary on graph paper or on any other graduated surface. The cell suspension is then adjusted to the desired density by removal or addition of buffer. This procedure is fast and can be used to standardize conditions with respect to dilution of isotope and changes in FA:albumin ratios brought about by the cell suspension. However, aliquots must be taken out for determination of intracellular water space or for DNA content for the packed cell volume in the assays. This can be done at a later time, but is important, in order to control for differences in cell size, and consequently in the number of cells per packed cell volume, between different preparations (e.g., when comparing cells from fasted vs fed animals or from control and diabetic, and so on). For details on determination of intracellular water volume, see the following subheading.

## 3. Methods

### 3.1. Assay of FA Transport by Adipocytes in Suspension

#### 3.1.1. FA Transport Assay (23°C)

1. Pipet an aliquot (30–50 μL) of the isotopic solution as a round drop in the bottom of a clear plastic tube.
2. Mix cell suspension 2–3× with the aid of the pipet, and rapidly pipet out an aliquot of cell suspension (30–50 μL) before cells begin to float. Place the tip of the pipet close to the drop of isotope.
3. With the aid of a timer, start uptake by rapidly ejecting the cell suspension aliquot onto the drop of isotope, and mix by shaking. Mix the solution gently for the duration of the assay time-point.
4. At times ranging between 2 s to 2 min, stop uptake by the addition of a large volume (about 50× that of assay mixture) of stop solution (cold buffer containing 200 μ*M* phloretin).
5. Separate cells from medium by filtration, using Gelman glass-fiber filters and a filtration apparatus (Hoefer Scientific manifold filtration apparatus). A single filtration apparatus can also be used. Filtration should be at very low vacuum

pressure (50 mmHg), so that addition of 1 mL stop solution should cover the filter for few a seconds before being filtered. Wet filter with 1 mL stop solution 20–30 s before addition of transport mixture. Add the transport mixture at the approximate rate of filtration, to minimize cell flotation and sticking to the walls of the filtering apparatus. Wash filter twice with 1 mL cold stop solution, then remove filter, and place in a scintillation vial containing 4 mL aqueous counting fluid for counting cell-associated radioactivity.

6. Controls for adsorption of FA to filters are done by adding to the prewet filters stop solution with isotope and an amount of buffer equivalent to that present in the cell suspension (35 µL for 50 µL 30% suspension), and containing the same amount of albumin.

7. Zero time controls are done routinely by pipeting cell suspension to premixed isotope and stop solution, and processing as for other samples. A match between zero and adsorption controls indicates that the stop is working well, and stopping uptake instantaneously.

### 3.1.2. Determination of Intracellular Water Space

Aliquots of cell suspension (30–50 µL) are pipeted into microcentrifuge tubes containing silicone oil and corn oil (about 100 µL each) and 140 µL ice-cold stop solution containing tritiated water and $^{14}C$-2-deoxyglucose are added. The cells are pelleted between the two oils by centrifugation for 30 s in a microfuge. The intracellular water space in the cell pellet is calculated as the water space minus the deoxyglucose space, which marks the extracellular space, since it is added in the presence of stop solution. For more details refer to Whitesell and Abumrad (*10*).

### 3.1.3. Determination of DNA per Packed Cell Volume

1. Wash adipocytes and resuspend them in phosphate-buffered saline, pH 7.4, (0.05 $M$ $NaH_2PO_4$, 2 $M$ NaCl).

2. Homogenize an aliquot of adipocyte suspension (100 µL of 10–30% [v/v]) by 5 – 10 pulse sonications (10 s each pulse, with a 20-s interval), using a probe sonifier at a power level of 40 W.

3. Add 2 m$M$ EDTA to the homogenate immediately after sonication, to prevent DNAase activity, and centrifuge the homogenate in a microcentrifuge at maximum speed for 12 min. Remove the clear liquid phase at the bottom, which contains the DNA, and transfer carefully to a new tube, with effort taken to avoid contamination by the top fat layer. Measurement of DNA content is then done (*11*).

### 3.1.4. Determination of FA Uptake Rates

Initial rates of FA uptake are determined from the best-fit curve for time-courses, using nonlinear regression analysis. The mathematical model used assumes two compartments of uptake, the first one reflecting entry of free label, and the second one reflecting the rate of formation of FA metabolites. In the case of incubations in which the FA is not processed metabolically, as when cells are kept without glucose and assayed at room temperature, or in the case of FA analogs that are slowly metabolized, only one compartment can be considered.

The following equation can be used to fit the curve:

$$S_1 = S_{14}[1 - e^{(-8t)}] + (kt)$$

$S_1$ represents uptake (nmol/mL packed cells) at time $t$, $S_{14}$ is FA taken up at steady state in the first compartment, with 8 as the fractional rate of approach to steady-state uptake, and $k$ is the rate of FA entering the second compartment. Initial rate of FA uptake is given by: $(S_{14})$ (8) + $k$. In the case of slow metabolism, one compartment can be reflected by $S_1 = S_{14}[1 - e^{(-8t)}]$.

### 3.2. Assay of FA Transport by Cultured Adipocytes (23° C)

1. Wash cells grown in 35-mm dishes with 1 ml KRH buffer, lacking albumin twice.
2. Add medium containing the FA (50–100 $\mu M$, 40,000–50,000 cpm/nmol) isotopic solution (0.7–1 mL, 4 × 10⁶ cpm/mL), using a 1-mL pipet with the plastic tip enlarged; the dish is swirled.
3. Stop transport by adding 3 mL cold stop solution, and ensure mixing by shaking the dish gently back and forth 6–8×.
4. Aspirate the mixture, and wash the dish twice with 1 mL cold stop solution. Lyse the cells in 1 mL cold 0.1 $N$ NaOH. After a 20 min incubation on ice, mix the lysate well using a 1-mL pipet, then take aliquots for scintillation counting, DNA measurements, and protein determination *(12)*.
5. For zero-time controls, add stop to cells, followed by isotope, and process as for regular samples. These controls should correct for extracellularly trapped counts, and will be subtracted from the uptake values. A zero time control can also be obtained from extrapolation of the early time-course by linear regression to the *y* axis (*see* **Notes 6** and **7**). Transport is expressed per protein or DNA content, or per cell number obtained from separate dishes following treatment with trypsin.

## 4. Notes

1. Adipocytes are usually prepared from the epididymal fat of male rats, although other fat depots can also be used. However, since there are suspected differences in the metabolic activity of various fat depots, care should be taken to match the tissues used. In the female, we use perirenal fat as the standard depot, because it has a large mass and can be easily harvested.
2. When a series of different FA:albumin ratios have to be used, it is helpful to prepare the total volume needed with the highest ratio. The mixture is then divided into aliquots, and the required amount of albumin is added to each aliquot, in order to obtain the desired molar ratio of FA: albumin. For example, if 20 mL each of FA:albumin ratios 0.2, 1, and 2 are needed, 60 mL ratio 2 is prepared, then divided into three 20-mL portions. Albumin is then added to two of these, to achieve molar ratios of 0.2 and 1. This method ensures that the same total FA and

the same specific activity are obtained for all three ratios, which minimizes variability, and can be crucial in kinetic studies in which comparison of uptake at the different ratios is required.

3. The FA concentration used should be based on the length of the incubation time and on the transport activity of the cell assayed. It is important to keep the FA taken up below 10–20% of that added to the medium, in order to avoid large changes in the FA:albumin ratio and the ubFA concentration during the uptake. Such change would be expected to be more significant if the FA:albumin ratio is high. The FA:albumin ratio should approximate the physiological range of 0.2–1. High ratios of 1–1.5 can be used, for example, when comparisons of maximal transport capacity between two tissues are desired. However, a passive diffusion component becomes a larger fraction of uptake at such ratios, and correction for this component will be required for estimating activity via the saturable protein-mediated component.

4. FA dissociation from albumin can be estimated to be about $2.8 \times 10^{-2}$/s, and should not rate-limit uptake unless either the FA–albumin ratio or the concentration of complex are too low. In general, it is best to maximize the concentration of FA–albumin complex or to increase the volume of isotope relative to cell suspension. To test whether dissociation may limit uptake, the following model can be used:
   a. FA:albumin $\rightleftharpoons$ ubFA + albumin.
   b. ubFA $\rightleftharpoons$ FA cell
   c. ubFA = unbound FA equilibrated with FA–albumin complex.
   d. $k_a$ = rate of association and $k_d$ = rate of dissociation and $k_u$ = rate of uptake.
   The reactions above give rise to a set of coupled differential equations:

$$dx/dt = k_d (Bo - x) - k_a xy \tag{1}$$

$$dy/dt = k_d (Bo - x) - k_a xy - k_u y \tag{2}$$

$$dz/dt = k_u y \tag{3}$$

$Bo$ = total albumin concentration; $x$, $y$, and $z$ are, respectively, the instantaneous values of free albumin, unbound, and cell-associated FA.

Utilizing a computer program to solve coupled first-order differential equations allows for the calculation of the instantaneous ubFA concentration during uptake studies. Based on these calculations, adjustments can be made in experimental conditions, to ensure that FA–albumin dissociation does not rate-limit uptake. In case it is found to limit uptake, the concentration of FA–albumin complex or the volume of isotope can be increased, to increase the number of dissociating complexes.

5. Cold buffer containing phloretin (a rapid, reversible inhibitor of FA transport) was formulated with studies in adipocytes. However, other inhibitors may be more effective in a particular cell type. A recent review (*1*) listed the compounds shown to inhibit FA transport in various cells. It is also possible that, with some cells or with certain procedures, cold buffer would be effective as a stop solution, and the addition of an inhibitor may not be required. This can be explored in the particular system used, by monitoring efflux of radioactivity from cells into buffer following a period of preloading.

6. Transport in cultured cells may occasionally require the use of extracellular markers, such as mannitol or L-glucose *(13)*. The use of these substances is recommended when very short time-points are needed, and when zero time values are more than half transport values. These markers can be obtained, labeled with an isotope that differs from that used in the case of the FA, and can be included in the transport buffer.

7. Occasionally, FA transport needs to be carried out on cultured cells in suspension following treatment with trypsin. For this, it is important to use a specific activity and a cell suspension aliquot that are extrapolated from those obtained with cells attached to dishes. If, at confluence, it is assumed that a 35-mm dish has about 2 million cells, an aliquot of cells in suspension after trypsin treatment, which contains a similar number of cells, should be used in order to obtain measurable rates.

## References

1. Abumrad, N., Harmon, C., and Ibrahimi, A. (1998) Membrane transport of long-chain fatty acids: evidence for a facilitated process. *J. Lipid Res.* **39,** 2309–2318.
2. Vorum, H., Brodersen, U., Kragh-Hansen, U., and Pedersen A. O. (1992) Solubility of long-chain fatty acids in phosphate buffer at pH 7.4. *Bochim. Biophys. Acta.* **1126,** 135–142.
3. Abumrad, N. A., Perkins, C. R., Park, J. H., and Park, C. R. (1981) Mechanism of long-chain fatty acids permeation in the isolated adipocyte. *J. Biol. Chem.* **256,** 9183–9191.
4. Sorrentino, D., Robinson, R. D., Kiang, C. L., and Berk., P. D. (1989) At physiologic albumin oleate concentrations oleate uptake by isolated myocytes, hepatocytes and adipocytes is a saturable function of the unbound oleate concentration. *J. Clin. Invest.* **84,** 1324–1333.
5. Spector, A A. (1975) Fatty acid binding to plasma albumin. *J. Lipid Res.* **16(3),** 165–179.
6. Richieri, G. V., Anel, A., and Kleinfeld, A. M. (1993) Interactions of long-chain fatty acids and albumin: determination of free fatty acid levels using the fluorescent probe ADIFAB. *Biochemistry* **32(29),** 7574–7580.
7. Richieri, G. V. and Kleinfeld, A. M. (1995) Unbound free fatty acid levels in human serum. *J. Lipid Res.* **36(2),** 229–240.
8. Storch, J., Lechenne, C. and Kleinfeld, A. M. (1991) Direct determination of fatty acid transport across the adipocyte plasma membrane using quantitative fluorescence microscopy. *J. Biol. Chem.* **266,** 13,473–13,476.
9. Knapp, F. F., Kropp, J., Goodman, M. M., Franken, P., Reske, S. N., Som, P., Biersack, H. J., Sloof, G. W., and Visser, F. C. (1993) Development of iodine125-methyl-branced fatty acids and their application in nuclear cardiology. *Ann. Nucl. Med.* **7,** SII-1–SII-14.
10. Whitesell, R. R. and Abumrad, N. A. (1986) Modulation of basal glucose transporter Km in the adipocyte by insulin and other factors. *J. Biol. Chem.* **261(32),** 15,090–15,096.
11. Labarca, C. and Paigen, K. (1980) Simple, rapid, and sensitive DNA assay procedure. *Analytic Biochem.* **102,** 344–352.

12. Ibrahimi, A., Sfeir, Z., Magharaie, H., Amri, E. Z., Grimaldi, P., and Abumrad, N. A. (1996) Expression of the CD36 homolog (FAT) in fibroblast cells: Effects on fatty acid transport. *Proc. Natl. Acad. Sci. USA* **93(7),** 2646–2651.
13. Abumrad, N. A., Forest, C. C., Regen, D. M., and Sanders, S. (1991) Increase in membrane uptake of long-chain fatty acids early during preadipocyte differentiation. *Proc. Natl. Acad. Sci. USA* **8(14),** 6008–6012.

# 13

## Assays of Glucose Entry, Glucose Transporter Amount, and Translocation

### Jean-François Tanti, Mireille Cormont, Thierry Grémeaux, and Yannick Le Marchand-Brustel

## 1. Introduction

Glucose enters the cell by a carrier-mediated, facilitated diffusion mechanism, which, in most tissues, exhibits no energy or counter-ion requirements. In adipose tissues and skeletal muscle, glucose entry is acutely regulated by insulin and other hormones *(1,2)*. Indeed, in those tissues, glucose transporter 4 (GLUT4) is the chief isoform which is, in basal conditions, retained in a specific intracellular storage compartment *(3)*. The GLUT4-containing vesicles are translocated to the plasma membrane in response to insulin, thus allowing for the massive entry of glucose into the cells *(1,2)*. Adipocytes also contain a small proportion of the ubiquituously expressed glucose transporter, GLUT1, which is at a similar level at the plasma membranes and inside the cell *(3)*. Because of this basal distribution, insulin effect on GLUT1 translocation is minor.

The isolated adipocyte is a very convenient system for measuring glucose transport in basal conditions. Further, it is both highly insulin sensitive ($EC_{50}$ approx 0.1 n$M$ insulin) and responsive (10- to 20-fold stimulation). This system can be used to look for an insulinomimetic effect of drugs (in a search for new therapeutic agents for diabetes) or to determine the ability of cells from patients to respond to insulin. More recently, the possibility of measuring GLUT4 translocation in adipocytes expressing an epitope-tagged GLUT4 transporter has been taken as a way of determining the possible molecular mechanism of insulin action.

This chapter describes the techniques allowing for the measurement of glucose uptake, using glucose itself or glucose analogs (2-deoxyglucose [DOG] and 3 O-methyl glucose [OMG]). The DOG method was first described *(4–6)*,

From: *Methods in Molecular Biology, vol. 155: Adipose Tissue Protocols*
Edited by: G. Ailhaud © Humana Press Inc., Totowa, NJ

and is still widely used. The principle of this technique is that labeled 2-DOG is transported into the cells with high affinity, and phosphorylated, but not further metabolized. Labeled 2-DOG phosphate is trapped in the cell, and the rate of uptake can be taken as a measure of unidirectional transport. By contrast, 3 OMG is not phosphorylated, and quickly equilibrates across the cell membrane *(7)*. The third (and easiest) method is based on the premise that glucose metabolism (glucose incorporation into lipids) provides a measurement of glucose transport at very low glucose concentration (<5 $\mu M$) *(8,9)*. The chapter then describes the method allowing for a direct measurement of GLUT translocation, using an epitope-tagged GLUT4 as a reporter gene *(10,11)*. Finally, because glucose transporter translocation can be directly followed by immunoblotting GLUT4 (and GLUT1) in the subcellular fractions obtained from control or treated adipocytes, the transporter immunodetection is briefly described. The sheet assay, which is used to visualize GLUT4 translocation in 3T3-L1 adipocytes, after cell sonication, is not described here, since it does not apply to normal adipocytes *(12)*.

## 2. Materials

1. Krebs-Ringer bicarbonate HEPES (KRBH) buffer: Prepare the following 10X stock solutions (store at 4°C):
   a. 1.2 $M$ NaCl, 40 m$M$ KH$_2$PO$_4$, 10 m$M$ MgSO$_4$, 7.5 m$M$ CaCl$_2$ (*see* **Note 1**).
   b. 100 m$M$ NaHCO$_3$.
   c. 300 m$M$ HEPES, pH 7.4.
      To prepare fresh KRBH buffer, mix 10 mL of each stock solution, and add water to 100 mL.
2. D-Glucose stock solution: Dissolve anhydrous glucose in water at a concentration of 10 mg/mL (store in aliquots at –20°C).
3. Adipocyte incubation buffer: To KRBH buffer, add glucose (200X dilution of the glucose solution) and 1% (w/v) bovine serum albumin (BSA) (*see* **Note 2**). Adjust the pH to 7.4.
4. Dulbecco's modified Eagle's medium (DMEM).
5. Laemmli buffer: 70 m$M$ Tris-HCl, pH 7.0, 3% sodium dodecyl sulfate (SDS), 11% glycerol *(13)*, final concentrations. Stock solutions of this buffer (2X or 4X) are usually prepared.
6. 1 $M$ HEPES, pH 7.4.
7. Potassium cyanide (KCN) stock solution: Dissolve KCN in water at a concentration of 200 m$M$, and store in aliquots at –20°C (*see* **Note 3**).
8. Insulin: Commercial insulin preparations (rapid type), used for diabetic patient care, can be used.
9. Antibodies (Abs) against Myc epitope (Santa Cruz) (*see* **Note 4**).
10. $^{125}$I-iodinated protein A (*see* **Note 5**).
11. Diisononylphthalate (oil of density 0.972, which allows for the separation of fat cells from the incubation medium) *(6)*.

12. Gentamicin solution at 10 mg/mL.
13. 2059 Falcon tubes (*see* **Note 6**).
14. 20- and 6-mL polyethylene scintillation vials (*see* **Note 6**).
15. Microtubes (400-μL, Beckman type) (*see* **Note 6**).
16. 15-mL polystyrene flat-bottomed tubes (*see* **Note 6**).
17. Nonpyrogenic syringes without needle.
18. 0.4-cm-gap electroporation cuvets.
19. Electroporator system (*see* **Note 7**).
20. 2-Deoxy-D-glucose and 3-O-methyl-D-glucose solution: Dissolve 2-deoxy-D-glucose or 3-O-methyl-D-glucose in water, at a concentration of 100 m$M$, and store in aliquots at –20°C.
21. Radiolabeled sugars:
    a. Radiolabeled deoxy-D-glucose, 2-[$^3$H(G)]: 259 GBq/mmol (7 Ci/mmol), 1 mCi/mL in EtOH:water (9:1); store at –20°C (*see* **Note 8**).
    b. Radiolabeled 3-O-methyl-D-$^3$H glucose: 2.8 TBq/mmol (75.2 Ci/mmol), 1 mCi/mL in EtOH:water (9:1) (*see* **Note 8**).
    c. Radiolabeled [3-$^3$H]-D-glucose: 647.5 Gbq/mmol (17.5 Ci/mmol), 1 mCi/mL in EtOH:water (9:1) (*see* **Note 8**).
22. Cytochalasin B solution: Dissolve cytochalasin B in dimethyl sulfoxide (DMSO) at a concentration of 1.5 m$M$, and store in aliquots at –20°C (*see* **Note 9**).
23. Toluene/butyl PBD scintillation cocktail: Dissolve 4 g butyl PBD (Serva 15075) per liter of toluene.
24. Commercial scintillation fluid accepting a small proportion of water (e.g., Ready Safe, Packard).
25. Phosphate-buffered saline (PBS).
26. Tris-buffered saline (TBS)-Tween: 50 m$M$ Tris-HCl, pH 7.4, 150 m$M$ NaCl, 0.1% Tween-20.
27. Fat-skimmed milk.
28. Rabbit anti-GLUT4 and anti-GLUT1 Abs: Abs directed against the C-terminus of GLUT4 and GLUT1, respectively (East Acres Biological, Southbridge, MA).
29. Horseradish peroxydase (HRP) conjugated Donkey antirabbit (Amersham).
30. Electrochemiluminescence (ECL) detection kit (Amersham).
31. Vinyl-covered exposure cassettes.
32. Sensitive X-ray film (Sigma, Amersham).
33. Nitrocellulose or polyvinylidene fluoride (PVDF) membranes (Millipore).

# 3. Methods

## 3.1. Measurements of Glucose Uptake by Freshly Isolated Adipocytes

### 3.1.1. DOG Uptake

1. Isolate adipocytes from epididymal fat pads of male Wistar rats (170–200 g) by collagenase digestion (*see* **Note 10**).
2. Wash the isolated adipocytes 2× by flotation in a 50-mL syringe with 30 mL KRBH buffer (which should be prewarmed at 37°C (*see* **Note 11**).

3. Resuspend adipocytes as a 30% (v/v) cell suspension (*see* **Note 12**) in incubation buffer without glucose.
4. Prepare a series of 15-mL flat-bottomed tubes with 100 µL assay buffer without glucose containing various insulin dilutions (0–100 n*M*).
5. Pipet in triplicates 100 µL adipocyte suspension, and put it in a water bath at 37°C (*see* **Note 13**). Incubate the cells for 10–20 min with the hormone.
6. Add in each tube, with a 15-s interval, 50 µL incubation buffer without glucose containing 0.5 m*M* deoxy-D-glucose and 0.5 µCi of $^3$H deoxy-D-glucose, and incubate for 2 min at 37°C, with gentle shaking (*see* **Note 14**).
7. Stop the transport by adding, with a 15-s interval, 10 µL 1.5 m*M* cytochalasin B.
8. Pipet 240 µL adipocyte suspension in a 400-µL microtube (Beckman type) already containing 100 µL diisononylphthalate.
9. Centrifuge 1 min at 6000*g* in a microcentrifuge at room temperature (RT) to separate the adipocytes from the medium (*see* **Note 15**).
10. Cut the tubes through the dinonylphthalate layer, and put the top part of the tube with the fat cake in 6-mL polyethylene scintillation vials.
11  Add 4 mL liquid scintillation (Ready Safe) in the vials, and vortex well.
12  Count the radioactivity associated with the cells in a β-counter, using the tritium program.
13. Substract the nonspecific transport from all the values (*see* **Note 16**).

### 3.1.2. 3-OMG Uptake

1. Proceed as described in **Subheading 3.1.1., steps 1–5**.
2. Add 50 µL incubation buffer, without glucose, containing 0.5 m*M* 3-*O*-methyl-D-glucose, 1 µCi 3-*O*-methyl-D-$^3$H glucose.
3. Incubate for 30 s for basal condition or 5 s for insulin condition (*see* **Note 17**) at 37°C, with gentle shaking.
4. Stop the transport with 10 µL 1.5 m*M* cytochalasin B.
5. Proceed as described in **Subheading 3.1.1., steps 8–13**.

### 3.1.3. Measurement of Glucose Incorporation into Lipids at Low Glucose Concentration

1. Proceed as described in **Subheading 3.1.1., steps 1** and **2**.
2. Resuspend adipocytes as a 10–15% (v/v) cell suspension (*see* **Note 12**) in assay buffer with 0.3 µ*M* glucose (1:183 dilution of the stock glucose solution).
3. Prepare a series of 20-mL polyethylene scintillation vials at 37°C, containing 50 µL of the various insulin concentrations (in assay buffer with 0.3 µ*M* glucose) and 50 µL assay buffer with 0.3 µ*M* glucose, 0.2 µCi [3-$^3$H]-D-glucose.
4. Pipet every 10 s, in triplicates for each insulin concentration, 900 µL adipocyte suspension into each vial, and incubate for 1 h at 37°C in a water bath, with gentle shaking (*see* **Note 13**).
5. Stop the reaction by adding 10 mL toluene/butyl PBD scintillation cocktail, vortex well and wait 2–12 h before counting in a β-counter, using the tritium program (*see* **Note 18**).

6. Subtract counts obtained from blank samples to all the values (*see* **Note 19**).
7. Do not forget to count an aliquot of the radioactive glucose solution (using a commercial scintillation fluid), which will allow expression of the results in absolute amounts of glucose taken up by the cells.

## 3.2. Determination of Glucose Transporter Translocation, Using an Epitope-Tagged GLUT4 in Transiently Transfected Rat Adipocytes

### 3.2.1. Transfection of Rat Adipocytes

1. Prepared adipocytes from epididymal fat pads of male Wistar rats (170–200 g) by collagenase digestion (*see* **Note 10**).
2. Wash isolated adipocytes twice in a 50-mL syringe with 30 mL prewarmed (37°C) incubation buffer without BSA and once with 30 mL prewarmed DMEM (*see* **Note 11**). Resuspend adipocytes as a 50% (v/v) cell suspension in DMEM.
3. Add 400 μL of the cell suspension (*see* **Notes 13** and **20**) in a 0.4-cm-gap electroporation cuvet, along with the plasmid cDNAs (0.5 μg of pCIS2 GLUT4-myc (*see* **Note 21**) and 9.5 μg of empty pCIS2), and perform electroporation with a double electric shock (800 V, 25 μF/200 V, 1050 μF) by using an Easyject electroporator system (*see* **Note 7**).
4. Immediately following the electric shock, put the cell suspension (400 μL) in a 2059 Falcon tube containing 1.5 mL DMEM, 25 m$M$ HEPES, pH 7.4, 5% (w/v) BSA, 100 μg/mL gentamicin (*see* **Note 22**), and incubate for 16 h (*see* **Note 23**) in a cell incubator at 37°C under an atmosphere of 5% $CO_2$/95% air.

### 3.2.2. Determination of Amount of Epitope-Tagged GLUT4 at the Cell Surface

1. Pool the tubes containing the transfected adipocytes (three tubes/conditions), and wash adipocytes 3× in a 5-mL syringe with 4 mL incubation buffer (prewarmed at 37°C).
2. Resuspend adipocytes in incubation buffer at a 10% (v/v) suspension (*see* **Note12**) and transfer adipocytes (3 mL) in 20-mL scintillation vials (*see* **Note 13**).
3. Incubate the cell suspension for 20–30 min at 37°C in a shaking water bath (50 cycles/min) in the absence or presence of insulin (100 n$M$ final concentration).
4. Add KCN (2 mM, final concentration), and wait for 5 min (*see* **Note 24**).
5. Transfer the vials containing the adipocyte suspension to a water bath at 25°C (all the following steps are performed at 25°C), and incubate with 1 μg/mL Abs to the myc epitope for 1 h, with gentle shaking.
6. Wash the cells 3× in a 5-mL syringe with 4 mL incubation buffer.
7. Resuspend the adipocytes with 450 μL incubation buffer (the total volume adipocyte + buffer should be 600 μL).
8. Pipet 200 μL in triplicate in 15-mL flat-bottomed tubes already containing [$^{125}$I]-iodinated protein A (200,000 cpm/tubes, 100 μL), and incubate for 1 h at 25°C, with gentle shaking.

9. Place 270 µL adipocyte suspension on 100 µL diisononylphthalate in a 400-µL microtube (Beckman type).
10. Centrifuge 1 min at 6000g in a microcentrifuge at RT, to separate cells from the medium (*see* **Note 15**).
11. Cut the tubes through the dinonylphthalate layer, and push the fat cell cake (with a yellow tip) into a 1.5-mL Eppendorf tube containing 300 µL of Laemmli buffer. Vortex, and boil for 30 min.
12. Put the tube in a γ-counter to count the radioactivity.
13. Centrifuge 1 min in a microcentrifuge at RT.
14. Transfer the aqueous phase, below the oil that is at the top of the tube, using a syringe, into new Eppendorf tubes.
15. Normalize the radioactivity by measuring the protein concentration in each sample by bicinchoninic acid assay (Pierce) (*see* **Note 25**).
16. Substract nonspecific binding from all the values (*see* **Note 26**).

## 3.3. Immunodetection of GLUT4

Translocation of GLUT4 can be determined following fractionation of adipocytes into plasma membrane and low-density microsomes (*see* Chapter 5). The amount of GLUT4 (and possibly GLUT1) in each fraction can be quantified by Western blotting. The immunoblotting is described for GLUT4, but is identical for GLUT1 immunodetection, except that a specific anti-GLUT1 Ab is used.

1. Separate proteins (50–80 µg) from each fraction on a 10% SDS-polyacrylamide gel electrophoresis (PAGE) *(13)*, and transfer proteins to membranes (nitrocellulose or PVDF membranes) (*see* **Note 27**).
2. Incubate the membrane in blocking buffer, with gentle shaking at RT for 1–2 h (*see* **Note 28**).
3. Incubate the membrane with gentle shaking overnight at 4°C with anti-GLUT4 Ab (diluted 1:500–1:1000 in blocking buffer) (*see* **Notes 28**).
4. To remove any unbound anti-GLUT4 Ab, wash the membrane 3× (10 min each) with washing buffer (*see* **Note 29**), with gentle shaking at RT.
5. Incubate the sheet with either [$^{125}$I]-protein A (500,000 cpm/mL blocking buffer) (*see* **Notes 5** and **28**) or with HRP-conjugated antirabbit immunoglobulin G (IgG) for ECL detection (freshly diluted to 1:5000 in TBS-Tween-20), with gentle shaking at RT for 1 h.
6. Discard the [$^{125}$I]-protein A or the secondary Ab and wash the sheet as in **step 4**.
7. When using [$^{125}$I]-protein A detection, make an autoradiography of the blot. For ECL detection, use the protocol of manufacturer.

## 4. Notes

1. To avoid calcium/phosphate precipitation, calcium chloride must be dissolved separately and added at the end to the salt solution.
2. BSA may contain some contaminants with insulinomimetic action. In accordance, highly purified BSA (Cohn fraction V) must be used, and several batches must be tested to select a batch that gives low basal values.

3. **CAUTION:** KCN is highly toxic, and must be handled with care.
4. Ig, from rabbit, against the Myc epitope must be used, and is commercially available (Santa Cruz). Do not use mouse Ig, Because protein A does not bind very well to mouse Ig.
5. [$^{125}$I]-iodinated protein A is commercially available (Amersham, ICN) or, alternatively, protein A can be labeled by the chloramine-T method.
6. Adipocytes are very sensitive to the nature of the plastic, and may break if wrong tubes or beakers are used. It is important to use the tubes and the vials mentioned in the materials.
7. An adequate electroporator is mandatory for successful transfection of adipocytes. An electroporator able to make double electric shock is required. Successful transfection of adipocytes has been obtained using the Easyject electroporator system *(11,14)*.
8. The EtOH–water stock solutions of the labeled compounds should be preferred to the aqueous solutions, which are more prone to microbial contaminations.
9. Cytochalasin B binds to glucose transporters, and inhibits the glucose transport *(3)*.
10. Collagenase dissociation of adipocytes is described in Chapters 5, 10, 15, and 24 of this book. Approximately 2 mL adipocyte suspension are usually obtained from rats weighing 170–200 g. The weight of the rat must not exceed 220 g, since adipocytes prepared from larger rats are more brittle and do not respond well to insulin stimulation.
11. Adipocytes contain a large amount of triglycerides, and float. To wash adipocytes, clamp a syringe, pour the adipocyte suspension, wait for sufficient time (2–5 min) to allow all the adipocytes to float at the top of the syringe, open the syringe to eliminate the washing solution, and repeat. Centrifugation should be avoided.
12. The adipocyte concentration must be between 10 and 30% (v/v), because a higher concentration could be responsible for a lower insulin response.
13. To accurately pipet adipocytes, the suspension should be kept homogenous by handshaking of the beakers or tubes during pipeting. Further, the tips of the automatic pipets should be cut to increase their diameter.
14. Uptake of DOG in these conditions is linear during the first 5 min following the addition of the medium containing labeled deoxy-D-glucose.
15. Because of its density, dinonylphthalate forms a layer between the adipocytes and the medium, and can thus be used to rapidly and efficiently separate adipocytes from the medium (centrifugation for 1 min at 3000–5000$g$ is sufficient).
16. Nonspecific transport is determined in the same conditions, except that cytochalasin B (10 µL 1.5 m$M$ stock solution) is added to the labeled medium before the cell addition.
17. Since 3-*O*-methyl-D-glucose can go in and out of the cells, it rapidly reaches an equilibrium. Thus, the uptake of 3-*O*-methyl-D-glucose is linear during a very short period of time (30 s for basal conditions, 5 s for insulin stimulation). A metronome can be used to time the assay.
18. Use a nonaqueous scintillation cocktail such as toluene/butyl PBD to extract radiolabeled lipids from the cells *(8)*. The cell suspension in the labeled medium

is carefully mixed with the scintillation fluid by vortexing, to allow for the lipid extraction in the toluene phase containing the scintillant. It is then necessary to wait for 2–12 h before counting the radioactivity, to ensure a complete separation between the aqueous phase and the toluene phase *(8)*. The aqueous phase, which contains the nonincorporated 3H-radiolabeled glucose, although present at the bottom of the scintillation vial, will not be counted. It should be noted that [$^{14}$C]-glucose cannot be used in this assay instead of [$^{3}$H]-glucose, because $^{14}$C is more energetic than $^{3}$H, and $^{14}$C-glucose present in the aqueous phase would be counted.

19. Blank values are obtained by adding toluene/butyl PBD scintillation cocktail to the cells, before the medium containing radiolabeled glucose

20. For each condition (i.e., basal or insulin) three 400-μL aliquots of adipocyte suspension are submitted to electroporation.

21. The sequence of GLUT4 has been modified in order to insert a Myc epitope tag in the first extracellular loop of GLUT4 *(10,11)*. Following insulin stimulation, the Myc epitope is exposed toward the extracellular medium, and the amount of GLUT4 at the cell surface can be estimated by the binding of an anti-Myc Ab *(10)*. pCIS2 is an expression vector with a cytomegalovirus promoter and enhancer *(15)*. All the described experiments *(10,11)* have been performed with this vector, but any vector with a cytomegalovirus promoter gives a high level of expression in adipocytes *(11,15,16)*.

22. It is crucial to add gentamicin to the medium to avoid bacterial contamination during the duration of the culture.

23. Adipocyte should not be maintained in culture for more than 16–24 h, since the insulin response falls dramatically thereafter.

24. KCN, which causes a depletion of cellular adenosine triphosphate *(2)*, is added to prevent the redistribution of GLUT4 during the time of the incubation with the Myc Abs and the protein A.

25. Five microliters of the sample are sufficient to measure the protein concentration with bicinchoninic acid (BCA reagent, Pierce), using the microassay protocol. For this volume of sample, SDS at the concentration used in the Laemmli buffer (3%) does not interfere with protein measurement.

26. Nonspecific binding of Abs obtained with cells transfected with empty pCIS2 alone, represents 20% of the total binding observed in cells transfected with pCIS GLUT4–Myc in the absence of insulin stimulation.

27. SDS-PAGE is not described here. The detailed technique has been described in **ref. *17***.

28. For [$^{125}$I]-protein A detection, use the following blocking buffer: PBS–5% BSA (w/v) or PBS–5% fat-skimmed milk. For ECL detection, use TBS-Tween-20 containing 5% fat-skimmed milk as a blocking buffer.

29. For [$^{125}$I]-protein A detection, use PBS–1% Nonidet-P40 (v/v). For ECL detection, use TBS-Tween-20.

## References

1. Rea, S. and James, D. E. (1997) Moving GLUT4. The biogenesis and trafficking of GLUT4 storage vesicles. *Diabetes* **46,** 1667–1677.

2. Simpson, I. A. and Cushman, S. W. (1986) Hormonal regulation of mammalian glucose transport. *Annu. Rev. Biochem.* **55,** 1059–1089.

3. Gould, G. W. and Holman, G. D. (1993) The glucose transporter family: structure, function and tissue-specific expression. *Biochem. J.* **295,** 329–341.

4. Rodbell, M. (1968) Metabolism of hormones on glucose metabolim and lipolysis. *J. Biol. Chem.* **239,** 375–380.

5. Olefsky, J. M. (1975) Effects of dexamethasone on insulin binding, glucose transport and glucose oxidation of isolated rat adipocytes. *J. Clin. Invest.* **56,** 1499–1508.

6. Livingston, J. N. and Lockwood, D. H. (1975) Effect of glucocorticoids on the glucose transport system of isolated fat cells. *J. Biol. Chem.* **250,** 8353–8360.

7. Vinten, J., Gliemann, J., and Østerlind, K. (1975) Exchange of 3-O-methylglucose in isolated fat cells. *J. Biol. Chem.* **251,** 794–800.

8. Moody, A. J., Stan, M. A., and Stan, M. (1974) A simple free fat cell bioassay for insulin. *Horm. Metabol. Res.* **6,** 12–16.

9. Kashiwagi, A., Verso, M. A., Andrews, J., Vasquez, B., Reaven, G., and Foley, J. E. (1983) In vitro insulin resistance of human adipocytes isolated from subjects with noninsulin-dependent diabetes mellitus. *J. Clin. Invest.* **72,** 1246–1254.

10. Quon, M. J., Butte, A. J., Zarnowski, M. J., Sesti, G., Cushman, S. W., and Taylor, S. I. (1994) Insulin receptor substrate 1 mediates the stimulatory effect of insulin on GLUT4 translocation in transfected rat adipose cells. *J. Biol. Chem.* **269,** 27,920–27,924.

11. Tanti, J.-F., Grémeaux, T., Grillo, S., Calleja, V., Klippel, A., Williams, L. T., Van Obberghen, E., and Le Marchand-Brustel, Y. (1996) Overexpression of a constitutively active form of phosphatidylinositol 3-kinase is sufficient to promote Glut 4 translocation in adipocytes. *J. Biol. Chem.* **271,** 25,227–25,232.

12. Robinson, L. J., Pang, S., Harris, D. S., Heuser, J., and James, D. E. (1992) Translocation of the glucose transporter (GLUT4) to the cell surface in permeabilized 3T3-L1 adipocytes: Effects of ATP, insulin, and GTPγS and localization of GLUT4 to clathrin lattices. *J. Cell Biol.* **117,** 1181–1196.

13. Laemmli, U. K. (1970) Cleavage of structural proteins during the assembly of the head of bacteriophage T4. *Nature* **227,** 680–685.

14. Cormont, M., Bortoluzzi, M.-N., Gautier, N., Mari, M., Van Obberghen, E., and Le Marchand-Brustel, Y. (1996) Potential role of Rab4 in the regulation of subcellular localization of Glut4 in adipocytes. *Mol. Cell. Biol.* **16,** 6879–6886.

15. Quon, M. J., Zarnowski, M., Guerre-Millo, M., De La Luz Sierra, M., Taylor, S. I., and Cushman, S. W. (1993) Transfection of DNA into isolated rat adipose cells by electroporation. Evaluation of promoter activity in transfected adipose cells which are highly responsive to insulin after one day in culture. *Biochem. Biophys. Res. Commun.* **194,** 338–346.

16. Tanti, J.-F., Grillo, S., Grémeaux, T., Coffer, P. J., Van Obberghen, E., and Le Marchand-Brustel, Y. (1997) Potential role of protein kinase B in glucose transporter 4 translocation in adipocytes. *Endocrinology* **138,** 2005–2010.

17. Smith, F. S. and Titheradge, M. A. (1998) Detection of NOS isoforms by Western-Blot analysis, in *Methods in Molecular Biology, Vol 100. Nitric Oxide Protocols* (Tithradge, M. A., eds.), Humana, Totowa, NJ, pp. 171–180.

# 14

## Methods to Study Phosphorylation and Activation of the Hormone-Sensitive Adipocyte Phosphodiesterase Type 3B in Rat Adipocytes

Eva Degerman, Svante Resjö, Tova Rahn Landström, and Vincent Manganiello

## 1. Introduction

Cyclic nucleotide phosphodiesterases (*PDE*s) include a large group of structurally related enzymes that are responsible for the hydrolysis of cyclic adenosine monophosphate (cAMP) and cyclic guanosine monophosphate (cGMP). These enzymes belong to at least nine related gene families (*PDE*s 1-9) *(1–5)*, which differ in their primary structures, affinities for cAMP and cGMP, responses to specific effectors, sensitivities to specific inhibitors, and regulatory mechanisms. The *PDE3* family *(6)* consists of two subfamilies, *PDE3A* and *PDE3B*, which exhibit tissue-specific distribution; grossly, *PDE3A* enzymes are expressed in the cardiovascular system, and PDE3B enzymes in insulin-sensitive cells, such as hepatocytes *(7)* and adipocytes *(6)*, and also in pancreatic β-cells *(8)*. One characteristic of PDE3s involves their phosphorylation and activation in response to insulin, as well as to agents that increase cAMP in adipocytes *(6)*, hepatocytes *(7)*, and platelets *(9–11)*, and in response to insulin-like growth factor-1 (IGF-)1 in pancreatic β-cells *(8)*.

The adipocyte, *PDE3B*, has been cloned, recombinant protein produced, and the organization of the gene partially elucidated *(6)*. PDE3B, which migrates as a 135-kDa protein during sodium dodecyl sulfate-polyacrylamide gel elcetrophoresis (SDS-PAGE), consists of a C-terminal catalytic domain, which is conserved among the different *PDE* families, and an *N*-terminal regulatory region. Regarding the *PDE3* family, the catalytic domains have high affinity for both cAMP and cGMP, with *Km* values in the range of 0.1–0.8 μ*M*, but

From: *Methods in Molecular Biology, vol. 155: Adipose Tissue Protocols*
Edited by: G. Ailhaud © Humana Press Inc., Totowa, NJ

$V_{max}$ values for cAMP are higher (four- to 10-fold) than for cGMP *(1,6)*. The *N*-terminal regulatory region contains a hydrophobic region important for membrane-association of the enzyme. Serine 302 is located downstream from this region. This site is phosphorylated in response to stimulation of adipocytes with insulin and isoproterenol *(6,12–14)*; phosphorylation is associated with activation of the enzyme.

In adipocytes, phosphorylation and activation of *PDE3B* is the major mechanism whereby insulin antagonizes catecholamine-induced lipolysis (**Fig. 1**). *PDE3B* activation results in increased degradation of cAMP, and thereby a lowering of the activity of cAMP-dependent protein kinase (PKA). The reduced activity of PKA leads to a net dephosphorylation and decreased activity of hormone-sensitive lipase and reduced hydrolysis of triglycerides *(15)*.

The mechanism whereby insulin-stimulation leads to phosphorylation and activation of *PDE3B* is only partly understood, but probably involves activation of phosphatidylinositol-3-kinase (PI3K) *(16)* and maybe activation of protein kinase B *(17–19)*. In rat adipocytes, lipolytic hormones and other agents that increase cAMP, including isoproterenol, also induce rapid phosphorylation on serine 302 of *PDE3B*, presumably catalyzed by PKA *(6,12–14)*. The phosphorylation is associated with activation of the enzyme, and probably represents feedback regulation of cAMP, presumably allowing close coupling of the regulation of steady-state concentrations of both cAMP and PKA, and thereby control of lipolysis. Regarding dephosphorylation and deactivation of *PDE3B*, protein phosphatase 2A was recently shown to act as a *PDE3B* phosphatase, both in vivo and in vitro *(20)*.

This chapter describes methods for studying phosphorylation and activation of *PDE3B* in rat adipocytes. This involves methods to prepare membrane-associated *PDE3B* from adipocytes, methods to measure *PDE3B* activity, and methods to isolate [$^{32}$P]-phosphorylated *PDE3B* from prelabeled adipocytes. We prepare rat adipocytes essentially according to the procedures of Rodbell *(21)* and Londos *(22)*, which are described in Chapters 5, 10, 15, and 24 of this book.

## 2. Materials

### 2.1. Preparation of PDE3B from Rat Adipocytes

1. Rat adipocytes isolated from epididymal fat pads, as described in Chapters 5, 10, 15, and 24 of this book.
2. Hematocrit tubes.
3. Hematocrit centrifuge.
4. Hematocrit sealing compound (Brand, Wertheim, Germany, cat. no. 749500).
5. Slide-caliper.
6. Adipocyte incubation buffer: 25 m$M$ HEPES, pH 7.5, 120 m$M$ NaCl, 4.74 m$M$ KCl, 1.19 m$M$ KH$_2$PO$_4$, 1.19 m$M$ MgSO$_4$, 2.54 m$M$ CaCl$_2$, 1% (w/v) bovine

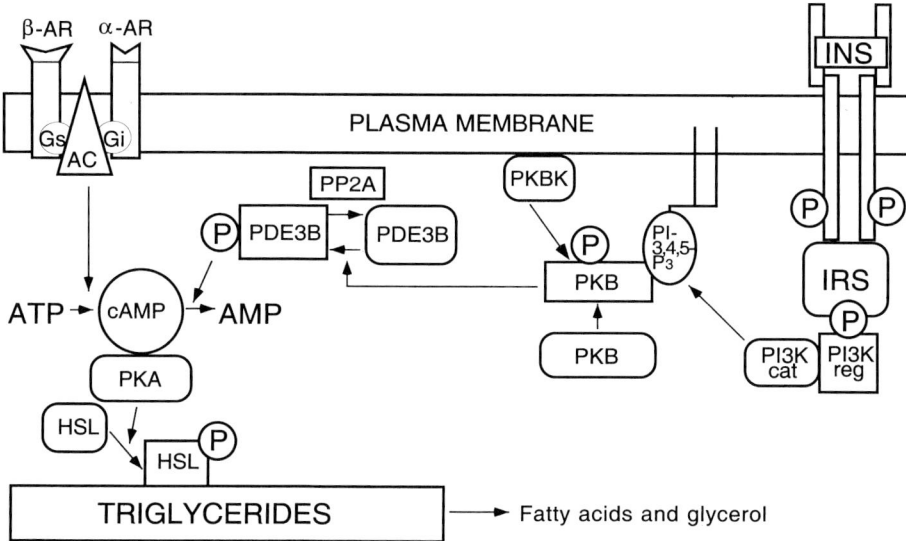

Fig. 1. Working hypothesis for mechanisms of hormonal regulation of lipolysis. Briefly, insulin stimulation of adipocytes results in phosphorylation and activation of *PDE3B* leading to reduction of cAMP, and thereby a lowering of the activity of cAMP dependent protein kinase A (PKA). The PKA mediated phosphorylation and activation of hormone-sensitive lipase (HSL) is thereby reduced, resulting in inhibition of lipolysis. The signaling pathway from the insulin receptor to *PDE3B* is only partially understood. Phosphatidylinositol 3-kinase (PI3K) is likely to be involved, since wortmannin, a selective PI3K inhibitor, blocks insulin-induced phosphorylation and activation of *PDE3B*. Recent data indicate that protein kinase B (PKB) is involved in the regulation of *PDE3B*. AR, adrenergic receptor; AC, adenylyl cyclase; Gs, stimulatory GTP-binding protein; Gi, inhibitory GTP-binding protein; P, phosphorylation sites; INS, insulin; IRS, insulin receptor substrate; PP2A, protein phosphatase 2A.

serum albumin (BSA) (Sigma, St. Louis, MO, cat. no. A-4503), 2 m$M$ glucose, 200 n$M$ adenosine (*see* **Note 1**).

7. Labeling buffer: Adipocyte incubation buffer containing 300–600 μ$M$ KH$_2$PO$_4$ (*see* **Notes 1** and **2**).

8. (–)-N$^6$-(R-phenyl-isopropyl) adenosine (PIA) (*see* **Note 1**).

9. Adenosine deaminase (Boehringer Mannheim, Mannheim, Germany, cat. no. 102,121) (*see* **Note 1**).

10. Insulin: 100× stock solution in adipocyte incubation buffer, prepared on the day of use, and stored at 4°C.

11. Isoproterenol: 100× stock solution in adipocyte incubation buffer, prepared less than 15 min before use.

12. [$^{32}$P]$_i$ (40 mCi/mL, Amersham, Uppsala, Sweden).

13. Hypodermic needles (18 gage) attached to a short piece of capillary tubing (polyethylene capillary tubing, Amersham, cat. no. 19-0040-01) (*see* **Fig. 2**).

Fig. 2. Hypodermic needle with a piece of polyethylene tubing attached in a 15-mL plastic tube.

14. 15-mL plastic tubes.
15. Adipocyte homogenization buffer: 50 m$M$ Tris EDTA-sucrose (TES), pH 7.4, 250 m$M$ sucrose, 1 m$M$ EDTA, 0.1 m$M$ EGTA. Add the following protease inhibitors on the day of use: 10 μg/mL antipain, 10 μg/mL leupeptin, and 1 μg/mL pepstatin A (*see* **Note 3**).
16. Swing out rotor centrifuge for 15-mL tubes.
17. Straight-necked scintillation vials for incubation of adipocytes (Poly Q vials, Beckman Instruments, Fullerton, CA) (*see* **Note 4**).
18. Homogenizers.
19. Heated shaking incubator (we use a Lab-Shaker LSR/L-V, Adolf Kühner, Birsfelden, Switzerland, with a stroke length of 5 cm).
20. Solubilization buffer: 50 m$M$ TES, pH 7.4, 250 m$M$ sucrose, 1 m$M$ EDTA, 0.1 m$M$ EGTA, 100 m$M$ NaBr, 50 m$M$ NaF, 20% (w/v) glycerol, 1% (v/v) $C_{13}E_{12}$ (nonionic alkyl polyoxyethylene glycol detergent from Berol Kemi AB, Stenungsund, Sweden). Store as stock solution at –20°C. Add the following protease inhibitors on the day of use: 10 μg/mL antipain, 10 μg/mL leupeptin, and 1 μg/mL pepstatin A (*see* **Note 3**).
21. PDE3B antibodies (anti-P$_{423-440}$ antiserum raised against a peptide corresponding to amino acids 423–440 of rat *PDE3B*) (*see* **Note 5**).
22. Immunoprecipitation wash buffer: 0.2 $M$ NaCl, 2.7 m$M$ KCl, 1.5 m$M$ KH$_2$PO$_4$, 4.3 m$M$ NaHPO$_4$, 0.1% (w/v) $N$-lauroyl sarcosine.
23. Protein A sepharose (Pharmacia) 50% slurry in 0.2 $M$ NaCl, 2.7 m$M$ KCl, 1.5 m$M$ KH$_2$PO$_4$, 4.3 m$M$ Na$_2$HPO$_4$. Store at 4°C.

24. SDS-sample buffer: 0.25 m$M$ Tris-HCl, pH 6.8, 4 m$M$ EDTA, 70 m$M$ SDS, 2% (w/v) β-mercaptoethanol, 25% (w/v) glycerol and bromophenol blue.
25. Equipment to perform SDS-PAGE.

## 2.2. PDE3B *Assay*

1. Diethylaminoethyl (DEAE)-Sephadex A-25 (Amersham, cat. no. 17-0170-01).
2. Small chromatography columns with the capacity to hold at least 1 mL gel and 10 mL eluant solution (Bio-Rad Poly-Prep columns, cat. no. 731-1550, Bio-Rad, Hercules, CA).
3. Column regeneration solution: 0.72 $M$ HCl.
4. [$^3$H]-cAMP, 30–50 Ci/mmol ([$^3$H]-cAMP dissolved in 1:1 EtOH:water, 1 mCi/mL, Amersham, cat. no. TRK498).
5. Cellulose thin-layer chromatography (TLC) plates, 20 × 20 cm (Merck, Whitehouse Station, NJ, cat. no. 1.05716).
6. TLC tank.
7. TLC buffer: 5:2 EtOH:ammonium acetate (0.5 $M$).
8. Scintillation cocktail for aqueous samples (ReadySafe, Beckman, Fullerton, CA).
9. PDE assay dilution buffer: 50 m$M$ TES, pH 7.4, 250 m$M$ sucrose, 1 m$M$ EDTA, 0.1 m$M$ EGTA. Store at –20°C.
10. Substrate buffer: 2× stock solution containing 100 m$M$ TES, pH 7.4, 500 m$M$ sucrose, 2 m$M$ EDTA, 0.2 m$M$ EGTA, 50 m$M$ MgCl$_2$, 3.0 m$M$ cAMP, 3.4 µg/mL ovalbumin. Store at –20°C.
11. Stop solution: 8 m$M$ cAMP, 5 m$M$ AMP, 0.25 $M$ HCl. Store at 4°C.
12. NaOH-Tris solution: Mix equal volumes of 0.5 $M$ Tris, pH 8.0, and 0.5 $M$ NaOH. Store at 4 °C.
13. Crotalus Atrox snake venom solution: 100 m$M$ Tris, pH 8.0, 1.88 mg/mL snake venom (Sigma, cat. no. V 7000), 1 m$M$ NaN$_3$. Store at –20°C in aliquots. Store thawed aliquots at 4°C.

## 3. Methods

## 3.1. *Determination of Adipocyte Concentration*

1. Gently mix the suspension of adipocytes by inverting the tube. Fill a hematocrit tube by dipping it quickly into the suspension, and seal it by stabbing it into hematocrit-sealing compound 3×. Repeat this with 2–3 tubes.
2. Centrifuge the tubes in a hematocrit centrifuge for 2 min.
3. After centrifugation, the adipocyte suspension separates into three fractions, one containing medium, one containing adipocytes, and sometimes one containing free fat. Use a slide-caliper to measure the total length of all fractions and the length of the adipocyte fraction. Calculate the proportion of adipocytes by dividing the length of the adipocyte fraction by the combined length of all the fractions.
4. The total amount of packed adipocytes can be calculated by multiplying the proportion of adipocytes in the suspension by the total volume of the suspension.
5. Using 37-d-old male Sprague-Dawley rats, approx 300 µL packed cells/rat are obtained.

## 3.2. Preparation of PDE3B from Rat Adipocytes Stimulated with Different Agents

1. To investigate six separate conditions, suspend freshly prepared adipocytes in adipocyte incubation buffer, to yield 13 mL 8–12% (v/v) cell suspension.
2. Incubate the suspension at 37°C in a shaking incubator (*see* **Note 6**) for 15 min, to allow the cells to recover after their preparation.
3. Prewarm the incubation vials to 37°C.
4. Transfer aliquots of the adipocyte suspension to the scintillation vials using a pipet (*see* **Note 7**). Use 1.5 mL suspension for each condition. Since the adipocytes rapidly float to the surface, it is important to move the container constantly while pipeting, in order to obtain a homogenous suspension. Leave the last 4 mL suspension in the container, because the composition of the last portion is not representative of the rest of the suspension.
5. Add hormones or other stimulants to the incubations, and incubate at 37°C in a shaking incubator at 120 bpm for the desired time (*see* **Note 8**).
6. Stop the incubations by adding 10 mL of room temperature (RT) adipocyte homogenization buffer.
7. Transfer the suspensions to 15-mL plastic tubes. To each tube, add a needle with a piece of polyethylene tubing attached (*see* **Subsheading 2.1., item 1.** and **Fig. 2**). Centrifuge the tubes for 15 s in a swing out rotor at approx 125$g$.
8. Attach a syringe to the needle, and remove the incubation buffer.
9. Resuspend the adipocytes in 1 mL (final volume) homogenization buffer containing 60 n$M$ calyculin A (*see* **Note 9**). Carefully rinse the walls of the tube to remove attached cells, and transfer the adipocytes to a homogenizer.
10. Homogenize the cells with 10 strokes at RT (*see* **Note 10**).
11. Immediately transfer the homogenates to centrifuge tubes placed on ice. The samples should be kept at 4°C in all following steps, unless stated otherwise.
12. Centrifuge the homogenates (33,000$g$, 45 min, 4°C).
13. Remove the floating fat cake with a small spatula, then remove the infranatant. To remove all traces of fat, wipe the test tube carefully with cotton-tipped applicators (Q-tips) without touching the membrane pellet.
14. Add 500–1000 µL adipocyte homogenization buffer containing 60 n$M$ calyculin A (*see* **Note 9**) to each pellet. Transfer to a homogenizer, and resuspend the crude membranes by homogenization with 10 strokes.

## 3.3. Preparation of PDE3B from [$^{32}$P]-Labeled Rat Adipocytes

1. To investigate six separate conditions, suspend freshly prepared adipocytes in labeling buffer to yield 13 mL 8–12% (v/v) cell suspension.
2. Add 0.5–1 mCi [$^{32}$P]$_i$/mL cell suspension (*see* **Note 11**), and incubate the cells for 75 min at 37°C with moderate shaking (*see* **Note 6**). Meanwhile, prewarm the incubation vials.
3. After labeling, divide the cells into 1.5-mL aliquots, using a pipet (*see* **Note 7**), add hormones or other stimuli (*see* **Note 8**), and incubate at 37°C in a shaking incubator at 120 bpm, for the desired time.

4. Stop the incubations by adding 7 mL of RT homogenization buffer. Transfer the suspensions to 15-mL plastic tubes. To each tube, add a needle with a piece of polyethylene tubing attached (*see* **Subheading 2.1., item 13** and **Fig. 2**). Centrifuge the tubes for 15 s in a swing-out rotor at approx 125*g*.

5. Attach a syringe to the needle, and remove the homogenization buffer.

6. Wash the cells once more in 10 mL homogenization buffer, and finally resuspend the adipocytes in 1 mL (final volume) homogenization buffer containing 60 n*M* calyculin A (*see* **Note 9**). Carefully rinse the walls of the tube to remove attached cells, and transfer the adipocytes to a homogenizer.

7. Homogenize the cells with ten strokes at RT (*see* **Note 10**), and immediately transfer the homogenate to centrifuge tubes placed on ice. During the following steps, the samples should be kept at 4°C.

8. Centrifuge the homogenate (33,000*g*, 45 min, 4°C).

9. Remove the floating fat cake with a small spatula, then remove the infranatant. To remove all traces of fat, wipe the test tube carefully with cotton-tipped applicators (Q-tips), without touching the membrane pellet.

10. Add 1000 µL solubilization buffer containing 60 n*M* calyculin A to each pellet (*see* **Note 9**). Transfer to a homogenizer, resuspend the crude membranes by homogenization with 10 strokes, and transfer to new centrifuge tubes.

11. After 1 h on ice, centrifuge the solubilized membranes (10,000*g*, 15 min, 4°C), using a tabletop centrifuge, transfer the supernatants to new tubes, and add anti-PDE3B antibodies (*see* **Note 5**).

12. After 4–16 h, add protein A-Sepharose (50% slurry) (*see* **Note 12**), and incubate the immunoprecipitate for another 20–30 min.

13. Wash the immunoprecipitates 4× with 1 mL immunoprecipitation wash buffer, and, finally, suspend the immunopellet in SDS-sample buffer.

14. Boil the immunoprecipitate for 10 min, centrifuge briefly to pellet the protein A-Sepharose, and subject the supernatant to SDS-PAGE, using 7% polyacrylamide in the separation gel. Dried gels are analyzed using digital imaging of [$^{32}$P] (Fujix Bas 2000).

## 3.4. Preparation of Ion Exchange Columns for PDE Assay

1. For 100 columns, suspend 40 g DEAE-Sephadex A-25 in 500 mL of water, and allow the gel to swell at RT overnight.

2. Aspirate the supernatant from the settled gel. Resuspend the gel in water, and allow it to settle for 30–60 min. Repeat this twice The purpose of this procedure is to remove the finest particles from the gel to increase the flow rate of the column.

3. Resuspend the gel in water, using equal volumes of water and swelled gel. Stir the suspension constantly to keep the gel particles suspended ,and transfer 2 mL slurry to each chromatography column.

4. Pack the columns by allowing the water to drain from the column.

5. Wash the columns once with 10 mL column regeneration solution, and twice with 10 mL water. The columns are now ready to use. Once the columns have been packed, they can be used for a large number of assays (several times a week for 6 mo to 1 yr).

### 3.5. Purification of [³H]-cAMP by TLC

1. Apply 400 μL of [³H]-cAMP (0.4 mCi) to a TLC plate in a streak 2.5 cm above the bottom edge of the plate. Leave space on both sides for cAMP standards (*see* **Fig. 3**).
2. As a standard, apply 10 μL 100 m*M* cAMP (dissolved in water) 2 and 4 cm from each edge (*see* **Fig. 3**).
3. Allow the applied material to dry, then run the plate in TLC buffer until the solvent has reached the upper edge of the plate (approx 12 h). Let the TLC plate dry overnight.
4. Use a UV-lamp to detect the cAMP standards, and scrape out the corresponding area that contains the [³H]-cAMP (*see* **Fig. 3**).
5. Transfer the cellulose to a sintered funnel, and elute the [³H]-cAMP with 4 × 2.5 mL water.
6. Store the eluate at –20°C in 1-mL aliquots.

## 3.6. Preparation of Substrate for PDE Assay

1. Wash a column with DEAE-Sephadex A-25, according to **Subheading 3.4., step 5**.
2. Apply 2 mL TLC-purified [³H]-cAMP to the washed column, and discard the flowthrough.
3. Wash the column with 10 mL water, and discard the water.
4. Elute the column with 4 mL 50 m*M* HCl.
5. Add 40 mL substrate buffer and water to the eluate, to a final volume of 80 mL.
6. Store the substrate at –20°C in 10-mL aliquots.

## 3.7. PDE Assay (see Note 13)

1. Use ion exchange columns prepared according to **Subheading 3.4.** Wet, resuspend, and pack the DEAE-Sephacel, by adding water, stirring the gel, and allowing the water to drain. This is not necessary, if the columns have been washed the same day.
2. Mix 10–30 μL membrane suspension containing *PDE3B* with *PDE* assay dilution buffer to a final volume of 200 μL. All reactions should be performed in duplicates or triplicates.
3. Prepare 2–3 control incubations for background evaluation (blanks) consisting of 200 μL *PDE* assay dilution buffer and 100 μL *PDE* assay substrate, and treat these the same way as the other samples.
4. Start the assay by the addition of 100 μL *PDE* assay substrate, vortex briefly, and place the samples at 30°C.
5. Incubate at 30°C for 6–8 min, and stop the reactions by adding 100 μL stop solution. If convenient, the samples can be stored at –20°C at this stage.
6. Add 100 μL NaOH-Tris solution, to neutralize the pH of the incubations.
7. Add 100 μL snake venom solution, and incubate for at least 20 min at 30°C, to completely dephosphorylate AMP to adenosine.
8. Place scintillation vials under the DEAE-Sephadex columns, and apply 500 μL sample to each column (see **Note 14** and **Fig. 4**).
9. Wait until the solutions have entered the columns, then wash the columns with 3.6 mL water. Collect the eluates in the scintillation vials.

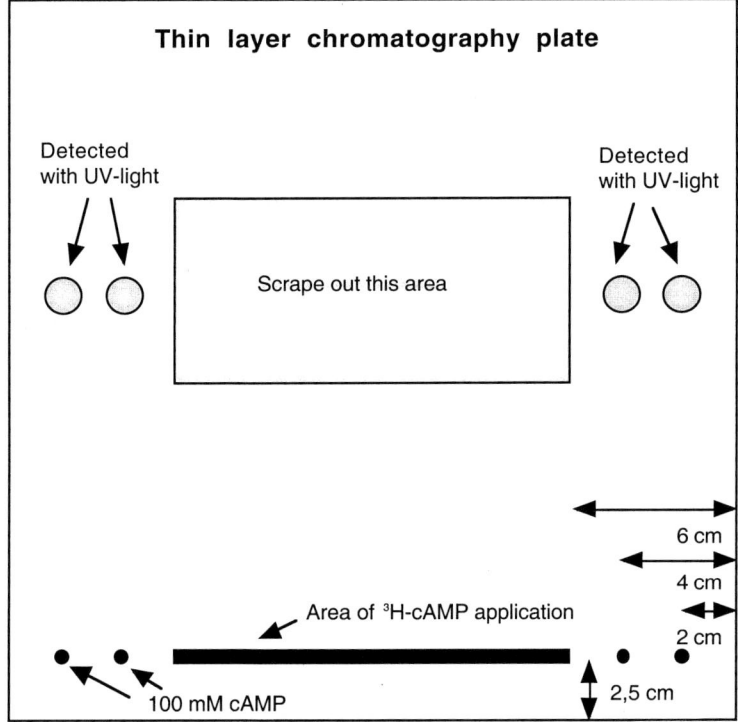

Fig. 3. Purification of $^3$H-cAMP, using TLC chromatography.

10. Add 10 mL scintillation cocktail to the vials, mix, and count in a scintillation counter.
11. Mix 100 μL *PDE* assay substrate with 4.1 mL water and 10 mL scintillation cocktail in a scintillation vial, and count in a scintillation counter. This sample is used to evaluate the theoretical maximal cpm in the assay.
12. Calculate *PDE* activity: Activity is expressed in units equaling ability to hydrolyze 1 pmol cAMP/minute/mL enzyme. The symbols used are: C, the amount of cpm in the sample; B, the amount of cpm in the blank; M, the amount of cpm in the maximum sample; $T_i$, the incubation time (in minutes) in **item 5**; and $V_e$, the volume of membrane suspension (in milliliters) in **item 2**. The *PDE* activity in the membrane suspension is then $([C - B]*150)/(M*[5/6]*T_i*V_e)$ (*see* **Note 15**).
13. Wash the columns once with 10 mL column regeneration buffer and twice with water, to make them ready to use again

## 4. Notes

1. If adipocytes are to be incubated for 3 h or more, the concentration of BSA in the adipocyte incubation buffer should be increased to 3%, to bind the fatty acids released from the adipocytes. The pH of the adipocyte incubation and labeling

cAMP + adenosine

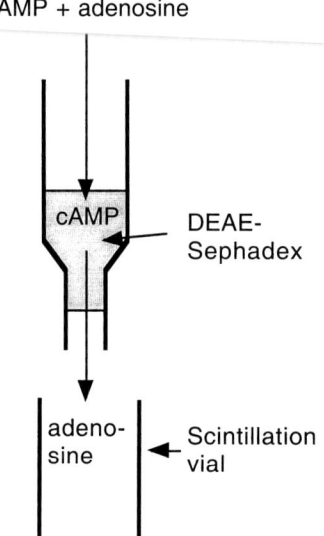

Fig. 4. *PDE* assay; separation of cAMP and adenosine using DEAE-Sephadex.

buffers should be adjusted, so that it is always 7.4 at the temperature at which the adipocytes are to be incubated. For example, the pH for a HEPES buffer, adjusted to 7.5 at RT is 7.4 at 37°C. The adipocyte incubation buffer contains 200 n$M$ adenosine, which is added to inhibit basal lipolysis. Since the adipocytes produce endogenous adenosine, cell lysis may influence the concentration of adenosine in the medium during the course of the experiment. This variation can be avoided by adding 100–200 n$M$ PIA, an adenosine analog insensitive to adenosine deaminase, followed by 0.5–1 U/mL adenosine deaminase, to hydrolyze the adenosine.

2. 300–600 µ$M$ KH$_2$PO$_4$ (instead of 1.19 m$M$, normally present in the standard medium) is used in the labeling medium, because decreasing the concentration below 600 µ$M$ has been shown to decrease rather than increase the specific activity of the intracellular adenosine triphosphate *(23)*.

3. The protease inhibitors should be prepared as 1000× stock solutions: Pepstatin 1 mg/mL in methanol; leupeptin and antipain 10 mg/mL in water. The stock solutions can be stored at –20°C for 6 mo.

4. Straight-necked scintillation vials can be used for 0.8- to 2.0-mL incubations. For larger volumes, other plastic containers have to be used. The volume of the container should be adjusted so that the suspension of adipocytes covers the bottom to a depth of 4–10 mm. **CAUTION:** Glass vessels must never be used to incubate adipocytes, since glass will cause lysis of the cells.

5. In addition to the anti-P$_{423–440}$ antiserum, antibodies directed against the *N*-terminal portion of *PDE3B* have also been successfully used for immunoprecipitaion of *PDE3B* in these laboratories *(24,25)*. All new batches of *PDE3B* antiserum must be titered to estimate the efficiency of the antibodies.

6. The speed of the shaking incubator must be adjusted, depending on the size of the adipocyte container. The suspension should be rocked, so that there is an even wave across the bottom of the container.

7. All pipet tips for volumes under 1000 µL should be cut a few mm from the tip, to produce a wider opening.

8. In order to achieve maximal phosphorylation and activation of *PDE3B*, rat adipocytes may be stimulated with insulin (1 n*M*, 10 min) or isoproterenol (300 n*M*, 10 min).

9. To maintain maximal phosphorylation and activation of *PDE3B*, the phosphatase inhibitor, calyculin A (60 n*M*), may be added to the homogenization/solubilization buffer before homogenization and resuspension/solubilization of the membranes.

10. The adipocytes should be homogenized at RT, since homogenization at 4°C may lead to trapping of proteins in the fraction containing the fat and subsequent loss of material, because a cold, fatty homogenate is difficult to remove from the homogenizer.

11. 0.1–0.5 mCi [$^{32}$P]$_i$/mL cell suspension results in phosphorylation of sufficient *PDE3B* to detect [$^{32}$P]-*PDE3B* by digital imaging. However, if further manipulation of the phosphorylated *PDE3B* is desired (such as tryptic phosphopeptide mapping or phosphoaminoacid analysis), the authors suggest the use of 1 mCi [$^{32}$P]$_i$/mL cell suspension.

12. Protein A-Sepharose must be titrated for every new batch of *PDE3B* antiserum. However, as a rule of thumb, a 1:1 ratio is suggested.

13. The assay works according to the following principles (*see also* **ref. 26**): cAMP is hydrolyzed to AMP by the *PDE* in the first incubation. The reaction is stopped by the addition of stop solution with low pH, and the proteins are denatured. The pH is neutralized, and snake venom is added. The venom contains an enzyme that dephosphorylates AMP to adenosine, but does not affect cAMP. During the (at least) 20-min incubation with venom, all the AMP is converted to adenosine. The resulting mixture of cAMP and adenosine is then applied to the ion exchange column. Because cAMP is charged and adenosine is uncharged, only cAMP will bind to the column. The adenosine will elute when the column is washed with water, and can then be counted in a scintillation counter. In adipocyte membranes prepared according to **Subheading 3.2.**, *PDE3B* makes up more than 90% of the total *PDE* activity. However, if there is a mixture of different *PDE* activities, the assay can be performed in the absence and presence of 3–10 µ*M* OPC3911, an example of an inhibitor specific for *PDE3*. The *PDE3* activity can then be calculated by subtracting the activity with OPC3911 from the activity without OPC3911.

14. To easily place scintillation vials under the columns, the columns should be placed in a rack designed to keep them in the correct position relative to each other and to the vials (*see* **Fig. 4**).

15. The formula for calculating *PDE* activity is determined as follows: The reaction mixture in the assay contains 150 pmol cAMP. The proportion of cAMP that has been hydrolyzed in a given sample can be calculated by dividing the cpm in the sample by the total cpm that would be generated if all the cAMP in the reaction mix was hydrolyzed. Because there is always a small number of cpm, even in a

sample without any added *PDE* (or in the presence of heat-inactivated *PDE*), the assay blank must first be subtracted from the sample cpm. The cpm that would be generated if all the cAMP were hydrolyzed is obtained from the maximum sample, but, since only 500 of 600 µL reaction mixture is applied to the column, the cpm in 100 µL substrate must be multiplied by 500/600 = 5/6. Thus, the amount of cAMP that has been hydrolyzed in a given sample = 150 pmol*([C – B]/(M*[5/6]). This is then divided by the incubation time, and the volume of sample is added to the assay, to yield: *PDE* activity (in pmol/min/mL) = ([C – B]*150)/(M*[5/6]*$T_i$*$V_e$).

## Acknowledgments

This work was supported by the Swedish Medical Research Council Projects 3362 for Per Belfrage and 12537 for Eva Degerman; the Swedish Diabetes Association; Albert Påhlsson's foundation, Malmö, Sweden; Novo Nordisk, Copenhagen, Denmark; the Crafoord Foundation, Lund, Sweden; and by a Center of Excellence grant from the Juvenile Diabetes Foundation and Knut and Alice Wallenberg Foundation.

## References

1. Manganiello, V. C., Murata, T., Taira, M., Belfrage, P., and Degerman, E. (1995) Perspectives in biochemistry and biophysics, Diversity in cyclic nucleotide phosphodiesterase isoenzyme families. *Arch. Biochem. Biophy.* **322**, 1–13.
2. Conti, M., Nemos, G., Sette, C., and Vicini, E. (1995) Recent progress in understanding the hormonal regulation of phosphodiesterases. *Endocrine Rev.* **16**, 370–389.
3. Beavo, J. (1995) Cyclic nucleotide phosphodiesterases: functional implications of multiple isoforms. *Physiol. Rev.* **75**, 725–748.
4. Fischer, D. A., Smith, J. F., Pillar, J. S., St. Dennis, S. H., and Cheng, J. B. (1998) Isolation and characterization of of PDE9A, a novel cGMP specific phosphodiesterase. *J. Biol. Chem.* **273**, 15,559–15,564.
5. Soderling, S. H., Bayuga, S. J., and Beavo, J. A. (1998) Identification and characterization of novel family of cyclic nucleotide phosphodiesterases. *J. Biol. Chem.* **273**, 15,553–15,558.
6. Degerman, E., Belfrage, P., and Manganiello, V. (1997) Minireview: Structure, localization and regulation of cGMP-inhibited phosphodiesterase (PDE3). *J. Biol. Chem.* **272**, 6823–6826.
7. Houslay, M. D. and Kilgour, E. (1990) Cyclic nucleotide phosphodiesterases in liver: A review of their characterization, regulation by insulin and glucagon and their role in controlling intracellular cyclic AMP concentration, in *Cyclic Mucleotide Phosphodiesterases: Structure, Regulation and Drug Action.* (Beavo, J. A. and Houslay, M. D., eds.), Wiley, Chichester, UK, pp. 185–224.
8. Zhao, A. Z., Zhao H., Teague, J., Fujimoto, W., and Beavo, J. A. (1997) Attenuation of insulin secretion by insulin-like growth factor 1 is mediated through activation of phosphodiesterase 3B. *Proc. Natl. Acad. Sci. USA* **94**, 3223–3228.

9. Grant P. G., Mannarino A. F., and Colman R. W. (1988) cAMP mediated phosphorylation of the low-Km cAMP phosphodiesterase markedly stimulates its catalytic activity. *Proc. Nat. Acad. Sci. USA* **85,** 9071–9075.

10. Lopez-Aparicio, P., Rascón, A., Manganiello, V. C., Andersson, K. E., Belfrage, P., and Degerman, E. (1992) Insulin induces phosphorylation and activation of the cGMP-inhibited cAMP phosphoodiesterase in human platelets. *Biochem. Biophys. Res. Commun.* **186,** 517–523.

11. Macphee, C. H., Reifsnyder, D. H., Moore, T. A., Levea, K. M., and Beavo, J. A. (1988) Phosphorylation results in activation of a cAMP phosphodiesterase in human platelets. *J. Biol. Chem.* **263,** 10,353–10,358.

12. Degerman, E., Smith, C. J., Tornqvist, H., Vasta, V., Manganiello, V., and Belfrage, P. (1990) Evidence that insulin and isoprenaline activate the cGMP inhibited low Km cAMP-phosphodiesterase in fat cells by phosphorylation. *Proc. Natl. Acad. Sci. USA* **87,** 533–537.

13. Smith, C. J., Vasta, V., Degerman, E., Belfrage, P., and Manganiello, V. (1991) Hormone-sensitive cyclic GMP-inhibited cyclic AMP phosphodiesterase in rat adipocytes. *J. Biol. Chem.* **266,** 13,385–13,390.

14. Rahn, T., Rönnstrand, L., Wernstedt, C., Leroy, M.-J., Tornqvist, H., Manganiello, V., Belfrage, P., and Degerman, E. (1996) Identification of the site in the cGMP inhibited phosphodiesterase phosphorylated in adipocytes in response to insulin and isoproterenol. *J. Biol. Chem.* **271,** 11,575–11,580.

15. Holm, C., Langin, D., Manganiello, V., Belfrage, P., and Degerman, E. (1997) Regulation of hormone-sensitive lipase activity in adipose tissue, in *Methods of Enzymology* (Rubin, B. and Dennis, E. A., eds.), Academic Press, pp. 45–67.

16. Rahn, T., Ridderstråle, M., Tornqvist, H., Manganiello, V., Fredrikson, G., Belfrage, P., and Degerman, E. (1994) Essential role of phosphatidylinositol 3-kinase in insulin-induced activation and phosphorylation of the cGMP-inhibited cAMP phosphodiesterase in rat adipocytes. *FEBS Lett.* **350,** 314–318.

17. Wijkander, J., Stenson Holst, L., Rahn, T., Resjö, S., Castan, I., Manganiello, V., Belfrage, P., and Degerman, E. (1997) Regulation of protein kinase B in rat adipocytes by insulin, vanadate and peroxovanadate. Membrane translocation in response to peroxovanadate. *J. Biol. Chem.* **272,** 21,520–21,526.

18. Wijkander, J., Rahn Landström, T., Manganiello, V., Belfrage, P., and Degerman, E. (1998) A possible role for protein kinase B but not mitogen-activated protein kinases and p70 S6 kinase in insulin-induced phosphorylation and activation of phosphodiesterase 3B in adipocytes. *Endocrinology* **139,** 219–227.

19. Ahmad, F., Cong, L-N., Stenson Holst, L., Wang, L-M., Pierce, J., Rahn Landström, T., Quon, M., Degerman, E., and Manganiello, V. (2000) Cyclic nucleotide phosphodiesterase 3B is a downstream target of protein kinase B and may be involved in regulation of effects of protein kinase B on thymidine incorporation of FDCPZ cells. *J. Immunol.* **164,** 4678–4688.

20. Resjö, S., Zolnierowicz, S., Manganiello, V., Belfrage, P., and Degerman, E. (1999) Phosphorylation and activation of phosphodiesterase 3B in adipocytes in response to phosphatase inhibitors. *Biochem. J.* **341,** 839–845.

21. Rodbell, M. (1964) Metabolism of isolated fat cells. *J. Biol. Chem.* **239,** 375–380.

22. Honnor, R. C., Dhillon, G. S., and Londos, C. (1985) cAMP-dependent protein kinase and lipolysis in rat adipocytes. *J. Biol. Chem.* **260,** 15,122–15,129.

23. Hopkirk, T. J. and Denton, R. M. (1986) Studies on the specific activity of [γ-$^{32}$P]-ATP in adipose tissue and other tissue preparations incubated with medium containing [$^{32}$P]phosphate. *Biochim. Biophys. Acta.* **885,** 195–205.

24. Rascón, A., Degerman, E., Taira, M., Meacci, E., Smith, C. J., Manganiello, V., Belfrage, P., and Tornqvist, H. (1994) Identification of the phosphorylation site in vitro for cAMP dependent protein kinase on the rat adipocyte cGMP-inhibited cAMP phosphodiesterase. *J. Biol. Chem.* **269,** 11,962–11,966.

25. Leroy, M-J., Degerman, E., Taira, M., Wang, L-H., Movsesian, M., Murata, T., Meacci, E., and Manganiello, V. (1996) Characterization of two recombinant PDE3 (cGMP-inhibited cyclic nucleotide phosphodiesterase) isoforms, RcGIP1 and HcGIP2, expressed in NIH 3006 murine fibroblasts and Sf9 insectscells. *Biochemistry* **35,** 10,194–10,202.

26. Kincaid, R. L. and Manganiello, V. C. (1988) Assay of cyclic nucleotide phosphodiesterase using radiolabelled and fluorescent substrate. *Methods Enzymol.* **159,** 457–471.

# 15

## Measurements of Glucose Conversion to its Metabolites

### Mario DiGirolamo

### 1. Introduction

Glucose is the main substrate for most tissues of the body, and provides energy for cellular respiration and metabolic activities. The fate of glucose end-products depends, in part, on the physiological demands of the tissue in question. Similarly, the type and function of glucose transporters varies with tissue and responsiveness to hormones.

In adipocytes, glucose is transported by glucose transporter I (GLUT1) under basal conditions. Insulin, the chief hormone stimulating glucose transport and metabolism, activates glucose-transporter 4 (GLUT4) and facilitates glucose translocation.

Glucose is phosphorylated intracellularly and converted to its main metabolites (approximate percent of total conversion under basal conditions in isolated adipocytes from a 150–220 g rat is in parenthesis): Glycogen (2–3%); $CO_2$ (25–30%); glyceride-glycerol (20–30%); glyceride-fatty acids (FAs) (20–25%); lactate (1%); and pyruvate (10%). With the addition of insulin, all products of glucose metabolism are enhanced, but the hierarchy of stimulation is distinct: glucose conversion to lactate > glyceride-FAs > $CO_2$ > glyceride-glycerol = pyruvate (see **Note 1** and **ref. 1**).

This chapter describes the measurement of adipocyte glucose utilization and conversion to its end products by a combination of $^{14}C$-glucose-uniformly labeled (UL) tracer isotope recovery into $CO_2$, glyceride-glycerol, and glyceride-FAs, and by enzymatic recovery of lactate and pyruvate into the incubation medium. Glycogen recovery is not done routinely, because, usually, glucose conversion to glycogen is less than 2–3% of glucose metabolized.

From: *Methods in Molecular Biology, vol. 155: Adipose Tissue Protocols*
Edited by: G. Ailhaud © Humana Press Inc., Totowa, NJ

Independent confirmation of glucose metabolized by the fat cells, with the method described here, shows that, in the rat, the five products measured account for approx 98% of all glucose metabolized (*see* **Note 2** and **ref.** *2*).

## 2. Materials

1. Incubation buffer, Krebs-Ringer bicarbonate (KRB) with 4% bovine serum albumin (BSA) should be prepared on the morning of the experiment. Reagents for KRB are prepared ahead of time, and are stored at 4°C. To make up one batch of KRB, add 100 mL 0.9% NaCl (0.154 $M$) to 4 mL 1.15% KCl (0.154 $M$), 3 mL 1.22% $CaCl_2$ (0.11 $M$), 1 mL 2.11% $KH_2 PO_4$ (0.154 $M$), 1 mL 3.82% $MgSO_4 \cdot 7 H_2O$ (0.154 $M$), and 21 mL 1.3% $NaHCO_3$ (0.154 $M$). $NaHCO_3$ needs to be gassed with 95% $O_2$–5% $CO_2$ for 1 h, when first prepared. Multiple batches (3–6) can be prepared simultaneously, depending on the size of the experiment. Considerable amounts of KRB–4% BSA are used in the washing of the isolated fat cell preparation and resuspension of the cells for incubation. Also, save 50 mL of the solution for preparation of standard curves for lactate and pyruvate. Once prepared, the KRB solution is gassed with 95% $O_2$–5% $CO_2$ (3 L/min) for 20–30 min.

   BSA (Sigma A-4503) is added to KRB at a concentration of 4% (40 mg/mL). After stirring the albumin in solution, pH 7.4 is achieved by addition of 10 $N$ NaOH. For best results in assays in which insulin action is measured, we test several batches of BSA until one is found that does not elevate basal glucose metabolism, and allows good insulin action (~200–300% increase in the glucose metabolism rate at 6 m$M$ glucose with 1 mU/mL insulin) in adipocytes from young, insulin-sensitive rats (*see* **Note 3**).

2. Collagenase from *Clostridium histolyticum* (Worthington, no. 4197 C/SI) is added to minced adipose tissue (AT) at a concentration of 1.5 mg/mL. A total of 20 mL is used to digest 5–10 g of tissue and isolate the adipocytes (*see* **Note 3**).

3. [14]C-glucose-UL is purchased from New England Nuclear (250 mCi/mmol) in 1 mCi amount (*see* **Note 4**). Approximately 0.5 µCi are added to 1 mL final KRB–4% BSA medium for the metabolic incubation. Cold glucose is present at 6 m$M$ concentration (6 µmol/mL) (*see* **Note 5**). Prior to the incubation, 50 µL radioactive cocktail, containing [14]C-glucose-UL and nonlabeled glucose, is measured in the liquid scintillation counter, to determine the specific activity (dpm/µmol glucose) of the incubation mixture.

4. Benzathonium hydroxide (1 $M$ solution in methanol) is purchased from Sigma (no. B2156), and is used to trap $CO_2$ on filter paper. Use cautiously.

5. Sulfuric acid (5 $N$ $H_2SO_4$) from Fisher Scientific. Use cautiously.

6. Filter paper (P8) from Fisher is used in strips of 2 × 8 cm.

7. Rubber stoppers and plastic center wells (Kontes, respectively, no. 882310-0000 and no. 882320-0000) are assembled prior to use, and rolled filter paper is added prior to the incubation.

8. Scintillation fluid, Ecoscint, from National Diagnostic.

9. Scintillation vials (20 mL, polyethylene) with foil-lined caps, from Packard. are used for cell incubation and FA extraction. For $CO_2$ counting, use plastic-lined caps. Also, borosilicate glass vials (20 mL) are used for triglyceride (TG) extraction and counting.
10. Porcine insulin, diluted to reach final concentration of 1 mU/mL in the incubation vials (a gift from E. Lilly, 28 U/mg).
11. Methylene blue. Dissolve 20 mg in 1 mL water; store at room temperature (RT).
12. Dole's extraction mixture: 40 parts isopropyl alcohol, 10 parts heptane, and one part 1 $N$ $H_2 SO_4$.
13. Materials and reagents for lactate and pyruvate assays (purchased from Sigma as kits) (*see* **Note 6**).
    a. Lactate: 8% Perchloric acid (PCA) (store at 4°C); lactate standard, Sigma 826-10 (store at 4°C); lactate dehydrogenase (LDH), Sigma 726-10 (store at 4°C); glycine-hydrazine buffer (containing 0.6 $M$ glycine and 0.5 $M$ hydrazine), Sigma 826-3 (store at 4°C); β-nicotinamide adenine dinucleotide (β-NAD⁺), Sigma N-7004 (store dessicated at –20°C); and 3-mL cuvets, VWR Scientific Products 58017-825.
    b. Pyruvate: 8% PCA (store at 4°C); LDH, Sigma 726-10 (store at 4°C); Trizma base, Sigma 726-4 (store at 4°C); β-NAD⁺, Sigma N-8129 (store desiccated at 4°C); and 3-mL cuvets, VWR 58017-825.

## 3. Methods

### 3.1. Preparation of Isolated Fat Cells

1. Fragments of adipose tissue (usually 50–100 mg in weight), totaling 3–5 g are removed from various sites of a given adipose depot, minced with scissors, and placed in 20 mL warm KRB medium with glucose and collagenase.
2. After 45–60 min of gentle shaking at 37°C in a Dubnoff metabolic shaker (60–80 strokes/min), the fat cells are isolated by passing the tissue cells and debris through a 250-μm Nitex mesh (Sefar America, Kansas City, MO) and collecting the cells in a 75-mL plastic tube. Three consecutive washings in KRB–4% BSA medium, without collagenase, then follow.
3. Letting the cells float by gravity or by gentle centrifugation (1 min at 1000$g$) allows separation of the cells from the medium and easy removal of the infranatant by a 50-mL syringe and a 18-gage needle connected to PE160 tubing, prior to adding fresh medium. After addition of fresh medium, the plastic vial is capped and gently inverted a few times, to allow washing of the cells (*1*).

### 3.2. Determination of Fat Cell Diameter

1. A 0.5-mL aliquot of the isolated fat cell suspension is added to 2–3 mL warm medium in a plastic scintillation vial, to which about 1 drop methylene blue is added.
2. After staining at 37°C for 3–5 min, aliquots of the stirred suspended cells are removed with a plastic pipetter, and placed on a siliconized glass slide inside a premade round silicone well, and covered with a glass cover slip.

3. After being brought into focus, fat cells are recognized by their spherical shape, stained nucleus, and cytoplasm. The cells are aligned on the caliper scale with systematic motion of the stage control knobs. The transverse diameter of 200–300 fat cells is measured and recorded in successive 7-μm multiples, to provide a frequency distribution of diameter in 9–15 categories of size *(3)*.

4. For more details of this optical method or an electronic method, *see* Chapter 4 in this book.

## 3.3. Calculation of Mean Diameter and Cell Volume

1. Data for diameter are converted to mean diameter and standard deviation (SD). The heterogeneity of the cell population is expressed as coefficient of variation (CV = SD/mean).

2. From the diameter data, the frequency distribution can also be calculated for fat cell surface area $(SA = \Pi D^2)$ and cell volume $[V = (\Pi D^3)/6]$, taking into account consideration the nonlinear transformation *(3)*.

3. For additional calculations of fat cell size and fat cell number, *see* Chapter 4 in this book.

## 3.4. Preparation of Incubation Buffer for $CO_2$ and TG

Samples are run in triplicate. Plastic scintillation vials are labeled with markers of different color to indicate content (e.g., basal, insulin, and so on).

1. For $CO_2$ and TGs, 1 mL incubation medium is used: 0.5 mL fat cell suspension, and 0.5 mL KRB–4% BSA, containing glucose and $^{14}$C-glucose.

   One milliliter of the final incubation will contain approximately 200,000 fat cells, 6 mM glucose (i.e., 6 μmol/mL) and 0.5 μCi $^{14}$C-glucose-UL (*see* **Note 7**). Thus, the fat cell suspension from which cells are distributed into the incubation vials should contain approx 400,000 fat cells/mL; and the buffer cocktail to distribute into the incubation vials ahead of time, should contain 12 μmol glucose/mL and 1.0 μCi $^{14}$C-glucose-UL. For any addition of hormones to samples, the hormone concentration needed for 1 mL incubation vial should be dispensed in 10 μL vol (*see* **Note 8**).

2. While the cell suspension is being washed after the collagenase incubation, the cocktail buffer is added, in 0.5 mL vol, to the incubation vials. Hormones, if appropriate, will also be added at this stage. The incubation vials are then placed in the Dubnoff metabolic shaker, in individual slots of a plastic or metal rack, and kept at 37°C.

3. The isolated fat cell suspension is added at timed intervals of 20–30 s. Once 0.5 mL fat cells is added, the vial is gassed with 95% $O_2$-5% $CO_2$ for 10–15 s and capped with a rubber stopper equipped with a central plastic well and $CO_2$ collecting paper (*see* **Note 9**). Then incubation for 60 min is started, and the vials are gently shaken at 60–80 strokes/min.

4. At the end of the incubation, 0.2 mL benzathonium hydroxide is added to the filter paper in the center well with a 21-gage needle and a 1-mL syringe, through the rubber stopper; 0.25 mL 5 *N* sulfuric acid is added to the medium, to destroy the cells and stop the metabolic incubation (*see* **Note 10**).

5. The vials are then returned to the metabolic shaker for 20 min to complete collection of $CO_2$ into the strip of filter paper wetted with benzathonium hydroxide. Twenty minutes later, the rubber stopper is removed, and the central well with the filter paper is cut and dropped into a scintillation vial containing 10 mL scintillation fluid. The plastic vials are capped with plastic-lined caps, and counted.
6. The cells left in the incubation vials are treated with the Dole's procedure (4), to extract the TGs and to determine the radioactivity content.

### 3.5. Preparation of Incubation Buffer for Recovery of Lactate and Pyruvate

For lactate and pyruvate, 3 mL incubation are used: 1.5 mL fat cell suspension and 1.5 mL of KRB–4% BSA containing glucose.

1. Prior to the final washings of the isolated fat cell preparation, after the collagenase incubation, the incubation buffer is added, in 1.5 mL vol, to the plastic scintillation vials kept at 37°C. Hormones, if appropriate, are also added at this stage (10 μL for 1 mL incubation buffer). The isolated fat cell suspension is added at timed intervals of ~15 s. Once 1.5 mLfat cells are added, the vials are gassed with 95% $O_2$_5% $CO_2$ for 10–15 s, then are capped with foil-lined caps. Incubation for 60 min is started, and the vials are gently shaken at 60–80 strokes/min.
2. At the end of the 60-min incubation, the content of each individual vial is poured into 6-mL polypropylene test tubes kept on crushed ice, and the incubation is terminated. After 3–5 min to allow the fat cells to rise to the surface of the medium, 1.5 mL infranatant is removed by a 3-mL syringe and a 21-gage needle connected to 5 cm PE 60 polyethylene tubing, and added to a test tube containing equal volume of 8% PCA. After vortexing the test tube for 30 s, the vial is allowed to rest on ice for 20–30 min prior to centrifugation for 20 min in a refrigerated centrifuge at 4°C (1000$g$). The supernatant is then stored at –20°C, prior to determination of lactate and pyruvate, which is usually done within 1 wk of the experiment.

### 3.6. Recovery of $^{14}$C-Glucose Radioactivity into $CO_2$

The scintillation vials, containing the center wells with filter paper and $CO_2$ trapped by benzathonium hydroxide, are counted for 10 min in a liquid scintillation counter.

The cpm obtained are converted to dpm (with the assistance of a quenching curve), and the total dpm radioactivity are related to the specific activity (dpm $^{14}$C-glucose/6 μmol glucose) of the incubation buffer.

### 3.7. Recovery of $^{14}$C-Glucose Radioactivity into TGs and FAs

The TGs in the incubated fat cells are extracted by the Dole's extraction procedure (4) described in brief below.

1. Add 5 mL Dole's extraction mixture to the medium and the cells in incubation vials, cap, and shake well.

2. Add 3 mL heptane and 3 mL $dH_2O$, cap, shake well, and let sit for 20 min.
3. Uncap the vials and remove 2 mL upper (heptane) phase (usually 4.3 mL) for hydrolysis of TGs into glycerol and FAs into a second set of 20-mL scintillation vials. Dry under a stream of air overnight in a fume hood.
4. While samples for hydrolysis are drying, remove 1 mL remaining upper phase into tared glass vials for measurement of TG weight and of radioactivity recovered into the TGs. Leave overnight in hood for drying, weigh, then add 10 mL scintillation fluid to scintillation vials, shake, and count for 10 min. This gives glucose converted to TGs.
5. Hydrolysis of TGs:
   a. To dried samples in plastic scintillation vials, add: 2.0 mL 95% EtOH and 0.5 mL 40% KOH. Place in 60°C water bath, 1 h, for hydrolysis. Vials should be capped, but not screwed tightly, before they are put in the water bath. Cool vials to RT, add 4 mL 3 $N$ HCl, and add 4 mL n-heptane, mix well, and let sit for 30 min.
   b. Measure upper (heptane) phase (usually 4.1 mL), then remove 2.0 mL upper phase to plastic scintillation vials for FA measurement, and dry overnight.
   c. To dried FA vials, add 10 mL scintillation fluid, shake, and count. This gives glucose converted to TG-FAs.
6. Glucose converted to TG-glycerol is estimated from glucose converted to TGs minus that converted to TG-FAs.

### 3.8. Lactate Assay

1. A standard curve of lactate (Sigma) concentration is prepared in KRB-BSA buffer from samples frozen on the experimental day. Label eight tubes for a standard curve, and add 3 mL KRB-BSA to each tube. Add 1 mL lactate (Sigma, 4,440 nmol) to first tube, and vortex. This now contains 1,110 nmol/mL. Remove 3 mL to add to the next tube. Follow this procedure to produce a serial dilution with corresponding concentrations of 555, 277.5, 136.7, 69.4, 34.7, 17.3, and 8.7 nmol/mL. Include a blank sample made from only KRB-BSA. Precipitate standards with 1:1 addition of PCA, according to the same method used for the samples.
2. To prepare the glycine-hydrazine cuvet buffer, dilute one part glycine-hydrazine solution to two parts $dH_2O$ (for a final concentration of 0.2 $M$ Glycine and 0.1667 $M$ Hydrazine). To this solution, add 1.667 mg β-NAD+ for each ml of the glycine-hydrazine solution. Prepare about 100 mL, or as much as needed for the number of samples to run.
3. To each cuvet, add 0.2 mL blank (in triplicate), standard (in duplicate), or unknown sample and 2.8 mL cuvet buffer. Invert using parafilm.
4. Preread each cuvet at 340nm in a standard spectrophotometer. Add 25 µL LDH to each cuvet and invert using parafilm. Let sit at RT for 45 min. Reread cuvet at 340nm. Record the first and the second reading.

### 3.9. Pyruvate Assay

1. A standard curve of pyruvate (Sigma) concentration is prepared in KRB-BSA buffer frozen from the experimental day. Label six tubes for standard curve, and

add 2.3 mL KRB-BSA to each tube. Add 2.3 mL pyruvate to first tube, vortex, and remove 2.3 mL add to the next tube. Follow this procedure to produce a serial dilution with corresponding concentrations of 227.1, 113.6, 56.8, 28.4, 14.2, and 7.1 nmol/mL. Include blank sample made from only KRB-BSA. Precipitate standards with 1:1 addition of PCA, according to the same method used for the samples.

2. Prepare Trizma solution by adding 0.45 mg β-NADH to 1 mL Trizma.
3. To separate cuvets add 1 mL blank (in triplicate), standard (in duplicate), or unknown sample, 0.5 mL Trizma with β-NADH, 0.5 mL Trizma, and 1 mL dH$_2$O. Invert using parafilm.
4. Preread each cuvet at 340nm. Add 25 μL LDH to each cuvet and invert using parafilm. Let sit at RT for 45 min. Reread cuvet at 340nm. Record each reading.

## 3.10. Calculations for the Amounts of Lactate and Pyruvate Recovered in the Incubation Medium

### 3.10.1. Lactate

In order to extrapolate the correct value for the lactate concentration per milliliter of sample, you need to express your standard curve values in terms of the original dilutions, since standard values and unknown values are treated similarly by 1:1 PCA dilution.

Generate a standard curve with lactate in nmol/mL on the $x$-axis and delta absorbance (minus blank) at 340 nm on the $y$-axis.

To correlate the absorbancies of data to this standard curve, use the equation $y = mx + b$. This is transformed to $x = (y-b)/m$, where $x$ = nmoles lactate/sample, $y = \delta$ of initial and final absorbancies at 340 nm (after subtraction of average blank absorbance), $b = y$-intercept of standard curve, and $m$ = slope of standard curve.

After finding the concentration of lactate in nmol/mL, relate this value to cells present in the incubation vial. Then normalize the lactate produced per $10^7$ fat cells as follows:

nmol lactate/$10^7$ fat cells = (nmol lactate in 1 mL × incubation volume × 10,000,000)/ (no. fat cells.incubation × 10,000,000)

To convert the answer to μmol, divide by 1000.

Remember that, in glucose metabolism, two moieties of lactate are produced for every one of glucose metabolized. So, in order to use the lactate values to calculate glucose metabolized to lactate, the final lactate value is divided by two.

To minimize calculations, the shorthand formula becomes:

(nmol/mL × incubation vol × 10,000,000)/(no. fat cells in incubation × 500)

This gives μmol lactate produced per $10^7$ fat cells.

### 3.10.2. Pyruvate

Similar steps are taken in calculating pyruvate production from glucose, with the appropriate pyruvate standard concentrations.

## 3.11. Calculations of Glucose Conversion to its Metabolites and Expression per 10[7] Fat Cells

The best way to set up calculations for the glucose converted to individual products, and to their sum, is to establish the following:

1. Number of fat cells in incubations vials. This will be determined by measuring mean diameter, calculating mean volume, and determining lipid content by the Dole's procedure (*see* Chapter 4 for cellularity measurements).
2. The amount of glucose in the incubation flask, usually 6 µmol.
3. The amount of radioactivity in dpm. Fifty microliters of the medium cocktail are counted in the liquid scintillation counter. The counts per minute (cpm) are converted to dpm by a standard quench correction curve, then the dpm present in the incubation vial are calculated.
4. The specific activity of the incubation mixture, i.e., the amount of radioactivity (dpm/µmol) glucose is calculated. From that point on, the total radioactivity (dpm) collected in each individual product ($CO_2$, TGs and glyceride-FAs) is converted to micromoles of glucose converted in 1 h of incubation by the number of fat cells present.
5. At the end, all calculations are normalized as µmol glucose converted to specified product by 10[7] fat cells in 1 h.

### 3.12. Glucose Conversion to $CO_2$

1. All $CO_2$ radioactivity, collected on the filter paper in the center well with benzathonium hydroxide, is transferred into a liquid scintillation vial. The vial is filled with 10 mL scintillation fluid, and counted for 10 min. We have found that the best and most stable count is obtained 3 d after the experiment (some chemiluminescence needs to settle, to avoid spurious results). The scintillation counter is set for [14]C-isotope counting.
2. The cpm obtained for each vial are converted to dpm by the channel ratio quench correction method, and averaged for the triplicate samples. Final calculation with specific activity calculated for a given experiment allows determination of µmol glucose converted to $CO_2$ by the cells present in the incubation vials, normalized to 10[7] fat cells.

### 3.13. Glucose Conversion to TGs

1. After the incubation is complete, the cell TGs are extracted by the Dole's procedure. One milliliter is removed from 4.3 mL supernatant organic phase, placed in a tared scintillation vial, to dry overnight under a fume hood. After weighing the dried vial containing TGs, 10 mL scintillation fluid is added. The capped vial is shaken, then counted in the scintillation counter for 10 min.
2. The cpm obtained for each vial are converted to dpm and averaged for the triplicate samples. Final calculation with appropriate specific activity allows determination of µmoles glucose converted to TGs by the cells present in the incubation vial. It is important to use the correction factor of 4.3, since only 1 out of 4.3 mL organic phase was dried and counted. Glucose conversion to TGs is then normalized to 10[7] fat cells.

### 3.14. Glucose Conversion to Glyceride-FAs

1. Two milliliters of the 4.3 mL extracted in the Dole's solution are dried, then the lipids are hydrolyzed.
2. Two milliliters of the 4.1 mL FA extract are evaporated, scintillation fluid is added, and the samples are counted for 10 min.
3. The cpm obtained for each vial are converted to dpm and averaged for the triplicate samples.
4. Final calculation with appropriate specific activity allows determination of µmol glucose converted by glyceride-FAs by the cells present in the incubation vial. Correction factors are $4.3/2 \times 4.1/2$. Glucose conversion to glyceride FAs is then normalized to $10^7$ fat cells.

### 3.15. Glucose Conversion to Glyceride Glycerol

This estimate is calculated as the difference between glucose converted to TGs and glucose converted to glyceride-FAs.

### 3.16. Glucose Conversion to Lactate and Pyruvate

Glucose conversion to lactate and pyruvate, by the cells present in the incubation medium, is calculated as described above. It is important to remember that lactate produced and pyruvate produced must be divided by 2, if the data are expressed as glucose conversion to lactate and pyruvate (*see* **Note 6**).

### 3.17. Sum of Glucose Converted to Products

The sum of all five products ($CO_2$, glyceride-glycerol, glyceride-FAs, lactate, and pyruvate) is expressed as µmol glucose converted by the cells present in incubation vials, or µmol converted by $10^7$ fat cells in 1 h. Once the total glucose metabolized is calculated, relative conversion to each individual product can also be calculated.

## 4. Notes

1. The major advantage of this method for measuring glucose utilization by isolated fat cells is that it keeps track of the amount of glucose converted to individual products. This is important, because various conditions (both in vivo and in vitro) affect the rate and pattern of glucose metabolism to a great extent *(5–7)*. For instance, starvation of the animals leads to a marked decrease of glucose conversion to glyceride-FAs and a marked increase in glucose conversion to lactate *(8,9)*. In vitro, addition of hormones, such as insulin or isoproterenol, affects not only the rate, but also the pattern of glucose metabolism *(10)*. Thus, it is convenient to measure, in a given experimental condition, not only the total amount of glucose metabolized per $10^7$ fat cells, but also the proportion of glucose metabolized, to each individual product.
2. To determine the completeness of glucose recovery into its metabolic products, we have compared adipocyte glucose utilization, by summing up the five prod-

ucts of glucose, to a method that measures (5-$^3$H)-labeled glucose conversion to tritiated water (*2*). The two methods were in excellent agreement (98 vs 100%), indicating that the summation of the metabolic products does account for nearly all of the glucose utilized by the fat cells.

3. In setting up metabolic incubations in which the response to insulin is measured, the selection of collagenase and of the particular batch of BSA is important, because, occasionally, these products contain materials that interfere with the expression of the insulin response. We usually obtain small samples of collagenase from Worthington from different lots available under the same catalog number. Similarly, 3–4 samples of albumin are obtained from Sigma, from different lots available. Some quick experiments are then run to find out which of the lots combine to give the usual insulin response for this laboratory at 6 m*M* glucose. The aim is a baseline glucose metabolism of 3–4 μmol/10$^7$ fat cells in 1 h, and an insulin response (expressed as percent increase over baseline values) of about 200–300% (*11*). When the right combination is found, approx 1 g collagenase and 1–2 Kg albumin is bought and stored for future use.

4. Occasionally, other forms of labeled glucose (such as $^{14}$C-glucose-1 or $^{14}$C-glucose-6) are used for specific purposes. The $^{14}$C-glucose-UL is the best to estimate the average amounts of glucose converted to specified products by the cells.

5. We usually employ a physiological concentration of glucose, 6 m*M*. If one wanted to vary the glucose concentration to 1, 3, or 12 m*M*, the method could easily accommodate this variations. The variation in the specific activity (dpm/μmol glucose) allows the appropriate calculation of the glucose metabolized into products (*11*).

6. In this method, isotope recovery is measured in $CO_2$ and the two moieties of TGs. Extraction of lipids leads to a precise recovery of the isotope in either the TGs or glyceride-FAs. For recovery of glucose converted to lactate and pyruvate, we have chosen to measure the medium content of these products enzymatically, rather that attempting to recover the glucose isotope into aqueous medium, in which separation of $^{14}$C-lactate and pyruvate from the $^{14}$C-glucose background would have been more complex and probably more imprecise. This step requires a conceptual and practical correction at the end. When lactate and pyruvate production are measured, this is expressed in nmol/10$^7$ fat cells. But if one wants to add these two products to the sum of glucose converted to all its products, the values for lactate and pyruvate must be divided by 2, because one molecule of glucose generates two molecules of lactate or pyruvate.

7. Schwabe et al. (*12*) have shown that the fat cell density, namely, the cellular concentration, influences the rate of adipocyte lipolysis (e.g., the more the density, the lesser the rate of lipolysis per cell). We have also found (*11*) that fat cell density influences the pattern of glucose metabolism (i.e., the relative conversion of glucose to individual products) and the response of the cells to insulin (*13*).

Consequently, it is advisable to carry out experiments with isolated fat cells at a predictable cell concentration. Fine and the author (*14*) have developed such a method, which is based on rapidly obtaining a lipocrit (packed cell volume) and a mean fat cell volume for the newly isolated fat cell preparation. From this, the fat cell density can be estimated, and adjustments can be made by adding buffer, to bring cell concentration to desired values. Ten minutes are usually sufficient to carry out this step prior to distributing the cells for metabolic incubations.

8. We dilute hormones, in particular, insulin, in KRB medium containing 4% BSA, to avoid losing hormone on the wall of the glass test tube.

9. Gassing the cells for at least 10 s prior to the incubation allows the buffer to equilibrate, and provides enough oxygen for the duration of the incubation. Hypoxia is known to elevate lactate production, and it is important to avoid spurious results caused by insufficient oxygen supply.

10. When, at the end of the metabolic incubation for $CO_2$, the rubber stopper is punctured with the needle and the syringe, to distribute benzathonium hydroxide into the rolled filter paper in the center well, the 0.2 mL should be dispensed slowly, to allow absorbency on the filter paper; excess spilling into the buffer would produce spurious results. Also, caution is needed when the needle touches the filter paper and is withdrawn. If the needle has lost sharpness, occasionally it hooks the paper and may bring it out of the center well. This must be avoided, because dropping the paper into the medium buffer would ruin collection of $CO_2$ radioactivity.

## References

1. DiGirolamo, M. and Fried, S. K. (1987) In vitro metabolism of adipocytes, in *Biology of the Adipocyte. Research Approaches*, (Hausman, G. J. and Martin, R., eds.), Van Nostrand Reinhold, New York, pp. 120–147.
2. Groff, J. L., Stugard, C. E., Mays, C. J., Koopmans, H. S., and DiGirolamo, M. (1992) Glucose metabolism in isolated rat adipocytes: Estimate of total recovery by the product summation method. *J. Lab. Clin. Med.* **119,** 216–220.
3. Cushman, S. W. and Salans, L. B. (1978) Determination of adipose cell size and number in suspensions of isolated rat and human adipose cells. *J. Lipid Res.* **19,** 269–273.
4. Dole, V. P. (1955) A relation between non-esterified fatty acids in plasma and the metabolism of glucose. *J. Clin. Invest.* **35,** 150–154.
5. Lockwood, D. H. and East, L. E. (1974) Studies of the insulin-like actions of polyamines on lipid and glucose metabolism in adipose tissue cells. *J. Biol. Chem.* **249,** 7717–7722.
6. Fried, S. K., Lavau, M, and Pi-Sunyer, F. X. (1982) Variations in glucose metabolism by fat cells from three adipose depots of the rat. *Metabolism* **31,** 876–883.
7. DiGirolamo, M. and Rudman, D. (1968) Variations in glucose metabolism and sensitivity to insulin of the rat's adipose tissue, in relation to age and body weight. *Endocrinology* **82,** 1133–1141.
8. Newby, F. D., Sykes, M. N., and DiGirolamo, M. (1988) Regional differences in adipocyte lactate production from glucose. *Am. J. Physiol.* **255,** E716–E722.
9. Thacker, S.V., Nickel, M., and DiGirolamo, M. (1987) Effects of food restriction on lactate production from glucose by rat adipocytes. *Am. J. Physiol.* **253,** E336–E342.
10. DiGirolamo, M. and Owens, J.L. (1976) Glucose metabolism in isolated fat cells: enhanced response of larger adipocytes from older rats to epinephrine and adrenocorticotropin. *Horm. Metab. Res.* **8,** 445–451.

11. DiGirolamo, M. (1981) Effects of variable glucose and fat-cell concentration on glucose metabolism and insulin responsiveness by adipocytes of different sizes. *Int. J. Obes.* **5,** 671–677.

12. Schwabe, U., Schönhöfer, P. S., and Ebert, R. (1974) Facilitation by adenosine of the action of insulin on the accumulation of adenosine 3',5'-monophosphate, lipolysis, and glucose oxidation in isolated fat cells. *Eur. J. Biochem.* **46,** 537–545.

13. DiGirolamo, M., Thacker, S. V., and Fried, S. K. (1993) Effects of cell density on in vitro glucose metabolism by isolated adipocytes. *Am. J. Physiol.* **264,** E361–E366.

14. Fine, J. B. and DiGirolamo, M. (1997) A simple method to predict cellular density in adipocyte metabolic incubations. *Int. J. Obes.* **21,** 764–768.

# 16

## Metabolite and Ion Fluxes and Ion Channels in Brown and White Adipocytes

### Jan Nedergaard

## 1. Introduction

Determination of the flux of substances (ions or metabolites) into or out of cells is a recurrent issue in cell biology. This demands the incubation of cells in a medium followed by separation of the cells from this medium. To perform this type of experiment in brown, and especially in white, adipocytes, is challenging, in that the intracellular water volume into which the metabolites/ions are transferred (or from which they are transferred) is small in comparison to total cell volume (because of the lipid droplets). Further, routine techniques to isolate cells from media, i.e., normal centrifugation, cannot be used, because adipocytes will float instead of sink, and will be found in a diffuse band at the top of the vial after centrifugation.

A solution to both these problems was introduced by Gliemann et al. *(1)*, originally for white adipocytes, and extensively utilized for these since then, and later also extended to brown adipocytes *(2–4)*. In this method, which is based on centrifugation with phthalates, the adipocytes are recovered in a concentrated volume, with almost no (labeled) medium being trapped, well separated from the bulk of the medium.

A similar problem is encountered if patch-clamp experiments are attempted, in order to study ion fluxes electrically. For practical reasons, the cell under study must be attached to a substrate to allow for this type of experiment, which is difficult with floating, lipid-filled cells. One suggestion is to select the cells based on their ability to float up to and adhere to a hydrophilic biofilm, as originally introduced by Siemen et al. *(5,6)* for brown adipocytes, and used routinely for these cells *(7,8)*, but the method also works for white adipocytes *(9)*.

From: *Methods in Molecular Biology, vol. 155: Adipose Tissue Protocols*
Edited by: G. Ailhaud © Humana Press Inc., Totowa, NJ

## 2. Materials
### 2.1. Ion and Metabolite Fluxes

1. Isolated brown or white adipocytes prepared by conventional techniques (*see* Chapters 5, 10, 15, and 24).
2. The relevant medium for the incubation (normally Krebs-Ringer bicarbonate buffer) (*see* Chapter 24).
3. Radioactively labeled substance under study, generally $^{14}$C-labeled if metabolite, or the radioactive ions directly for ion flux studies (e.g., $^{45}$Ca$^{2+}$ or $^{22}$Na$^{+}$ or $^{36}$Cl$^{-}$ or $^{86}$Rb$^{+}$ [instead of K$^{+}$]). High specific activities are necessary; thus, for ions, it is normally necessary to exchange the unlabeled component in the buffer entirely with the labeled.
4. [$^{3}$H]inulin, to be added to the incubation buffer as a measure of buffer entrapment (*see* **Note 1**).
5. Dibutyl phthalate and bis(2-ethylhexyl)phatalate, in a 3:5 mix (*see* **Note 2**).
6. Tabletop centrifuge.

### 2.2. Ion Channels

1. Isolated brown or white adipocytes (*see* Chapters 5, 10, 15, and 24).
2. Cell culture medium (*see* Chapter 18).
3. Multiwell culture plates.
4. Biofoil-25 (Heraeus, Hanau, Germany), cut into small (about 1 cm) squares.
5. Patch-clamp equipment.

## 3. Methods
### 3.1. Ion Fluxes

The method is sketched in **Fig. 1**, illustrated for examination of adrenergically induced efflux of $^{45}$Ca$^{2+}$ from brown fat cells.

1. Prepare the intended series of incubation tubes with 1–2 mL incubation medium, radioactively double-labeled (i.e., with substance under study and [$^{3}$H]inulin).
2. Prepare a parallel series of tubes, each containing 0.8 mL phthalate oil mix.
3. Start the experiment by adding the adipocytes to the incubation medium, and incubate as required for the experiment.
4. Stop the incubation by adding the incubate on top of the phthalate oil mix. Immediately centrifuge the tubes in a tabletop centrifuge for about 1 min, at about 1000$g$.
5. The cells are now found on top of the oil layer, and can be removed with a pipette (*see* **Note 3**).
6. Count the radioactivity, and compensate for that coming from trapped medium, using the value for [$^{3}$H]inulin.

### 3.2. Ion Channels

1. Add the cell culture medium to the wells.
2. Add the isolated cells to the medium in the wells.

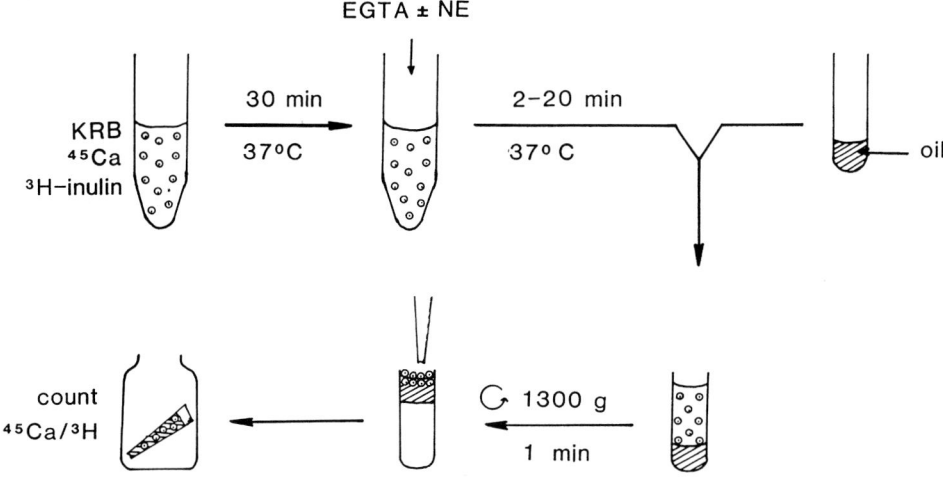

Fig. 1. The principle for the phthalate procedure, here illustrated for studies of $Ca^{2+}$ efflux. As seen, adipocytes are first preincubated in a medium containing radioactively labeled $Ca^{2+}$ and inulin (as a medium marker). The efflux is then monitored after addition of an extracellular $Ca^{2+}$ chelator. At the end of the incubation, the solution is poured on top of the phthalate oil solution and rapidly spun in a tabletop centrifuge, collected and the radioactivity counted.

3. Place pieces of the Heraeus biofilm on top of the cell culture medium in the wells; take care to place the roughened hydrophilic side of the biofilm downwards.
4. Leave the plates in a cell culture incubator for several hours, or better, overnight. During this time, the fat-filled cells will float and adhere to the biofilm.
5. Remove the biofilm with a pair of forceps, turn it around, and place it on the bottom of a small medium-filled Petri dish (i.e., with the hydrophobic part downwards). The film will stick to the bottom of this Petri dish.
6. The adhered cells can now be used for patch-clamp studies, using routine patch-clamp techniques (*see* **Note 4**).

## 4. Notes

1. If the radioactivity signal of the compound under study has an energy spectrum too close to that of tritium, $^{14}C$-labeled inulin can be used instead.
2. This proportion works well with hamster brown adipocytes. The two phthalates have different densities (the dibutyl, 1.045, and the *bis*[2-ethylhexyl], 0.985), and for different adipocyte preparations, it may be necessary to adjust the proportions between these two components.
3. It may sometimes be advantageous to add buffer to the top of the oil after centrifugation and in this way liberate the cells from the oil.
4. Alternatively, the adipocytes can be made to adhere to the bottom of Petri dishes, by ensuring that they are lipid-depleted at the time of preparation. This may be

accomplished by exposing newborn rat pups to 5°C for about 1 h before the cells are prepared *(10)*; whether this physiological stimulation in itself alters the cells is not known. Alternatively, cell culture systems of adipocytes *(4)* or of adipocyte-like cells may be used, but, again, the results may not be fully representative of those obtained with *in situ* differentiated cells.

## References

1. Gliemann, J., Österlind, K., Vinten, J., and Gammeltoft, S. (1972) A procedure for measurement of distribution spaces in isolated fat cells. *Biochim. Biophys. Acta* **286,** 1–9.
2. Connolly, E., Nånberg, E., and Nedergaard, J. (1986) Norepinephrine-induced Na⁺ influx in brown adipocytes is cyclic AMP-mediated. *J. Biol. Chem.* **261,** 14,377–14,385.
3. Connolly, E. and Nedergaard, J. (1988) β-adrenergic modulation of $Ca^{2+}$ uptake by isolated brown adipocytes. Possible involvement of mitochondria. *J. Biol. Chem.* **263,** 10,574–10,582.
4. Dasso, L., Connolly, E., and Nedergaard, J. (1990) $\alpha_1$-Adrenergic stimulation of $Cl^-$ efflux in isolated brown adipocytes. *FEBS Lett.* **262,** 25–28.
5. Siemen, D. and Reuhl, T. (1987) Non-selective cationic channel in primary culture cells of brown adipose tissue. *Pflügers Arch. Eur. J. Physiol.* **408,** 534–536.
6. Weber, A. and Siemen, D. (1989) Permeability of the non-selective channel in brown adipocytes to small cations. *Pflügers Arch.* **414,** 564–570.
7. Koivisto, A. and Nedergaard, J. (1995) Modulation of calcium-activated non-selective cation channel activity by nitric oxide in rat brown adipose tissue. *J. Physiol. (Lond)* **486,** 59–65.
8. Koivisto, A., Klinge, A., Nedergaard, J., and Siemen, D. (1998) Regulation of the activity of 27 pS nonselective cation channels in excised membrane patches from rat brown-fat cells. *Cell. Physiol. Biochem.* **8,** 231–245.
9. Koivisto, A., Dotzler, E., Ruß, U., Nedergaard, J., and Siemen, D. (1993) Nonselective cation channels in brown and white fat cells, in *Nonselective Cation Channels. Pharmacology, Physiology and Biophysics*, (Siemen, D. and Hescheler, J., eds.), Birkhäuser Verlag, Basel, pp. 201–211.
10. Lucero, M. T. and Pappone, P. A. (1989) Voltage-gated potassium channels in brown fat cells. *J. Gen. Physiol.* **93,** 451–472.

# 17

# Culture of Adipose Tissue and Isolated Adipocytes

## Susan K. Fried and Naima Moustaid-Moussa

## 1. Introduction
### 1.1. Overview
#### 1.1.1. Rationale for Use of Organ or Primary Cell Culture

Adipose tissue (AT) function is regulated by both short-term and long-term mechanisms. Long-term adaptations to hormonal, nutritional, and developmental factors take place over a time course of hours to days. For example, chronic alterations in the hormonal milieu (analogous to those that may occur during over- or underfeeding) have long-term effects on the cellular capacity for lipogenesis and lipolysis (*1–3*), as well as on the production of adipocyte secretory proteins (*4,5*). The mechanisms underlying these metabolic adaptations may involve pre- or posttranscriptional events that change the amount of enzyme protein or the number or activity of specific receptors or transporters. Elucidation of the mechanisms of long-term adaptations requires methods that maintain the function for AT ex vivo for periods up to 1 wk.

There are two principal methods used for studies of the long-term regulation of adipocyte metabolism: organ culture (OC) of AT fragments, and primary culture of fat cells isolated by collagenase digestion of AT.

#### 1.1.2. Organ Culture

Organ culture, as a term applied to the adipose organ, refers to intact fragments of AT placed in a complete culture medium that contains nutrients and electrolytes (glucose, amino acids, vitamins, minerals) buffered to pH 7.4. As is discussed in **Note 2**, the medium can be supplemented with serum, but this is not essential. The major strength of this method is the good maintenance of gene expression and adipocyte function within AT placed in OC for up to 2 wk,

From: *Methods in Molecular Biology, vol. 155: Adipose Tissue Protocols*
Edited by: G. Ailhaud © Humana Press Inc., Totowa, NJ

as a number of investigators have reported *(6–8)*. Furthermore, fat cells isolated from AT fragments in OC remain responsive to acute hormonal effects after culture *(9,10)*, allowing studies of mechanisms involved (e.g., effect of long-term culture with insulin or growth hormone on responsiveness to the acute effect of insulin on glucose transport).

OC has been used to assess the long-term effects of hormones on the metabolism of human *(7,11)*, rat *(12–15)*, ovine *(8,16)*, bovine *(17–19)*, and porcine *(10,20)* AT. In AT of all of these species, the long-term effects of hormones added to the culture medium appear to reflect known in vivo effects, lending confidence that the OC system is a useful research tool. In addition to studies of adipocyte glucose metabolism discussed below, OC has been used to study the long-term regulation of key adipocyte genes including lipoprotein lipase (LPL) *(3,11,21)*, leptin *(4,22,23)*, peroxisome proliferator activated receptor γ *(24)*, and plasminogen activator inhibitor-1 *(5,25)*.

### 1.1.3. Primary Fat Cell Culture

Fat cell culture refers to the maintenance of collagenase-isolated adipocytes from animal fat pads in primary culture. One advantage of this method, compared to OC, is that only adipocytes are present. Primary culture of fat cells is different from primary culture of newly differentiated fat cells derived from stromal precursor cells. The latter generally remain multilocular (multiple lipid droplets) in monolayer culture, and there is some evidence that these cells may not fully differentiate *(26)*. Thus, the chief advantage of primary culture of isolated adipocytes is the ability to study fully differentiated, unilocular adipocytes from fat pads of animal or human ATs.

## 1.2. Organ Culture of AT: Theoretical Considerations

AT contains many cell types, including adipocytes, endothelial cells, preadipocytes, and fibroblasts. Thus, OC preserves paracrine interactions among cells that can influence adipocyte metabolism. For example, cytokines are highly expressed in the AT of obese individuals and have profound effects on adipocyte insulin sensitivity and lipid metabolism *(27–30)*. We found that the majority of interleukin-6 (IL6) *(31)* and tumor necrosis factor-α (TNFα) (Fried, S. K., unpublished data) in human AT is found in the stromal fraction, not in the adipocytes themselves. Thus, the preservation of complex interactions between different cell types in AT may be critical to the function of AT. OC allows these paracrine effects to factor into the analysis of long-term regulation of AT function in vitro.

Although it can be argued that delineating paracrine interactions is critical to understanding the physiology of AT, there is no question that the complexity of the system renders mechanistic understanding difficult. The presence of multiple cell types complicates the interpretation of the data. Most importantly, with OC, it cannot be determined whether effects of hormones or substrates added to the culture medium can be explained by direct actions on the adipocyte itself, or require mediation by other cell types. Nevertheless, depending on one's objective, the ability to elicit physiologically relevant responses to hormones in OC may outweigh other concerns. Different systems (OC, isolated fat cell culture, primary culture of newly-differentiated adipocytes, established cells lines) offer complementary approaches.

In addition to paracrine factors, it seems possible that the extracellular matrix present within AT is also critical for maintaining the structure and function of the adipocytes during long-term culture in vitro. This may be the reason for improved maintenance of metabolic gene expression of isolated adipocytes cultured on matrigel (an artificial extracellular matrix) *(32)*. It remains to be established whether the extracellular matrix surrounding adipocytes has tissue- or depot-specific properties *(4,22,23)*.

## 1.3. Evidence for the Utility of AT OC

A number of investigators *(30,33)* have succeeded in eliciting effects of hormones or cytokines using OC of AT in cases in which cultured, isolated adipocytes were unresponsive. Although many laboratories report difficulties maintaining cultures of isolated fat cells, AT fragments placed into OC remain viable. The reasons underlying this marked contrast have not been established, however, as discussed above, structural or paracrine factors may be involved.

Smith *(2)* found that, after AT had been cultured 2 wk in the absence of hormones, fat cell size had slightly decreased, and the change was greatest in the largest cells. Smith went on to utilize his method to demonstrate that insulin had a long-term effect to increase glucose metabolic capacity *(2,6,34)*; long-term culture with insulin increases basal lipolysis measured in a subsequent 2-h incubation *(6,34)*; growth hormone, catecholamines, and glucocorticoids (GCs) were important long-term regulators of human AT metabolism *(3,35)*.

We have confirmed that adipocytes from human AT cultured for only 1 wk do not decrease in size, and remain acutely responsive to the ability of insulin to increase glucose metabolism *(9)* and glucose transport (Fried, S. K., unpublished observation). Acute insulin responses are best maintained when both insulin and moderate concentrations of GCs are present in the culture medium *(9)*. The authors have also recently shown that for AT of obese subjects, insulin, and GCs can maintain the initial level of expression of leptin mRNA and leptin secretion into the medium over 1 wk in culture *(4)*. In contrast, with AT from lean subjects, culture with insulin and GCs actually increases leptin secreted into the medium above initial levels over initial (Fried, S. K., unpublished observation). Thus, the long-term effects of hormones added to OC depend on subject factors existing prior to culture (e.g., lean, normoinsulinemic vs obese, hyperinsulinemic, and so on). In addition, depot-specific characteristics of omental and subcutaneous (sc) fat do persist during OC *(4,11)*. Whether depot differences depend on paracrine interactions within the AT fragments remains to be established. Improved techniques for the maintenance of isolated adipocyte cultures can help address these issues.

## 1.4. Overview of AT OC Methodology

After initial studies using samples of sc fat *(36)*, Slavin *(37,38)* was probably the first to use OC to study AT function. Smith *(6,7)* first proposed the use of OC for studies of long-term regulation of human AT metabolism. He sandwiched approximately three 5-mg (3 × 3 × 1 mm) fragments of sc human AT between glass cover slips (10× 40 mm), and placed them in Leighton tubes in 1.5 mL medium. The gas phase was air. The culture media were not replenished during culture. He compared results, using standard Hank's or Eagle's with Parker medium 199 (each supplemented with 300 mg/L HEPES and containing 5.5 m$M$ glucose). Only those cultured in M199 maintained the rounded appearance of the adipocytes, and allowed viable isolated adipocytes to be prepared after culture. Smith found that addition of serum at concentrations greater than 5% causes the outgrowth of fibroblast-like cells from the tissue fragments. Because serum was not necessary for cell viability, it was logical to use better-defined serum-free conditions *(39)*.

Although the technique of Smith was successful in maintaining AT metabolism for 1 wk, the method was cumbersome. We modified this method to accommodate culture of much larger amounts of tissue (up to 12 g), which are required for extraction of RNA and studies of adipocyte metabolism, enzyme activities, hormone binding, and so on. The major modification in the protocol was to eliminate the cover slips. Rather, the minced tissue was allowed to float freely in a petri dish. This procedure for human AT is similar to that used by Vernon *(8,16)* for ovine AT. The authors culture a total of 300–500 mg of AT (minced into 5–10 mg fragments) in 15 mL medium in a 100-mm dish. The high ratio of media to tissue appears to be critical for best maintaining adipocyte function (as judged by LPL activity) (Fried, S. K., unpublished observation). Culture media containing fresh hormones (when appropriate) are replaced every 2–3 d. The media are always refreshed the day prior to an experiment, to standardize conditions and minimize variability in culture conditions.

## 1.5. History and Rationale for Isolation of Human Adipocytes

Rodbell was the first, in 1964 *(40)*, to describe isolation of adipocytes from AT, using collagenase. These adipocytes were metabolically active following the digestion. Several studies since then used this system to investigate acute effects of hormones and other factors of fat cell metabolism: insulin stimulation of glucose transport *(41)*, as well as insulin sensitivity, insulin resistance, and insulin action *(42,43)*. Although freshly isolated adipocytes can be readily used to study hormonal or nutritional regulation of metabolism and gene expression, this does not allow studies on long-term effects of theses factors, and the responses obtained may reflect cell response to the in vivo neurohormonal conditions and effects of specific circulating factors. Establishment of long-term culture of fat cells was therefore necessary to alleviate these problems. In 1984, Marshall et al. *(44)*, developed a primary culture of rodent adipocytes, and demonstrated responsiveness of these cells to insulin after up to 72 h in culture. Several years later, Briquet-Laugier et al. *(45)* described a longer-term culture of rat adipocytes. Cells isolated from both lean and obese Zucker rats remained viable and metabolically active in culture for 9 d. The authors used a similar protocol to maintain human adipocytes in long-term culture, as described below.

## 2. Materials

### 2.1. AT OC

1. Laminar flow hood.
2. Tissue culture incubator, 5% $CO_2$ atmosphere.
3. Transport buffer for fat aspirations (phosphate buffered saline [PBS] [0.15 *M* NaCl, 10 m*M* $KH_2PO_4$] supplemented with 5 m*M* glucose, 100 n*M* $N^6$-phenylisopropyladenosine [PIA], and 50 µg/mL gentamicin). Prepare solution sterile filter, and store at 4°C. Add fungizone (Gibco-BRL, Grand Island, NY) (Amphotericin B [0.5 µg/mL]) just prior to use, if needed. $10^{-4}$ *M* PIA stock (19.6 mg in 5 mL propylene glycol brought to 500 mL with saline) can be stored frozen in aliquots at –80°C).
4. M199 (Gibco-BRL; liquid, bicarbonate buffered, supplemented with glutamine and 25 m*M* HEPES) and 50 µg/mL gentamicin.
5. 50-cc polypropylene sterile test tubes with caps.
6. Petri dishes (polystyrene, 20 × 100 mm).
7. Sterile pipets.
8. Sharp, long-handled scissors (15–18 cm) (autoclaved).
9. Forceps (angled, 16 cm) and/or perforated, round spoon (5 mm deep, 14.5 cm) (Fine Science Tools, Foster City, CA) (autoclaved).
10. Nylon mesh (pore size 250 µ; Sefar American, Kansas City, MO) shaped into a cone, secured with staples, and affixed to a plastic funnel with laboratory tape, and autoclaved wrapped in aluminum foil.
11. Insulin (recombinant human [e.g., Humulin, E. Lilly, Indianapolis, IN], 100 U/mL) dexamethasone (DEX) (10 mg/mL solution in sodium phosphate, Elkins-Sinn, Cherry Hill NJ).
12. Glass bead sterilizer for quick resterilization of forceps, and so on, as needed.
13. EtOH (95%).

### 2.2. Isolated Fat Cell Culture

1. Laminar flow hood approved for use of biohazard materials.
2. Incubator: 5% CO2 atmosphere.
3. Medium: Dulbecco's modified Eagle's medium (DMEM) (Gibco), which contains 5.5 m$M$ or 25 m$M$ glucose and 20 m$M$ HEPES,
4. Fetal bovine serum (FBS).
5. Bovine serum albumin (BSA).
6. Type I collagenase (Gibco).
7. (–)-N$^6$-phenylisopropyl adenosine or adenosine.
8. Antibiotics: penicillin, streptomycin, and gentamicin.
9. Insulin (Humulin, Eli Lilly, but cells are also responsive to bovine insulin).
10. Sterile pipets.
11. Sterile nylon mesh (350 µm) and filtration system (flask and funnel to support the filter) (spectrum medial).
12. 150 mm sterile Petri dishes, to prepare the tissue.
13. Sterile Falcon tubes (250 mL) or cell culture dishes.
14. Scissors and forceps (autoclaved).
15. EtOH (95%).
16. Bleach (cleaning/decontamination) of the work area.

## 3. Methods

### 3.1. AT OC (see Notes 1–13)

#### 3.1.1 Obtaining Tissue

##### 3.1.1.1. SURGICAL SAMPLES

Obtain surgical samples of AT from different sites (usually abdominal sc, omental) during the first 30 min after the induction of anesthesia. Place samples (1–3 g/tube) in a 50-cc plastic tube containing ~10 mL M199 (supplemented as above) at room temperature (RT). Coarsely mince the tissue to prevent tissue hypoxia. Cap the tube, and transport it to the laboratory (at 24°C), ideally within 30 min (longer times are possible without adverse effects for long-term cultures).

##### 3.1.1.2. FAT ASPIRATION SAMPLES

Samples of sc fat can also be obtained by needle aspiration (*36,46*). In this case, the tissue is already fragmented, and does not require further mincing. We aspirate the tissue into a 60-cc syringe, then transfer it to a bottle containing 400 mL transport buffer at 24°C. This transfer can be accomplished using sterile technique in a clean room, if a laminar flow hood is not available nearby the procedure room. The PIA is added to the transfer buffer, to maintain low rates of lipolysis during transport to the laboratory. The large volume of PBS helps wash the tissue free of blood.

## 3.1.2. OC Procedures

1. Tissue processing: All of the following procedures (except weighing) are carried out under a laminar flow hood, using sterile technique. Immediately upon arrival, mince tissue finely into 5–10 mg fragments by placing sharp scissors into the 50-cc tube (containing tissue and M199) and manually mincing the tissue using two gloved hands on the scissors. It takes approx 10 min to obtain sufficiently small fragments. Once the tissue is finely minced it can remain at RT for up to 1 h (while the other tubes of tissue are being minced). Do not place the tissue at 37°C, to avoid increasing its metabolic rate at this juncture.

2. Dilute hormones that will be needed, and label Petri dishes as needed. Use serial dilutions to obtain 100× working concentrations of hormones and drugs. When a solvent is used for hormone or drug stocks, this is added to controls at matching dilutions. These are then added to the volume of media required to culture the amount of tissue available.

3. Wash tissue free of lipid droplets and blood by pouring the contents of the tube through a nylon mesh (affixed to a funnel), which is placed on top of a ~500-mL bottle. Several tubes containing minced tissue can be combined on the funnel. Pour at least 300 mL sterile 0.9% saline (or PBS), at 37°C, over the tissue on the funnel. This removes free lipid and many red blood cells. Place the entire funnel, with washed tissue, in a sterile Petri dish. Scoop up tissue with a forceps or perforated spoon, and place into a tared Petri dish. Remove any large blood clots with a forceps.

4. Close the dish to maintain sterility, and then weigh. Thus, the potential number of dishes can be calculated according to the total tissue weight (~0.5 g wet weight/ dish). The tissue is not quite dry so the actual tissue weight is overestimated by about 10–20%.

5. Aliquot to the dishes, using the forceps or perforated spoon, based on the calculation above (~0.5 g/dish). Add media (15 mL with appropriate hormone additions, if using 100-mm Petri dishes), without excessive delay. Maintain cultures in a 37°C incubator under an atmosphere of 5% $CO_2$–95% air.

6. Replenish culture media at least every 2–3 d. This removes products of AT metabolism that accumulate in the medium (e.g., fatty acid [FA], cytokines) and ensures that hormones are maintained close to the desired concentration (e.g., insulin becomes degraded/denatured during long-term incubations at 37°C). In most instances, we change the media the day prior to the experiment. The procedure to change the media is as follows: Place tray with the Petri dishes in the hood so that it is at a slight angle (toward the experimenter). Insert a 10-mL pipet below the floating fat, and slowly aspirate the media. During the procedure, it is important to suction slowly, to avoid removing the AT, although some minimal loss of tissue is inevitable. The last milliliter of media can be left in the dish. As media is removed from a tray of several Petri dishes, add fresh media (with appropriate hormones), without excessive delay.

## 3.2. Isolated Fat Cell Culture Procedure (see Notes 14–20)

1. Obtain sc abdominal fat, following abdominal surgeries or from liposuctions. Fat is removed at the time of the surgery in a sterile environment and immersed in

Hank's balanced salt solution (HBSS) supplemented with penicillin (100 U/mL), streptomycin (100 μg/mL), and gentamicin (50 μg/mL). Process within 1 h following the surgery.

2. Wash tissue several times with HBSS (Gibco, MD), then remove the majority of connective tissue and blood clots using a scissors or scalpel and forceps. Remove any skin attached to abdominoplasty samples.

3. Mince tissue into small fragments, and digest with type I collagenase (1 mg/mL, Gibco) in a shaking water bath at 37°C for 30–60 min in a polypropylene flask. Mix every 15 min during the digestion by pipeting up and down using a 10-mL pipet.

4. The subsequent procedures are performed at RT. Filter cells through a sterile nylon filter (350 μm mesh) into a test tube. Centrifuge the suspension at 500*g* for 1 min to separate the pelleted stromal-vascular fraction from the floating adipocyte fraction. Wash adipocytes 3× with HBSS at RT, then resuspend (approx 2–3 mL of cells over 20 mL media) in DMEM, supplemented with HEPES (15 m*M*), glucose (5.5 or 25 m*M*), BSA (1%), 50 n*M* adenosine, and antibiotics; 1% FBS (standard medium). Culture cells in suspension in sterile polypropylene tubes (250-mL Falcon tubes; alternatively culture dishes or flasks can be used), in a humidified incubator at 37°C under 5% $CO_2$–95% air.

5. Exchange the media after the first 24 h, then every other day (and 1 d before harvesting cells for assays). First warm media to RT. Because adipocytes float, exchange media by introducing the pipet through the floating cell layer into the bottom of the tube, and pipet out the media. Then add fresh media in a similar way. It is inevitable to lose some cells during this process.

6. Incubate cells overnight in serum-free medium, prior to insulin (20 n*M*) or other hormonal treatment for the 24 h prior to an experiment.

## 4. Notes

1. For exchange of media (refeeding), it is not possible to use a vacuum system and Pasteur pipets, as with traditional cell culture, because the pressure is too great, and cannot be adequately controlled.

2. Use of serum in OC: The culture conditions and techniques used by different investigators vary. Many investigators use serum-free, albumin-free conditions, others include FBS (0.5–1%). There are very limited data in the literature regarding the relative merits of the different approaches. Smith, using human AT *(1,7)*, and Falconnier *(17)* using bovine AT, showed that addition of serum to the culture medium decreases the expression of lipogenic enzymes. Halleux et al. *(22)* use DMEM, supplemented with 10% FBS, for culture of human AT. They also used a higher tissue concentration than we do (0.7–1 g tissue in 10 mL) in a 100-mm Petri dish. Cultures were kept for only 48 h, so it is unclear whether it is suitable for longer-term cultures. Casabiell et al. *(23)* used small pieces of human AT (total 300–400 mg), placed in 2.5 mL medium in DMEM + 0.5% fetal calf serum.

   It should be noted that the addition of serum adds a level of complexity to the culture system. Serum is undefined, contains myriad growth factors, and shows variability from lot to lot. In contrast, serum-free medium provides a defined system that supports maintainance of the function of adipocytes. Differing results between studies may be attributable to variations in protocols for OC. Initial studies using OC should carefully compare various protocols with respect to the end points of interest. In addition, a potential problem in the interpretation of data on serum-containing cultures is that, as noted by Cigolini and Smith *(3)*, stromal fibroblasts (which have the potential to convert to adipocytes) proliferate if the serum concentration in the medium is more than 5%.

3. Use of albumin: Bernstein *(14)* showed that OC of rat AT in media consisting of Krebs buffer and 1 or 4% BSA, causes insulin resistance of glucose transport and metabolism. Charcoal treatment of the albumin and an atmosphere of oxygen, instead of air, improved insulin sensitivity of the cultured rat AT, but the improvement was not attributable to an alteration in FA release *(13)*. They did not use an atmosphere of 5% $CO_2$ and a bicarbonate buffer (needed to support FA synthesis). Addition of albumin as a FA acceptor during OC does not appear to be necessary (at least in the dilute cultures that we utilize). Apparently, under these conditions (high ratio of medium to tissue), the positively charged ions and amino acids in the medium suffice as FA acceptors. We have not measured FA release into the medium, to test this hypothesis. However, the accumulation of unesterified FAs in the tissue would be expected to be deleterious and to increase fat cell fragility, which has not been observed. Furthermore, Smith *(39)* found that M199 without albumin is adequate for studies of human adipocyte metabolism. The authors' unpublished observation is that addition of albumin or serum decreases the activity of LPL in AT cultured with insulin or insulin plus DEX.

4. AT from different species: For ruminant AT, most investigators add acetate (7 m*M* acetate may be optimal) as substrate *(16,18,20)*.

5. Time course: It is important to carefully assess the time course of the alterations in AT function in OC. The time course may vary with the end point assessed *(4,30)*, by species *(17)*, and according to subject factors (e.g., lean vs obese, fasted vs fed donors). While initially testing the OC system for human AT from obese subjects, we noted that LPL activity declined from control (initial) values during the first 3–4 d of culture, regardless of the hormonal additions to the culture medium. But, thereafter, during d 5–8, LPL activity increased to control values in cultures containing insulin, and rose to levels that exceeded initial, in cultures containing insulin and DEX. In contrast to the characteristic time course of hormonal effects on LPL during 1 wk of culture, we found that leptin secretion and expression are responsive to insulin and GCs during the first few days of culture; in fact, responses to insulin plus GC are evident after 5 h *(6)*. The reasons for the differences in the time-course of the expression of these genes are not established. One possible explanation for initial fall, then rise of LPL expression, in tissue cultured with insulin plus GC is the production of cytokines within the tissue during the early days of culture *(31)*.

Ottoson et al. *(21)* cultured human AT for 2 d with insulin alone, then added cortisol (with insulin) for an additional 48 h. Under these conditions, the cortisol effect on LPL activity is readily observed. An advantage of this technique is that interindividual variability caused by subject factors are diminished during the initial culture period, allowing good reproducibility in responses. It is feasible to culture adipose tissue for 7 d in the presence of insulin alone, and still observe a robust response to delayed addition of GC effects on LPL and leptin mRNA levels (4). Taken together, these data argue strongly that function of AT is maintained during OC, and that time-courses may be best studied after an initial period of adaptation.

6. Quantity of tissue per dish: We have found that placing more than 0.5 g tissue/ dish (15 mL) decreases the maximum LPL activity that can be induced by insulin and DEX and probably also affects other aspects of adipocyte function.

7. Alternate culture dishes: It is possible to use 6-, 12-, or 24-well plates, when it is desired to assess a large number of conditions and when the amount of tissue needed for assay is not large. For example, 50 mg tissue can be placed in 1.5–2 mL medium in a 6-well plate. It is important to keep the proportion of tissue to medium, at approx 30 mg/mL.

8. Choice of culture conditions (hormones): The hormonal additions to the culture medium should be adapted to the purpose of each experiment. Maintenance of in vivo levels of adipocyte gene expression (e.g. LPL [11], leptin [4,22], GLUT4 [22]), and unpublished preliminary observation) requires the presence of insulin and/or GC (and perhaps other additions, depending on the topic of investigation). Thus, for example, to test experimental maneuvers that are expected to decrease LPL activity, it is necessary to first culture under conditions that will produce a high and stable level in controls. The concern is that a sufficiently high baseline is needed to prevent a floor effect. For example, the suppressive effect of TNF-$\alpha$ or catecholamines on LPL activity cannot be observed in tissue that has been previously cultured (or co-cultured) without hormones, because the activity falls to very low levels in the control (no hormone condition). Thus, we culture AT for 6–7 d with insulin plus GC to achieve high LPL expression, and to minimize interindividual variability between patients, prior to tissue sampling. We may then test effects of hormones expected to decrease LPL expression. For manipulations that are expected to enhance anabolic gene expression (e.g., insulin sensitizers), consideration should be given to culturing the tissue with submaximal levels of insulin, with or without GC. For studying regulation of a novel process or gene of interest, it is important to test effects against a background of various combinations of hormones. Though this greatly increases the number of culture conditions, it is crucial to establish a milieu that is physiologically relevant.

9. Dose-dependence of hormonal responses during OC: Responses to hormones in OC occur at concentrations within the physiological range. For insulin, 50% maximal responses to insulin are obtained at concentrations that are clearly within the physiological range (0.7 n$M$). Additionally, cultured AT responds to a dose of 2.5 n$M$ DEX (corresponding to a physiologically-relevant cortisol concentration), with maximal responses in sc AT at 25 nM DEX (and 250 n$M$ in omental AT) (11). Similar results have been obtained for other end points, e.g., PAI-1 (5).

10. Assessment of fat cell and tissue viability: It is important to demonstrate that basic aspects of adipocyte metabolism are preserved during OC (e.g., rates of

glucose transport and conversion to products, rates of lipolysis, preservation of acute hormonal responses, and enzyme activities, such as LPL). For metabolic studies, it is preferable to prepare collagenase-digested fat cells from the cultured AT. Good viability of adipocytes prepared from cultured AT can be obtained, particularly if adenosine (200 n*M*) is added to collagenase digestion, wash buffers, and cell incubations (when appropriate) *(47)*. These methods are outlined in Chapters 12 and 15.

11. Several lines of evidence indicate that apoptosis of fat cells can occur in vitro after addition of a high concentration of TNF-α (in the absence of other hormones and in the presence of serum) *(48)*. In future studies, it will be important to assess the effect of experimental manipulations on apoptosis of fat cells (and other cell types within AT fragments) that occur during OC under specific conditions.

12. Size of fragments: It is critical to mince the tissue finely (5–10 mg), because Smith *(2)* observed necrosis of internal portions of the tissue within 24 h, if the pieces exceeded 4–5 mm in diameter. Our experience regarding LPL activity also shows that culture of small fragments show highest activities.

13. Possibility of differentiating of preadipocytes during OC: Few studies have addressed the possibility that preadipocytes present in the tissue could differentiate during culture and contribute alterations in gene expression. To avoid this issue, cultured AT can be subjected to collagenase digestion, to prepare isolated fat cell and stromal fractions *(11)*. This will allow conclusions on whether adipocyte gene expression is altered by manipulations during OC, which is usually the primary question being addressed in an AT OC experiment. Any discrepant findings between adipocyte and total tissue gene expression could be explained by the differentiation of preadipocytes in the tissue that are not recovered with either the floating fat cell or pelleted stromal fraction. Very small fat cells do not have sufficient lipid to float, and have too much to pellet: Hence, they remain in the internatant *(49)*. If differentiation is a concern, it can be addressed by immunohistochemical staining or *in situ* hybridization using fixed AT. We have found that alterations in AT lipoprotein lipase mRNA after 7 d of OC of human AT in serum-free media reflect alterations at the level of the adipocyte *(11)*. This parameter should be monitored for the target of interest and the specific techniques of OC that are employed.

14. Routinely, cells are used for experiments within 1 wk of surgery. As indicated by various tests of viability and metabolic activity, cells can be maintained (depending on the original quality of the tissue: *see* **Note 17**) for 1–2 wk. Most of the cell breakage occurs during the first 48 h of culture (as evidenced by oil release into the medium).

15. Evidence of viability: Trypan blue exclusion test is conducted in all cultures to confirm cell viability. In addition, enzyme activities (fatty acid synthase [FAS] and glycerol-3-phosphate dehydrogenase [G3PDH]), glucose consumption (measured in culture media), cell size, apoptosis, and DNA content can be monitored *(50)*.

16. Human subjects: As discussed in **Subheading 1.3.**, and depending on the parameter studied, there may be differences in the response of cells from lean vs obese

individuals. We have found differences in the response of adipocytes from lean and obese to insulin-stimulated lipogenesis in the presence of DEX (Fried, S. K., unpublished data). Therefore, it is important to obtain information regarding the patient's medical record (body mass index, or diabetes and other metabolic disorders). We have not observed, however, any statistical difference in the response of adipocytes from lean and obese patients to insulin *(50,51)*. However, Briquet-Laugier et al. *(52)* have reported that the difference in the lipogenic capacity between adipocytes from lean and obese Zucker rats is maintained in vitro after several days of culture, reflecting a difference in the intrinsic property between lean and obese adipocytes. We usually maintain human adipocytes in culture for 2–3 d prior to addition of any hormones, to eliminate any differences contributed by in vivo circulating factors, and have observed differences between freshly isolated cells and cultures cells in certain genes studied: Angiotensinogen mRNA levels are variable in freshly isolated adipocytes or AT (very low to very high levels, depending on the patient); however, once cells are cultured, the level of angiotensinogen expression (in 1% FBS and without hormone addition) is comparable among adipocytes from various patients *(53)*. Similarly, the FAS mRNA content is variable among adipocytes from different patients. However, when cells are cultured in the absence of insulin, the FAS mRNA content and activity are dramatically decreased; but these levels can be reinduced by insulin *(8)*.

17. Liposuction vs abdominoplasty: We have found that cells obtained from liposuctions are more fragile, probably because of the pressure of the suction system. In some cases, cells do not survive more than a couple of days. Cells obtained from abdominal surgeries, in which a large piece of fat tissue is removed, remain viable for several days, and most of our studies are performed on these cells. This also varies from one patient to another; most samples are from Caucasian females, and no relationship has been observed between cell survival or breakage and the age of the patient (Fried, S. K., unpublished data).

18. Insulin: Not only can adipocytes can be maintained for longer times in the presence of insulin, but lipogenic enzyme activities (such as FAS and G2PDH) remain at induced levels in the presence of insulin *(50)*. This effect is reversible because removal of insulin causes a rapid decrease in these enzymes, and addition of insulin to previously starved cells stimulates lipogenic gene activities *(51)*.

19. FBS, BSA, and adenosine: Although FBS is required for differentiation of most preadipocytes, it is not required to maintain viability of isolated mature cells. However, for both rat and human adipocytes, lipogenic enzyme activities and glucose consumption is twice as high in the presence of serum, compared to the absence of serum. We do not, however, routinely test the effects of different sera batches on the adipocyte function, and have not systematically compared the effect of BSA on cultured human adipocyte metabolism, as previously reported for rat adipocytes. Cells appear to survive longer, break less, and release less lipids into the medium in the presence of adenosine (Fried, S. K., unpublished observations).

20. Adipocyte attachment and dedifferentiation: It is difficult to account for this dedifferentiation when cells are maintained in tubes, but it can be easily observed

when adipocytes are cultured in flasks. When we first cultured adipocytes in tubes for 24 h then transferred them to culture flasks or dishes, cells were observed to attach to the flask base and appear as mature adipocytes. This is not the result of preadipocyte contamination, because adipocytes were isolated, then cultured for 1 d in tubes, and the low serum concentration would not allow rapid attachment and differentiation of these cells. After several days in culture, these adipocytes become smaller, and some of them multilocular, then progressively dedifferenti-ate into preadipocytes-like cells.

## Acknowledgments

We gratefully acknowledge Dr. Marcelle Lavau for teaching us about the art of experimenting with fat cells and adipose tissue. We also thank Colleen Russell for reviewing the manuscript.

## References

1. Smith, U. (1974) Studies of human adipose tissue in culture. 3. Influence of insulin and medium glucose concentration on cellular metabolism. *J. Clin. Invest.* **53,** 91–98.
2. Smith, U. (1972) Studies of human adipose tissue in culture. I. Incorporation of glucose and release of glycerol. *Anat. Rec.* **172,** 597–602.
3. Cigolini, M. and Smith, U. (1979) Human adipose tissue in culture: VII. Studies on the insulin-antagonistic effect of glucocorticoids. *Metabolism* **28,** 502.
4. Russell, C. D., Petersen, R. N., Rao, S. P., Ricci, M. R., Prasad, A., Zhang, Y., Brolin, R. E., and Fried, S. K. (1998) Leptin expression in adipose tissue from obese humans: depot-specific regulation by insulin and dexamethasone. *Am. J. Physiol.* **275,** E507–E515
5. Morange, P. E., Aubert, J., Peiretti, F., Lijnen, H. R., Vague, P., Verdier, M., Negrel, R., Juhan-Vague, I., and Alessi, M. C. (1999) Glucocorticoids and insulin promote plasminogen activator inhibitor 1 production by human adipose tissue. *Diabetes* **48,** 890–895.
6. Smith, U. (1973) Studies of human adipose tissue in culture II. Effects of insulin and of medium glucose on lipolysis and cell size. *Anat. Record* **176,** 181–183.
7. Smith, U. (1971) Morphologic studies of human subcutaneous adipose tissue in vitro. *Anat. Rec.* **169,** 97–104.
8. Robertson, J. P, Faulkner, A., and Vernon, R. G. (1982) Regulation of glycolysis and fatty acid synthesis from glucose in sheep adipose tissue. *Biochem. J.* **206,** 577–586.
9. Appel, B. and Fried, S. K. (1992) Effects of insulin and dexamethasone on lipo-protein lipase in human adipose tissue. *Am. J. Physiol.* **262,** E695–E699
10. Wang, Y., Fried, S. K., Petersen, R. N., and Schoknecht, P. A. (1999) Somatotropin regulates adipose tissue metabolism in neonatal swine. *J. Nutr.* **129,** 139–145.
11. Fried, S. K., Russell, C. D., Grauso, N. L., and Brolin, R. E. (1993) Lipoprotein lipase regulation by insulin and glucocorticoid in subcutaneous and omental adi-pose tissues of obese women and men. *J. Clin. Invest.* **92,** 2191–2198.
12. Livingston, J. N., Purvis, B. J., and Lockwood, D. H. (1978) Insulin-dependent regulation of the insulin-sensitivity of adipocytes. *Nature* **273,** 394–396.

13. Bernstein, R. S. (1982). Improved insulin responsiveness in rat adipose tissue pieces cultured with charcoal-treated albumin and oxygen. *J. Lipid Res.* **23,** 360–363.

14. Bernstein, R. S. (1979) Insulin insensitivity and altered glucose utilization in cultured rat adipose tissue. *J. Lipid Res.* **20,** 848–856.

15. Maloff, B. L., Levine, J. H., and Lockwood, D. H. (1980)Direct effects of growth hormone on insuln action in rat adipose tissue maintained in vitro. *Endocrinology* **538,** 544.

16. Vernon, R. G., Barber, M. C., and Finley, E. (1991). Modulation of the activity of acetyl-CoA carboxylase and other lipogenic enzymes by growth hormone, insulin and dexamethasone in sheep adipose tissue and relationship to adaptations to lactation. *Biochem. J.* **274,** 543–548.

17. Faulconnier, Y., Thevenet, M., Flechet, J., and Chilliard, Y. (1994) Lipoprotein lipase and metabolic activities in incubated bovine adipose tissue explants: effects of insulin, dexamethasone, and fetal bovine serum. *J. Anim. Sci.* **72,** 184–191.

18. Faulconnier, Y., Ferlay, A., and Chilliard, Y. (1997). Insulin and/or dexamethasone regulation of lactate production and its relationship to glucose utilization by ovine and bovine adipose tissue explants incubated for 7 days. *Reprod. Nutr. Dev.* **37,** 401–410.

19. Miller, M. F., Cross, H. R., Lunt, D. K., and Smith, S. B. (1991) Lipogenesis in acute and 48-hour cultures of bovine intramuscular and subcutaneous adipose tissue explants. *J. Anim. Sci.* **69,** 162–170.

20. Walton, P. E. and Etherton, T. D. (1986) Stimulation of lipogenesis by insulin in swine adipose tissue: antagonism by porcine growth hormone. *J. Anim. Sci.* **62,** 1584–1595.

21. Ottoson, M., Vikman-Adolfsson, K., Enerback, S., Olivecrona, G., and Bjorntorp, P. (1994). The effects of cortisol on the regulation of lipoprotein lipase activity in human adipose tissue. *J. Clin. Endocrinol. Metab.* **79,** 820–825.

22. Halleux, C. M., Servais, I., Reul, B. A., Detry, R., and Brichard, S. M. (1998) Multihormonal control of ob gene expression and leptin secretion from cultured human visceral adipose tissue: increased responsiveness to glucocorticoids in obesity. *J. Clin. Endocrinol. Metab.* **83,** 902–910.

23. Casabiell, X., Pineiro, V., Peino, R., Lage, M., Camina, J., Gallego, R., Vallejo, L. G., Dieguez, C., and Casanueva, F. F. (1998). Gender differences in both spontaneous and stimulated leptin secretion by human omental adipose tissue in vitro: dexamethasone and estradiol stimulate leptin release in women, but not in men. *J. Clin. Endocrinol. Metab.* **83,** 2149–2155.

24. Rieusset, J., Andreelli, F., Auboeuf, D., Roques, M., Vallier, P., Riou, J. P., Auwerx, J., Laville, M., and Vidal, H. (1999) Insulin acutely regulates the expression of the peroxisome proliferator- activated receptor-gamma in human adipocytes. *Diabetes* **48,** 699–705.

25. Alessi, M. C., Peiretti, F., Morange, P., Henry, M., Nalbone, G., and Juhan-Vague, I. (1997). Production of plasminogen activator inhibitor 1 by human adipose tissue: possible link between visceral fat accumulation and vascular disease. *Diabetes* **46,** 860–867.

26. Lu, Z. D., Pineyro, M. A., Kirkland, J. L., Li, Z. H., and Gregerman, R. I. (1988) Prostaglandin-sensitive adenylyl cyclase of cultured preadipocytes and mature adipocytes of the rat: probable role of Gi in determination of stimulatory or inhibitory action. *J. Cell. Physiol.* **136,** 1–12.

27. Hotamisligil, G. S., Shargill, N. S., and Spiegleman, B. M. (1993). Adipose expression of tumor necrosis factor-a: Direct role of obesity-linked insulin resistance. *Science* **259,** 87–91.

28. Hotamisligil, G. S., Murray, D. L., Choy, L. N., and Spiegelman, B. M. (1994) Tumor necrosis factor alpha inhibits signaling from the insulin receptor. *Proc. Natl. Acad. Sci. USA* **91,** 4854–4858.

29. Kern, P. A., Saghizadeh, M., Ong, J. M., Bosch, R. J., Deem, R., and Simsolo, R. B. (1995). The expression of tumor necrosis factor in human adipose tissue. Regulation by obesity, weight loss, and relationship to lipoprotein lipase. *J. Clin. Invest.* **95,** 2111–2119.

30. Fried, S. K. and Zechner, R. (1989) Effects of cachectin/tumor necrosis factor on human adipose tissue lipoprotein lipase activity, mRNA levels, and biosynthesis. *J. Lipid Res.* **30,** 1917–1923.

31. Fried, S. K., Bunkin, D. A., and Greenberg, A. S. (1998) Omental and subcutaneous adipose tissues of obese subjects release interleukin-6: depot difference and regulation by glucocorticoid. *J. Clin. Endocrinol. Metab.* **83,** 847–850.

32. Hazen, S. A., Rowe, W. A., and Lynch, C. J. (1995) Monolayer cell culture of freshly isolated adipocytes using extracellular basement membrane components. *J. Lipid Res.* **36,** 868–875.

33. Viguerie-Bascands, N., Saulnier-Blache, J.-S., Dandona, P., Dauzats, M, Davieaud, D., and Langin, D. (1999) Increase in uncoupling protein-2 mRNA expression by BRL49653 and bromopalmitate in human adipocytes. *Biochim. Biophys. Res. Commun.* **256,** 138–141.

34. Smith, U., Bostrom, S., Johansson, R., and Nyberg, G. (1976) Human adipose tissue in culture V. Studies on the metabolic effects of insulin. *Diabetologia* **12,** 137–143.

35. Smith, U., Isaksson, O., Nyberg, G., and Sjorstrom, L. (1976) Human adipose tissue in culture. IV. Evidence for the formation of a hormone antagonist by catecholamines. *Eur. J. Clin. Invest.* **6,** 35–42.

36. Hirsch, J. and Goldrick, R. B. (1964) Serial studies on the metabolism of human adipose tissue. I. Lipogenesis and free fatty acid uptake and release in small aspirated samples of subcutaneous fat. *J. Clin. Invest.* **43,** 1776–1792.

37. Slavin, B. G. and Elias, J. J. (1970) Morphologic changes in white and brown adipose tissue by insulin, thyroxin and cortisol in organ culture. *Anat. Rec.* **167,** 213–218.

38. Slavin, B. G. and Elias, J. J. (1969) The influence of pituitary hormones and norepinephrine on the size of adipose cells in organ culture. *Anat. Rec.* **164,** 141–151.

39. Smith, U. (1970) Insulin responsiveness and lipid synthesis in human fat cells of different sizes: effect of the incubation medium. *Biochim. Biophys. Acta* **218,** 417–423.

40. Rodbell, M. (1964) The metabolism of isolated fat cells. I. Effects of hormones on glucose metabolism and lipolysis. *J. Biol. Chem.* **239,** 375–380.

41. Cushman, S. W. and Wardzala, L. J. (1980) Potential mechanism of insulin action on glucose transport in the isolated rat adipose cell. Apparent translocation of intracellular transport systems to the plasma membrane. *J. Biol. Chem.* **255,** 4758–4762.

42. Klein, H. H., Freidenberg, G. R., Matthaei, S., and Olefsky, J. M. (1987) Insulin receptor kinase following internalization in isolated rat adipocytes. *J. Biol. Chem.* **262,** 10,557–10,564.

43. Traxinger, R. R. and Marshall, S. (1989) Recovery of maximal insulin responsiveness and insulin sensitivity after induction of insulin resistance in primary cultured adipocytes. *J. Biol. Chem.* **264,** 8156–8163.

44. Marshall, S., Garvey, W. T., and Geller, M. (1984) Primary culture of isolated adipocytes: a new model to study insulin receptor regulation and insulin action. *J. Biol. Chem.* **259,** 6376–6384.

45. Briquet Laugier, V., Quignard-Boulangé, A., Dugail, I., Argeson, A. C., LeLiepvre, X., and Lavau M. (1991) White adipocytes from young obese Zucker rates maintain excessive lipogenic enzyme activity in long term culture, in *Obesity in Europe 91* (Ailhaud, G., Guy-Grand, B., Lafontan, M., and Ricquier, D., eds.), Libbey, London, pp. 95–99.

46. Dowling, H. J., Fried, S. K., and Pi-Sunyer, F. X. (1995) Insulin resistance in adipocyte of obese women: effects of body fat distribution and race. *Metabolism* **44,** 987–995.

47. Honnor, R. C., Dhillon, G. S., and Londos, C. (1985) cAMP dependent protein kinase and lipolysis in rat adipocytes. I. cell preparation, manipulation, and predictability in behavior. *J. Biol. Chem.* **260,** 15,122–15,129.

48. Prins, J. B., Niesler, C. U., Winterford, C. M., Bright, N. A., Siddle, K., O'Rahilly, S., Walker, N. I., and Cameron, D. P. (1997) Tumor necrosis factor-alpha induces apoptosis of human adipose cells. *Diabetes* **46,** 1939–1944.

49. Francendese, A. and DeMartinis, F. D. (1985) Very small fat cells. II. Initial observations on basal and hormone-stimulated metabolism. *J. Lipid Res.* **26,** 149–157.

50. Moustaid, N., Jones, B. H., and Taylor, J. W. (1996) Insulin increases lipogenic enzyme activity in human adipocytes in primary culture. *J. Nutr.* **126,** 865–870.

51. Claycombe, K. J., Jones, B. H., Standridge, M. K., Guo, Y., Chun, J. T., Taylor, J. W., and Moustaid-Moussa, N. (1998) Insulin increases fatty acid synthase gene transcription in human adipocytes. *Am. J. Physiol.* **274,** R1253–R1259.

52. Briquet-Laugier, V., Dugail, I., Ardouin, B., Le Liepvre, X., Lavau, M., and Quignard-Boulange, A. (1994) Evidence for a sustained genetic effect on fat storage capacity in cultured adipose cells from Zucker rats. *Am. J. Physiol.* **267,** E439–E446.

53. Jones, B. H., Standridge, M. K., Taylor, J. W., and Moustaid, N. (1997) Angiotensinogen gene expression in adipose tissue: analysis of obese models and hormonal and nutritional control. *Am. J. Physiol.* **273,** R236–R242.

# 18

## Cultures of Adipose Precursor Cells from Brown Adipose Tissue and of Clonal Brown-Adipocyte-Like Cell Lines

### Barbara Cannon and Jan Nedergaard

## 1. Introduction

The possibility of maintaining viable cells in an incubator for prolonged periods, and studying dynamic changes in them in an environment that can be manipulated at will, is an attractive experimental approach. For studies of brown adipose tissue (BAT) recruitment, it provides a powerful approach for investigating the processes of proliferation, differentiation, and apoptosis and their control.

The methods that have been developed for culturing brown adipocytes have derived from those earlier developed for white adipocytes (*1*), based on the recognition that adipose tissues contain a potential mitotic compartment, which can allow for hyperplastic growth (for white adipose tissue, *see* **refs.** *2,3*; for BAT, *see* **refs.** *4–7*). Most conveniently, tissue pieces are treated with collagenase to degrade the extracellular matrix (ECM) and thus release all cells. The result is a mixture of cells, termed the stromal-vascular fraction. This contains all cells unable to float in an aqueous medium, e.g., stem cells, preadipocytes, fibroblasts, erythrocytes, endothelial cells, and so on (*8*). This rather crude mixture is filtered through a series of filters in an attempt to increase homogeneity (particularly from preparation to preparation), and to remove mature adipocytes. This heterogeneous mixture is then inoculated into culture flasks or wells, and allowed to develop in a nutrient medium.

The choice of the medium is important, not only for the chemical nutrients present, but also because of the presence or not of serum, with its concomitant growth factors and the quality of the serum. Often discussed is the cellular

From: *Methods in Molecular Biology, vol. 155: Adipose Tissue Protocols*
Edited by: G. Ailhaud © Humana Press Inc., Totowa, NJ

composition of the cell cultures, since the initial material appears heterogeneous. In our experience, this worry is valid, but is not an insurmountable problem. Under suitable conditions, the great majority of cells surviving in the culture will show the potential to develop into genuine brown adipocytes. The problem is qualitatively insignificant when studies of (brown) adipocyte markers are performed, because these are by definition cell-type-specific, but its significance may increase if more general cellular proteins are investigated.

That multilocular adipocytes can be defined as brown adipocytes can only be accepted if the cells are shown to be able to express the brown-fat-specific uncoupling protein, UCP1, either spontaneously or after stimulation with the physiological agonist of BAT norepinephrine (the only clearly acceptable exception to this is cell cultures from mice with a homologous ablation of the *UCP1* gene *[9]*, and, even there, an unfunctional *UCP1* mRNA fragment can be detected).

In attempts to overcome the potential problems of heterogeneity of the stromal-vascular fraction, and also to have continual access to cells, without the use of further animals, clonal cell lines have also been created. These provide uniform populations of immortalized cells, which are superior to the primary cultures in homogeneity, but which may have properties that are not representative of the cells from the tissue, e.g., proliferation control is clearly altered as the cells are immortalized. Furthermore, certain genes normally considered characteristic of mature adipocytes (e.g., $\beta_3$-adrenergic receptors) may not be expressed *(10)*. Some of the cell lines are also aneuploid *(11)*.

Detailed here is the method used for the most frequently studied system: brown preadipocytes from young mice. Other animal sources, and brief information on the clonal cell lines, are presented in tabular form (**Tables 1** and **2**).

## 2. Materials

1. Male mice, about 3- to 4-wk-old, NMRI outbred strain.
2. HEPES-buffered solution: 123 m$M$ NaCl, 5 m$M$ KCl, 1.3 m$M$ CaCl$_2$, 5 m$M$ glucose, 1.5% crude serum albumin, and 100 m$M$ N-2-hydroxyethylpiperizine-$N'$-2-ethanesulfonic acid (HEPES) (pH-adjusted with NaOH to 7.4). The buffer is sterile-filtered (Sarstedt 0.2 μm) before use, after addition of collagenase.
3. Crude collagenase type II (Sigma).
4. Crude bovine serum albumin (fraction V).
5. 250-μm nylon filter (Nytal).
6. 25-μm nylon filter (Nytal).
7. Sterile tubes, 15 and 50 mL, and sterile syringes.
8. Low-speed centrifuge.
9. Cell culture medium, consisting of Dulbecco's modified Eagles's medium (DMEM), supplemented with 10% newborn calf serum (*see* **Note 1**), 4 n$M$ insulin, 10 m$M$ HEPES, 4 m$M$ glutamine, 50 IU of penicillin, 50 μg streptomycin,

**Table 1**
**Primary Cell Culture Systems**

| Source | NE-induced UCP1 expression | Comments | Refs. |
|---|---|---|---|
| Newborn mouse | Not studied | | *(16)* |
| Young mouse | Yes | Most studied system, that described in detail above | *(17–22)* |
| Fetal rat | Initially high UCP1 expression spontaneously declines. Maintained by insulin or insulin-like growth factor 1 | cAMP inhibits proliferation | *(23–25)* |
| Newborn rat | | $\beta_3$-stimulated glucose transport | *(26,27)* |
| Young rat | Yes. Response to $\beta_3$-agonists: most successful with 6-d treatment; 10% FCS for 1 d, then serum-free medium | Probably needs thyroxine *See* **Note 7** on collagen | *(12,15, 28,29)* |
| Djungarian hamster | Yes. Response to $\beta_3$-agonist | Insulin and $T_3$ important in serum-free conditions | *(30,31)* |
| Syrian hamster | | No successful preparation reported | |
| Newborn lamb | DEX needed for UCP1 expression | DEX in serum-free conditions 3-wk animals give white adipocytes | *(32)* |
| Newborn man | Not successful | | *(33,34)* |

and 25 µg sodium ascorbate/mL. Alternatively, a fully defined medium may be used (*see* **Note 2**).

10. 25-cm$^2$ tissue culture flasks, alternativel, multiwell culture plates (6, 12, or 24, according to experimental requirements).

11. Cell culture incubator, maintained at 37°C, with an atmosphere of 8% $CO_2$ in air.

## 3. Methods

The main steps in this method are summarized in **Fig. 1**.

1. Male mice (*see above*) are maintained at room temperature (22°C) for 2–3 d after purchase. An NMRI outbred strain is suitable for regular culturing. On the day of preparation, the mice are killed by $CO_2$ anesthesia.

2. Dissect out the interscapular, axillary and cervical brown adipose tissue depots under sterile conditions into DMEM.

**Table 2**
**Brown Adipocyte-Like Cell Lines (All of Mouse Origin)**

| Cell line | Origin/derived from | NE-induced UCP1 | | Notes | Present source | Examples of refs. |
|---|---|---|---|---|---|---|
| | | expression | $\beta_3$-receptors | | | |
| BFC-1 | Mouse brown adipocytes | No | Yes (BRL EC$_{50}$ 0.5 n$M$ for lipolysis) | | Forest, Meudon | (11,14) |
| BFC-1b/BFC-1β | Subclone(s) of BFC-1 | No | Not tested | | Forest, Meudon | (35,36) |
| B3, B7 B9, B13 | From hibernoma from urinary protein promoter/SV40 T-antigen transgenic mouse | Yes | No. $\beta_3$-mediated stimulation of UCP1 expression; very low $\beta_3$-mRNA expression | NE stimulates proliferation through $\beta_1$-receptors | Kozak, Baton Rouge | (37–39) |
| HIB 1B | From hibernoma from aP2 promoter/SV40 T-antigen transgenic mouse | Yes | BRL response much lower than isoprenaline, thus unclear if | | Spiegelman, Harvard | (40) |

216

| Cell line | Origin | Differentiation | Thermogenic response | Remarks | Source | Ref. |
|---|---|---|---|---|---|---|
| HIB 1B/8 MB 4.9.2 | Subclone of HIB-1B From hibernoma from *ras* promoter/ SV40 T-antigen transgenic mouse | Yes Not studied | through β₃-receptors; very low β₃-mRNA expression | | Spiegelman, Harvard Lorenzo, Madrid | (41) (42,43) |
| T37i | From hibernoma from P1-mineral-ocorticoid receptor gene promoter/ SV40 T-antigen transgenic mouse | Yes (=iso-prenaline plus retinoic acid) | Not investigated | | Lombès, Paris | (44) |
| HB2 | From p53-knock-out mice | Yes (after prolonged passage, requires troglitazone also) | Presumably, CL-316243 gives response | Troglitazone also for lipid accumulation | Saito, Hokkaido | (45) |

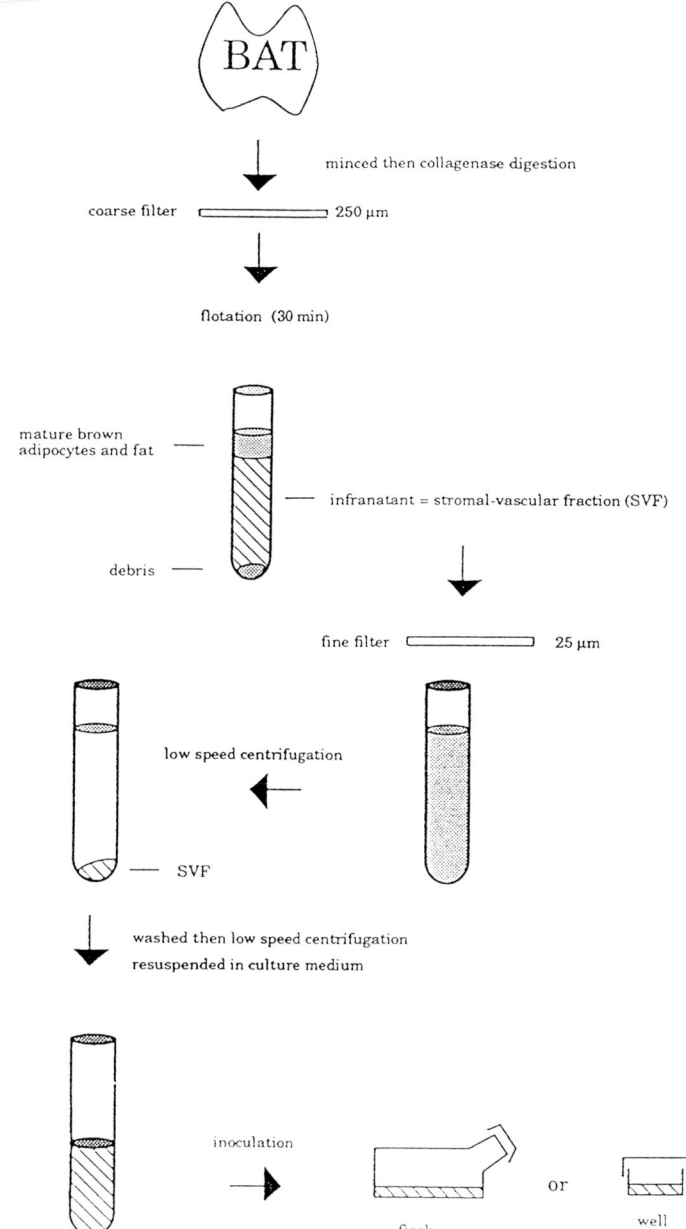

Fig. 1. Main steps in the preparation procedure.

3. Mince the tissue carefully and put in the HEPES-buffered solution (pH 7.4) detailed above, containing 0.2% (w/v) crude collagenase type II (Sigma) (*see*

**Note 3**) . Routinely, pooled tissue (about 200 mg/mouse) from six mice may be digested in 12 mL (2 mL/mouse) of the HEPES-buffered solution in a sterile 50-mL Falcon tube. Digest the tissue for 30 min at 37°C and vortex every 5 min. Pour the digest through a 250-μm nylon filter into a 15-mL sterile tube.

4. Put the solution on ice for 15–30 min, to allow the mature brown fat cells and lipid droplets to float. Filter the infranatant (using a sterile syringe) through a 25-μm nylon filter, and collect it in a 15-mL sterile tube. Now collect the actual precursor cells by centrifuging this tube for 10 min at 700$g$. Resuspend the pellet in DMEM and centrifuge once more at 700$g$ for 10 min, to wash the cells.

5. Finally, resuspend the pellet in 0.5 mL cell culture medium for each mouse dissected.

6. The cell suspension is cultivated in 25-cm$^2$ tissue culture flasks (0.5 mL is used with 4.5 mL cell culture medium), or in 6-well culture plates (0.2 mL with 1.8 mL cell culture medium in each well, or equivalent) (*see* **Note 4**).

7. Place the flasks or wells, at 37°C, in a water-saturated atmosphere of 8% $CO_2$ in air, in a cell culture incubator.

8. On d 1 (the day after isolation), wash the cells with prewarmed (37°C) DMEM, and add fresh prewarmed medium. On d 3, and every other day thereafter, change the medium (without washing). The medium is not changed on the day of an experiment.

9. Follow the development of the cell culture microscopically. It is expected that, after 5–7 d, nearly all cells will contain a ring of small droplets surrounding the nucleus (**Fig. 2**) (*see* **Notes 5–7**).

## 4. Notes

1. Fetal calf serum (FCS) is used for clonal cell lines and for cultures of fetal rat cells. In our experience, it appears to prevent differentiation in the primary cultures from mouse.

2. Concerning the composition of the culture medium: a high insulin concentration promotes lipid accumulation and the presence of ascorbate seems to promote plating, perhaps by its action on stimulation of ECM production. Fully defined media have not been developed for culture from plating to differentiation. However, such media can be used after an initial period in the serum-containing medium, to promote plating and initial proliferation. On d 1–3, the medium can be exchanged for one consisting of equal parts of DMEM/Ham's F12, and containing 0.5% (w/v) fatty-acid-free bovine serum albumin. Insulin, HEPES, glutamate, ascorbate, and antibiotics are present at the same concentrations as in the serum-supplemented medium. Triiodothyronine ($T_3$) is added daily to a concentration of 1 n$M$. In this medium, the cells are essentially quiescent, and these cultures can therefore be used to investigate the influence of hormones, neurotransmitters, and growth factors on proliferation, and subsequently on differentiation (Golozoubova et al., to be submitted). As indicated in **Table 2**, further additions may be required for the clonal cell lines to promote successful differentiation, e.g., troglitazone, dexamethasone (DEX). Additions, such as IBMX, insulin, and $T_3$, are also common, as with white adipocyte cell lines.

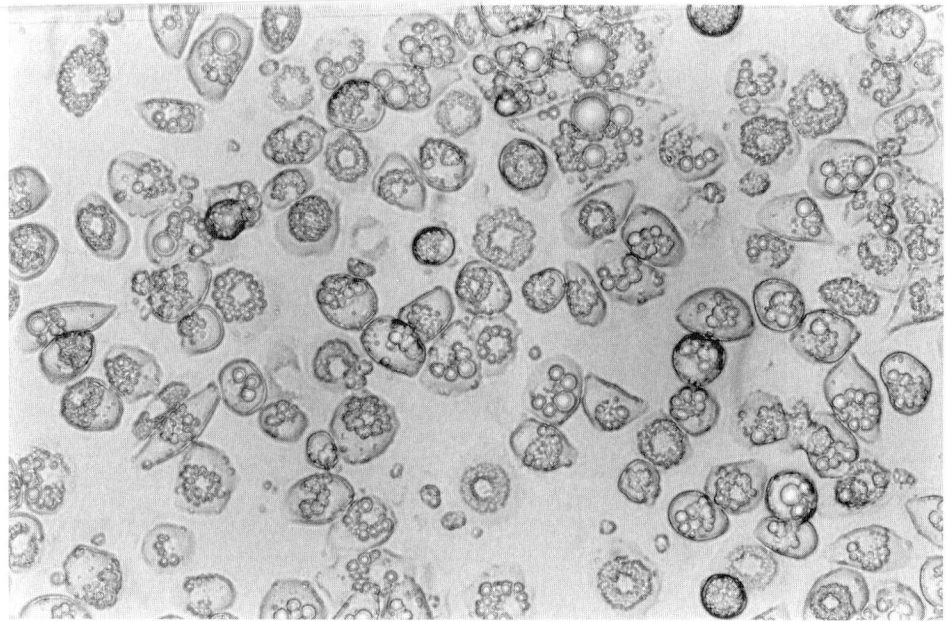

Fig. 2. Differentiated brown adipocytes in culture.

3. The exact concentrations and incubation times may differ between different investigators but the actual concentrations, and so on, are probably not critical.
4. Cell density of cell suspension that is inoculated. The cells may be counted in a Bürker chamber before inoculation. Erythrocytes from the tissue can interfere with cell counting, and can be removed by hypo-osmotic shock (1 mL DMEM + 2 mL H$_2$O for 20 s at 4°C, followed by addition of 5 mL DMEM and rewashing) *(12)*. In routine preparations, we do not find cell counting necessary.
5. Multilocularity. Adipose conversion starts to occur when the cells approach confluence. The development of triglyceride droplets is easily visible. The droplets are always small and multilocular in appearance when they first appear, and this is also true for cultures of white adipocytes (and also for cultures of other cell types that can demonstrate an ability to accumulate lipid without being true adipocytes). This feature can therefore not in itself be used to define brown adipocyte cultures. After prolonged culture, some differences between parallel-cultured white and brown adipocytes can nonetheless be discerned. The brown adipocyte cultures retain a higher degree of multilocularity; the white adipocyte cultures tend toward paucilocularity and unilocularity *(12)*.
6. ECMs have only rarely been tested: Collagen I, collagen IV, laminin, and fibronectin have been tested *(see* **Note 7***) (13)*.
7. Gene expression: There may be difficulties to express UCP1 in some species. DEX is needed for differentiation in cultures of some species. In newborn rat cultures, it has not been easy to obtain a good UCP1 response *(14,15)*, which

probably explains the low use of rats. On collagen matrices (I and IV), there is spontaneous UCP1 expression (norepinephrine [NE] not tested) *(13)*. Three-dimensional collagen I matrix gave both spontaneous expression and good NE stimulation *(13)*.

## Acknowledgments

The authors would like to thank members of the department for valuable discussions.

## References

1. Björntorp, P., Karlsson, M., Pertoft, H., Pettersson, P., Sjöström, L., and Smith, U. (1978) Isolation and characterization of cells from rat adipose tissue developing into adipocytes. *J. Lipid Res.* **19,** 316–324.
2. Hausman, G. J., Campion, D. R., and Martin, R. J. (1980) Search for the adipocyte precursor cell and factors that promote its differentiation. *J. Lipid Res.* **21,** 657–670.
3. Cryer, A. (1985) Biochemical markers of adipocyte precursor differentiation, in *New Perspectives in Adipose Tissue: Structure, Function and Development*, (Cryer, A. and Van, R. L. R., eds.), Butterworth, London, pp. 383–405.
4. Cameron, I. L. and Smith, R. E. (1964) Cytological responses of brown fat tissue in cold-exposed rats. *J. Cell Biol.* **23,** 89–100.
5. Hunt, T. E. and Hunt, E. A. (1967) A radioautographic study of proliferation in brown fat of the rat after exposure to cold. *Anat. Rec.* **157,** 537–546.
6. Bukowiecki, L. J., Géloen, A., and Collet, A. J. (1986) Proliferation and differentiation of brown adipocytes from interstitial cells during cold acclimation. *Am. J. Physiol.* **250,** C880–C887.
7. Rehnmark, S. and Nedergaard, J. (1989) DNA synthesis in mouse brown adipose tissue is under β-adrenergic control. *Exp. Cell Res.* **180,** 574–579.
8. Néchad, M. and Olson, L. (1983) Development of interscapular brown adipose tissue in the hamster. II. Differentiation of transplants in the anterior chamber of the eye: Role of the sympathetic innervation. *Biol. Cellulaire* **48,** 167–174.
9. Cannon, B., Matthias, A., Golozoubova, V., Ohlson, K. B. E., Andersson, U., Jacobsson, A., and Nedergaard, J. (1999) Unifying and distinguishing features of brown and white adipose tissues: UCP1 versus other UCPs, in *Progress in Obesity Research 8*, (Ailhaud, G. and Guy-Grand, B., eds.), John Libbey, London, pp. 13–26.
10. Klaus, S., Choy, L., Champigny, O., Cassard-Doulcier, A.-M., Ross, S., Spiegelman, B., and Ricquier, D. (1994) Characterization of the novel brown adipocyte cell line HIB 1B. Adrenergic pathways involved in regulation of uncoupling protein gene expression. *J. Cell Sci.* **107,** 313–319.
11. Forest, C., Doglio, A., Ricquier, D., and Ailhaud, G. (1987) A preadipocyte clonal line from mouse brown adipose tissue. Short- and long-term responses to insulin and β-adrenergics. *Exp. Cell Res.* **168,** 218–232.
12. Néchad, M., Kuusela, P., Carneheim, C., Björntorp, P., Nedergaard, J., and Cannon, B. (1983) Development of brown fat cells in monolayer culture. I. Morphological and biochemical distinction from white fat cells in culture. *Exp. Cell Res.* **149,** 105–118.

13. Hikichi, Y., Sugihara, H., and Sugimoto, E. (1993) Differentiation of brown adipose cells in three-dimensional collagen cell culture. *Path. Res Pract.* **189**, 73–82.
14. Forest, C., Doglio, A., Casteilla, L., Ricquier, D., and Ailhaud, G. (1987) Expression of the mitochondrial uncoupling protein in brown adipocytes. Absence in brown preadipocytes and BFC-1 cells. Modulation by isoproterenol in adipocytes. *Exp. Cell Res.* **168**, 233–246.
15. Kuusela, P., Rehnmark, S., Jacobsson, A., Cannon, B., and Nedergaard, J. (1997) Adrenergic stimulation of lipoprotein lipase gene expression in rat brown adipocytes differentiated in culture: mediation via $\beta_3$- and $\alpha_1$-adrenergic receptors. *Biochem. J.* **231**, 759–767.
16. Masuno, H., Schultz, C. J., Park, J.-W., Blanchette-Mackie, E. J., Mateo, C., and Scow, R. O. (1991) Glycosylation, activity and secretion of lipoprotein lipase in cultured brown adipocytes of newborn mice. Effect of tunicamycin, monensin, 1-deoxymannojirimycin and swainsonine. *Biochem. J.* **277**, 801–809.
17. Rehnmark, S., Néchad, M., Herron, D., Cannon, B., and Nedergaard, J. (1990) $\alpha$- and $\beta$-Adrenergic induction of the expression of the uncoupling protein thermogenin in brown adipocytes differentiated in culture. *J. Biol. Chem.* **265**, 16,464–16,471.
18. Kopecky, J., Baudysová, M., Zanotti, F., Janíková, D., Pavelka, S., and Houstek, J. (1990) Synthesis of mitochondrial uncoupling protein in brown adipocytes differentiated in cell culture. *J. Biol. Chem.* **265**, 22,204–22,209.
19. Bronnikov, G., Houstek, J., and Nedergaard, J. (1992) $\beta$-Adrenergic, cAMP-mediated stimulation of proliferation of brown fat cells in primary culture. Mediation via $\beta_1$ but not via $\beta_3$ receptors. *J. Biol. Chem.* **267**, 2006–2013.
20. Thonberg, H., Zhang, S.-J., Tvrdik, P., Jacobsson, A., and Nedergaard, J. (1994) Norepinephrine utilizes $\alpha_1$-and $\beta$-adrenoreceptors synergistically to maximally induce c-*fos* expression in brown adipocytes. *J. Biol. Chem.* **269**, 33,179–33,186.
21. Alvarez, R., de Andrés, J., Yubero, P., Vinas, O., Mampel, T., Iglesias, R., Giralt, M., and Villarroya, F. (1995) A novel regulatory pathway of brown fat thermogenesis: Retinoic acid is a transcriptional activator of the mitochondrial uncoupling protein gene. *J. Biol. Chem.* **270**, 5666–5673.
22. Bronnikov, G., Bengtsson, T., Kramarova, L., Golozoubova, V., Cannon, B., and Nedergaard, J. (1999) $\beta_1$ to $\beta_3$ switch in control of cAMP during brown adipocyte development explains distinct $\beta$-adrenoceptor subtype mediation of proliferation and differentiation. *Endocrinology* **140**, 4185–4197.
23. Sugihara, H., Miyabara, S., Yonemits, N., and Ohta, K. (1983) Hormonal sensitivity of brown fat cells of fetal rats in monolayer culture. *Exp. Clin. Endocrinol.* **82**, 309–319.
24. Lorenzo, M., Roncero, C., Fabregat, I., and Benito, M. (1988) Hormonal regulation of rat foetal lipogenesis in brown-adipocyte primary cultures. *Biochem. J.* **251**, 617–620.
25. Teruel, T., Valverde, A. M., Navarro, P., Benito, M., and Lorenzo, M. (1998) Inhibition of PI 3-kinase and RAS blocks IGF-I and insulin-induced uncoupling protein 1 gene expression in brown adipocytes. *J. Cell. Physiol.* **176**, 99–109.

26. Shimizu, Y. and Shimazu, T. (1994) Effects of wortmannin on increased glucose transport by insulin and norepinephrine in primary culture of brown adipocytes. *Biochem. Biophys. Res. Commun.* **202,** 660–665.

27. Nikami, H., Shimizu, Y., Sumida, M., Minokoshi, Y., Yoshida, T., Saito, M., and Shimazu, T. (1996) Expression of β3-adrenoceptor and stimulation of glucose β3-agonists in brown adipocyte primary culture. *J. Biochem.* **119,** 120–125.

28. Champigny, O., Holloway, B. R., and Ricquier, D. (1992) Regulation of UCP gene expression in brown adipocytes differentiated in primary culture. Effects of a new β-adrenoceptor agonist. *Mol. Cell. Endocrinol.* **86,** 73–82.

29. Chaudhry, A. and Granneman, J. G. (1999) Differential regulation of functional responses by β-adrenergic receptor subtypes in brown adipocytes. *Am. J. Physiol.* **277,** R147–R153.

30. Klaus, S., Cassard-Doulcier, A.-M., and Ricquier, D. (1991) Development of *Phodopus sungorus* brown preadipocytes in primary cell culture: Effect of an atypical beta-adrenergic agonist, insulin, and triiodothyronin on differentiation, mitochondrial development, and expression of the uncoupling protein UCP. *J. Cell Biol.* **115,** 1783–1790.

31. Klaus, S., Ely, M., Encke, D., and Heldmaier, G. (1995) Functional assessment of white and brown adipocyte development and energy metabolism in cell culture. Dissociation of terminal differentiation and thermogenesis in brown adipocytes. *J. Cell Sci.* **108,** 3171–3180.

32. Casteilla, L., Nouguès, J., Reyne, Y., and Ricquier, D. (1991) Differentiation of ovine brown adipocyte precursor cells in a chemically defined serum-free medium. Importance of glucocorticoids and age of animals. *Eur. J. Biochem.* **198,** 195–199.

33. Cigolini, M., Cinti, S., Brunetti, L., Bosello, O., Osculati, F., and Björntorp, P. (1985) Human brown adipose cells in culture. *Exp. Cell Res.* **159,** 261–266.

34. Kuusela, P., Herva, R., Hirvonen, J., Nedergaard, J., and Cannon, B. (1985) Differentiation in vitro of preadipocytes originated from human brown fat. *Acta Physiol. Scand.* **124 (Suppl. 542),** 399–399.

35. Abumrad, N. A., Forest, C. C., Regen, D. M., and Sanders, S. (1991) Increase in membrane uptake of long-chain fatty acids early during preadipocyte differentiation. *Proc. Natl. Acad. Sci. USA* **88,** 6008–6012.

36. Sasaki, A., Sivaram, P., and Goldberg, I. J. (1993) Lipoprotein lipase binding to adipocytes: evidence for the presence of a heparin-sensitive binding protein. *Am. J. Physiol.* **265,** E880–E888.

37. Kozak, U. C., Held, W., Kreutter, D., and Kozak, L. P. (1992) Adrenergic regulation of the mitochondrial uncoupling protein gene in brown fat tumor cells. *Mol. Endocrinol.* **6,** 763–772.

38. Kozak, U. C. and Kozak, L. P. (1994) Norepinephrine-dependent selection of brown adipocyte cell lines. *Endocrinology* **134,** 906–913.

39. Rohlfs, E. M., Daniel, K. W., Premont, R. T., Kozak, L. P., and Collins, S. (1995) Regulation of the uncoupling protein gene (*Ucp*) by β1, β2, and β3-adrenergic receptor subtypes in immortalized brown adipose cell lines. *J. Biol. Chem.* **270,** 10,723–10,732.

40. Ross, S. R., Choy, L., Graves, R. A., Fox, N., Solevjeva, V., Klaus, S., Ricquier, D., and Spiegelman, B. M. (1992) Hibernoma formation in transgenic mice and isolation of a brown adipocyte cell line expressing the uncoupling protein gene. *Proc. Natl. Acad. Sci. USA* **89**, 7561–7565.

41. Klaus, S., Champigny, O., Cassard-Doulcier, A.-M., Choy, L., Spiegelman, B., and Ricquier, D. (1994) Cell lines derived from brown fat tumour of transgenic mice: new tools for the study of brown adipocyte development, in *Obesity in Europe 1993*, (Ditschuneit, H., Gries, F. A., Hauner, H., Schusdziarra, V., and Wechsler, J. G., eds.), John Libbey, London, pp. 65–72.

42. Benito, M., Porras, A., and Santos, E. (1993) Establishment of permanent brown adipocyte cell lines achieved by transfection with SV40 large T antigen and *ras* genes. *Exp. Cell Res.* **209**, 248–254.

43. Navarro, P., Valverde, A. M., Benito, M., and Lorenzo, M. (1999) Activated Ha-ras induces apoptosis by association with phosphorylated Bcl-2 in a mitogen-activated protein kinase-independent manner. *J. Biol. Chem.* **274**, 18,857–18,863.

44. Zennaro, M.-C., Le Menuet, D., Viengchareun, S., Walker, F., Ricquier, D., and Lombès, M. (1998) Hibernoma development in transgenic mice identifies brown adipose tissue as a novel target of aldosterone action. *J. Clin. Invest.* **101**, 1254–1260.

45. Irie, Y., Asano, A., Canas, X., Nikami, H., Aizawa, S., and Saito, M. (1999) Immortal brown adipocytes from p53-knockout mice: differentiation and expression of uncoupling proteins. *Biochem. Biophys. Res. Commun.* **255**, 221–225.

# 19

## Cultures of Adipose Precursor Cells and Cells of Clonal Lines from Animal White Adipose Tissue

**Raymond Négrel and Christian Dani**

## 1. Introduction

Preadipocyte cellular models have proven to be useful tools for the study of adipose cell development in vitro *(1–5)*. Among these models are preadipocyte clonal lines from various origins and primary cultures of fibroblast-like adipose precursor cells present in the stromal-vascular fraction of adipose tissue, which can be isolated from several depots and various species, including the human (*see* Chapter 20). The availability of these cellular models has allowed crucial advances with respect to the characterization of the main steps of the differentiation process from the preadipocyte to the adipocyte in (1) delineating the kinetics of expression of adipose specific markers, (2) cloning some of these markers and studying their hormonal control, and (3) giving evidence of new secreted compounds from adipose cells *(6–13)*. In addition, the availability of multipotent or totipotent stem cells *(1,14,15)* gives the opportunity of approaching the events that govern the process of determination of progenitors to the adipose lineage. These stem cell systems are thus expected to allow characterization of still missing key genes, as well as marker(s) of the unipotent adipoblast before its commitment to a preadipocyte.

This chapter focuses on each type of these cellular models. Emphasis is given to growth and differentiation of preadipocyte clonal lines including 3T3-L1, 3T3-F442A, Ob17, and its subclone, Ob1771 *(16–19)*. A simple method to differentiate rat adipose precursor cells in primary culture is also described *(20,21)*. Furthermore, a procedure for growing cells of embryonic stem (ES) cell lines derived from the inner cell mass of 3.5-d blastocysts *(22,23)* in the absence of feeder-layer *(14,15)*, and to induce their differentiation into adipocytes by means of embryoid body (Ebs) formation, is presented *(24)*.

From: *Methods in Molecular Biology, vol. 155: Adipose Tissue Protocols*
Edited by: G. Ailhaud © Humana Press Inc., Totowa, NJ

Whereas induction of adipose cell differentiation in Ebs still requires a serum-containing medium, primary adipose precursor cells, as well as cells of preadipocyte clonal lines (such as Ob17 and its subclone, Ob1771, which have been the most extensively studied), can be grown and induced to differentiate under serum-free culture conditions. Therefore, this chapter describes culture conditions that allow one to induce growth of rodent adipose precursor cells and their subsequent differentiation in serum-containing media, on the one hand, and in serum-free media, on the other hand.

## 2. Materials

### 2.1. Animals

Four- to 8-wk-old male Wistar rats, but younger or older animals of other strains may be used (*see* **Note 1** and **ref. 25**).

### 2.2. Plastic Culture Materials

Plastic culture dishes or multiwell plates (from Corning or Greiner for ES cells), sterile plastic tubes, pipets and syringes, sterile nylon sheets (250-, 80-, 40-, and 25-μm mesh), and plastic funnels for filtration.

### 2.3. Chemicals and Biochemicals of Cell Culture Grade

1. Phosphate-buffered saline (PBS) calcium- and magnesium-free: 0.17 $M$ NaCl, 3.4 m$M$ KCl, 4 m$M$ Na$_2$H PO$_4$, and 2.4 m$M$ KH$_2$PO$_4$, pH 7.4.
2. Trypsin solution 1 for subculture of cells: commercial 0.25% trypsin-EDTA solution, stored frozen in aliquots, and ready for use after thawing (Gibco-BRL, 453000).
3. Trypsin solution 2 for dissociation of ES cells: a commercial 2.5% solution of trypsin in normal saline (Gibco-BRL 250900) is diluted 1:100 in PBS containing 1 m$M$ EDTA and 1% chicken serum.
4. Culture media: Dulbecco's modified Eagle's medium (DMEM) containing 1 g/L glucose and 110 mg/L sodium (Na) pyruvate, Ham's F12 medium, and Glasgow MEM/BHK21.
5. Selected batches of fetal bovine serum (FBS) (*see* **Note 2a,b**).
6. Miscellaneous cell culture products: bovine serum albumin (BSA) fraction V, and collagenase A from *Clostridium histolyticum* are dissolved into culture medium at the time of use. 2% gelatin, stored at 4°C, is diluted to 0.1% with PBS, to prepare an appropriate coating solution. Biotin, pantothenic acid, ascorbic acid, Na selenite, glutamine, methylisobutylxanthine (Mix), human transferrin, bovine insulin (*see* **Note 3** and **refs. 26,27**), basic fibroblast gowth factor (bFGF), growth hormone (GH) (*see* **Note 4**), bovine fetuin (Pedersen preparation, Sigma F2379; *see* **Note 5**), and a rat submaxillary gland extract (SMGE) (*see* **Note 6** and **ref. 28**) are stored as frozen aqueous concentrated stock solutions. Arachidonic acid (AA), carbaprostacyclin (cPGI$_2$), which is a stable analog of prostacyclin (PGI$_2$), dexamethasone (DEX), as well as other glucocorticoid hormones and

triodothyronine ($T_3$), are stored as 100- to 1000-fold concentrated alcoholic stock solutions. Therefore, after dilution into culture media, the concentration of alcohol never exceeds 1%. Leukemia inhibitory factor (LIF) which is commercially available (named Esgro by Gibco), can also be produced in-house (*see* **Note 7**; *29*). *All trans* retinoic acid (RA) is diluted in the dark into dimethyl sulfoxide, to prepare 10 m*M* stock solution stored at –20°C. Subsequent dilutions of RA performed in alcohol are used for one experiment only.

## 2.4. Clonal Lines

3T3-L1 and 3T3-F442A from H. Green, Harvard Medical School, Boston, MA; Ob1771, a subclone of Ob17, from R. Négrel et al., and ES cell lines from American Type Culture Collection, or from C. Dani et al.

## 2.5. Serum-Containing Media for Growth and Differentiation

1. For growth and differentiation of adipose precursor cells (primary cells or clonal lines): DMEM containing 1 g/L glucose and 110 mg/L Na pyruvate, supplemented with 3.7 g/L Na bicarbonate, 33 µ*M* biotin, 17 µ*M* pantothenate, antibiotics (62 mg/L penicillin and 50 mg/L streptomycin or 10 mg/L tetracyclin), and 10% FBS (selected batch; *see* **Note 2**). This serum-supplemented DMEM allows the growing of stock cultures of subconfluent cells, and is also used to grow cells to confluence. To induce terminal differentiation of confluent preadipocytes, this culture medium is usually supplemented with 17 n*M* insulin and 2 n*M* $T_3$, and is changed every other day. The concentration of insulin may also be increased to 170 n*M* (*see* **Note 3**). Furthermore, this culture medium can be enriched with additional adipogenic effectors (*see* **Note 8**; *30–35*).
2. For growth of pluripotent ES cells: Glasgow MEM/BHK21 supplemented with 0.23% Na bicarbonate (from a 7.5% stock solution), nonessential amino acids 1X (from a 100X stock solution), 2 m*M* L-glutamine, 1 m*M* pyruvate, 0.1 m*M* 2-mercaptoethanol and 10% FBS (selected batch; *see* **Note 2**). This medium, referred to as "complete Glasgow MEM," is supplemented with LIF (1000 U/mL Esgro, final concentration), which is required to maintain the pluripotentiality of ES cells. LIF is diluted from a stock solution into this culture medium just before use (*see* **Note 7**).
3. For induction of Ebs to differentiate into adipocytes: complete Glasgow MEM, freshly supplemented with 85 n*M* insulin and 2 n*M* $T_3$. This medium is referred to as "differentiation medium."

## 2.6. Serum-Free Media for Growth of Primary Cells or Clonal Lines (4F Medium)

1. DMEM-F12 complete basal medium: a mixture of DMEM and Ham's F12 medium (1/1; v/v) containing 15 m*M* Na bicarbonate, 15 m*M* HEPES buffer, pH 7.4, 33 µ*M* biotin, 17 µ*M* pantothenate, and antibiotics as above (*see* **Subheading 2.5., item 1**).
2. 4F medium: DMEM-F12 complete basal medium, freshly supplemented with (5 µg/mL, i.e., 850 n*M*) bovine insulin (*see* **Note 3**), 10 mg/mL human transferrin, 10 µg/mL bFGF, and 1–2 µg/mL of SMGE (*see* **Note 6**).

## 2.7. Serum-Free Medium for Differentiation of Rat Adipose Precursor Cells in Primary Culture (ITT medium)

ITT medium: DMEM-F12 complete basal medium is supplemented with (5 µg/mL, i.e., 850 n$M$) bovine insulin (*see* **Note 3**), 10 µg/mL human transferrin, and 200 p$M$ T$_3$ (*see* **Note 9**).

## 2.8. Serum-Free Media for Differentiation of Ob17 and Ob1771 Preadipose Cells

5F medium: DMEM-F12 complete basal medium is supplemented with 20 n$M$ selenite, 100 µ$M$ ascorbate, 5 µg/mL (850 n$M$) bovine insulin (*see* **Note 3**), 10 µg/mL human transferrin, 200 p$M$ T$_3$, 2 n$M$ GH (*see* **Note 4**) and 500 µg/mL bovine fetuin (*see* **Note 5**). This medium allows to maintain Ob17 and Ob1771 cells at the stage of preadipocytes (i.e., expressing early markers of differentiation only) for at least 2 wk (*13,17–19*). 5F medium, supplemented with 200 n$M$ cPGI$_2$ (*18*), induces terminal differentiation of confluent Ob17 or Ob1771 preadipocytes (*see* **Notes 10** and **11**; *36–39*).

## 3. Methods

### 3.1. Growth and Differentiation of Rat Adipose Precursor Cells in Primary Culture

#### 3.1.1. Isolation and Plating of Adipose Precursor Cells from Rat Adipose Tissue

1. Adipose depots (periepididymal, retroperitoneal or subcutaneous) are dissected under sterile conditions, freed as much as possible from blood capillaries, and bathed in serum-free DMEM containing antibiotics (*see* **Subheading 2.5., item 1**).
2. Minced tissue is then digested in this serum-free medium (3 mL/g issue) containing 2 mg/mL collagenase and 20 mg/mL BSA for 40 min at 37°C, under mild controlled agitation.
3. After a final agitation by means of several aspirations in a plastic pipet, cells and tissue remnants are poured, at room temperature (RT) into 3 vol culture medium supplemented with 8% FBS. The cell suspension is filtered through a 250-µm nylon screen.
4. The filtrate is centrifuged at 250$g$ for 8 min, in order to collect a first pellet of stromal-vascular cells.
5. The floating fat cell layer is resuspended in serum-containing medium as above, washed through aspirations in a plastic pipet, and centrifuged as above.
6. The first and second pellets are pooled after resuspension in the same medium, and further filtered through a 25-µm nylon screen (*see* **Note 12**).
7. After centrifugation as above, the pellet is resuspended in a small volume (1–5 mL) of 10% FBS-containing DMEM, and aliquots of the cell suspension are counted either with a Coulter counter or by using a cell-counting chamber (*see* **Notes 13–15**).

8. Isolated stromal-vascular cells suspended in 10% FBS- containing DMEM are usually plated at a density of $3.10^4$ cells/cm$^2$, either in 35-mm culture dishes or in 24-multiwell culture plates, and maintained at 37°C in a humidified atmosphere of 5% $CO_2$ in air for 16–20 h.

### 3.1.2. Culture of Primary Adipose Precursor Cells

Stromal-vascular cells, plated as indicated above, are thoroughly, but carefully washed twice, for at least 1 h, with complete basal DMEM-F12 medium, and fed with the same medium enriched with ITT medium. This culture medium is renewed every 3 d (*see* **Notes 16** and **17**; *40,41*).

## 3.2. Culture of Cells from Preadipocyte Cell Lines in Serum-Containing Media

1. Cells of 3T3-L1, 3T3-F442A, Ob17, or Ob1771 clonal lines are grown in biotin- and pantothenate-enriched DMEM, supplemented with 10% FBS (*see* **Subheading 2.5., item 1**). The inoculation density of freshly dissociated subconfluent cells, using trypsin solution 1 (see **Note 18** and **Subheading 2.3., item 2**), is usually $5 \times 10^3$ cells/cm$^2$ for stock culture maintenance in 100 mm culture dishes, and $8.5 \times 10^3$ cells/cm$^2$ to inoculate experimental culture dishes or wells. Cells are maintained at 37°C in a 5% $CO_2$ humidified atmosphere, and the culture medium is changed every other day. Confluence is reached within 4–5 d. At this stage, growth-arrested preadipocytes are able to express early markers of differentiation only, such as lipoprotein lipase and A2COL6/pOb 24 (*6,8,13*).
2. To induce confluent preadipocytes to terminally differentiate, this culture medium is usually supplemented with 17 n$M$ insulin and 2 n$M$ $T_3$, and changed every other day, but other agents can be added concomitantly to insulin and $T_3$ during the first 2–3 d, to accelerate and/or amplify the differentiation process (*see* **Note 8**). The concentration of insulin may also be increased to 170 n$M$ (*see* **Note 3**).

## 3.3. Culture of Cells from Preadipocyte Clonal Lines in Serum-Free Culture Media

1. Trypsinized subconfluent cells, suspended in 10% FBS-supplemented medium, are seeded at a cell density of $8.5 \times 10^3$ cells/cm$^2$ (usually in 24-multiwell tissue culture plates; 2 cm$^2$/well), and maintained at 37°C in a humidified atmosphere of 5% of $CO_2$ in air for 16–20 h.
2. Cells are then thoroughly washed twice with DMEM-F12 complete basal medium (*see* **Subheading 2.6., item 1**).
3. Feed cells with 4F medium (*see* **Subheading 2.6., item 2**), which allows them to reach confluence within 3–4 d.
4. After washing twice with DMEM-F12 complete basal medium, confluent cells are fed with 5F medium (*see* **Subheading 2.8.**), supplemented with 200 n$M$ cPGI$_2$, to induce maximal adipose differentiation (*see* **Notes 8–10**).
5. Perform medium change at d 3 and 7 with fresh prewarmed 5F medium enriched with cPGI$_2$. Differentiated cells may be used for experiments at d 10, or later.

### 3.4. Routine Culture of Pluripotent ES Cells and Differentiation of Ebs

#### 3.4.1. Routine Culture of ES Cells

CGR8 or E14TG2a ES cells, which are feeder-layer independent ES cells, can be grown and maintained in a multipotent, undifferentiated state, providing they are exposed to LIF (*see* **Note 7**).

1. 25 cm$^2$ flasks previously coated with gelatin (*see* **Note 19**) are used to inoculate $10^6$ ES cells kept in 10 mL prewarmed complete Glasgow MEM medium containing 1000 U/mL LIF as a single cell suspension (*see* **Note 20**), after trypsin dissociation using trypsin solution 2 (*see* **Subheading 2.3., item 3**).
2. Change medium every day.
3. Trypsinize the cultures 2–3 d later (*see* **Note 18**): Aspirate the medium, and wash twice with 3–5 mL PBS. Aspirate off the PBS, and add 1 mL trypsin solution 2. Ensure that the trypsin covers the cell monolayer, and incubate at 37°C, in 5% $CO_2$, for 2–5 min (*see* **Note 20**).
4. Under an inverted microscope, check that cells are correctly dissociated. Add 5 mL complete medium, and suspend the cells by vigorous pipeting. Transfer the cells to a sterile tube, and centrifuge at 250g for 5 min at RT.
5. Resuspend the cell pellet with 5 mL complete medium by pipeting up and down 2–3×. Count cells.
6. Seed cells as in **step 1**, to maintain a culture of pluripotent stem cells.

#### 3.4.2. Differentiation of Ebs into Adipocytes

Multilineage differentiation of ES cells is initiated by aggregation. The aggregates Ebs. To induce adipocyte lineage, the hanging-drop method for the formation of Ebs is used.

1. Feed ES cells with complete medium 2 h before subculture.
2. Trypsinize as described in **Subheading 3.4.1., step 3**.
3. Resuspend the cell pellet into 10 mL complete Glasgow MEM without LIF, but supplemented with penicillin and streptomycin as in **Subheading 2.5., item 1** and, after cell-counting, adjust the suspension at a concentration of $5 \times 10^4$ cells/mL.
4. Place 20-µL aliquots of this suspension (containing 1000 cells) onto the lids of bacteriological grade dishes. We use a multipipetor with sterile combitip dispenser (Eppendorf). Approximately 80 drops can be easily fitted on the lid of a 90-mm Petri dish. This is defined as d 0 of Eb formation.
5. Invert the lid. and place it over the bottom of a bacteriological Petri dish filled with 8 mL PBS, which is essential to prevent evaporation from the hanging drops. Be sure that the bottom of the dish is covered by the liquid. If not, add a few drops of serum in PBS. When the lid is inverted, each drop hangs and the cells fall to the bottom of the drop, where they aggregate into a single clump.
6. Two d later, remove the lid, invert it, and collect drops containing Ebs in a conical sterile tube. Leave 5 min at RT, to allow aggregates to sediment. Aspirate the supernatant, and resuspend the pellet in 4 mL complete Glasgow MEM, supple-

mented with $10^{-6}$ *M* RA (*see* **Note 21**). Transfer the suspension into 60-mm bacteriological-grade Petri dishes (*see* **Note 22**).

7. Incubate for 3 d in the presence of RA with change of medium every day.

8. At d 5 after Eb formation, remove RA from fresh complete Glasgow MEM, and leave Ebs in suspension in this medium for 2 more d (*see* **Note 23**).

9. At d 7, plate 1–4 Ebs/well in 24-well plates (*see* **Note 24**) containing 1 mL complete medium.

10. The day after plating, change the complete medium for differentiation medium (*see* **Note 25**). Change medium every other day.

11. After 10–20 d in the differentiation medium, 50–70% of Eb outgrowths should contain adipocyte colonies (*see* **Note 26**).

## 4. Notes

1. Consistent with the fact that a pool of dormant preadipocytes subsists all life long, but decreases with age, both in rodents and humans *(25)*, the extent of differentiation may be lower when using old animals.

2. FBS samples are tested for their ability to support growth and differentiation of Ob1771 preadipocytes, as well as growth and differentiation of ES cells before selection.

   a. In the case of Ob1771 cells, the criterion of selection is a good terminal differentiation under both serum-supplemented (*see* **Subheading 3.2.**) and serum-free culture conditions (*see* **Subheading 3.3.**), after three passages in each batch of serum tested.

   b. ES cell cultures may contain a proportion of differentiated cells that have lost their pluripotency. It is crucial to minimize the proportion of differentiated cells. This is achieved by addition of an appropriate concentration of LIF in a high-quality culture medium, i.e., an adequate batch of serum. Identification of pluripotent stem cells is difficult, unless one is familiar with the appropriate cellular morphology. Pluripotent stem cells are small, and have a large nucleus containing prominent nucleoli structures and minimal cytoplasm. Pluripotent stem cells, compared to differentiated cells, grow rapidly and divide approximately every 10–15 h. The authors' experience in selecting a batch of serum able to support growth of stem cells is to plate $10^6$ cells/25 cm$^2$ flask in 10% of each set of FBS, and to subculture cells every 2 d for four passages. For a high-quality serum, a flask should yield $5$–$10 \times 10^6$ cells at each passage. Furthermore, no toxicity of the selected serum should be observed at a 30% concentration. The batch of serum selected for differentiation of Ob1771 cells is used for differentiation of ES cells.

3. Insulin (as well as insulin-like growth factor-1 [IGF-1]) is prepared as 1-mg/mL small aliquots in *N*/100 HCl, stored at –20°C. After thawing, they can be stored at 4°C for at least 2 wk. In the presence of IGF-1 (i.e., in bovine-serum-supplemented media), insulin acts as an hormone in the physiological range of concentration: 1.7 to 17 n*M* are thus concentrations able to positively modulate adipose cell differentiation. Nevertheless, insulin added at higher concentrations

is able to bind to the IGF-1 receptor, and to mimic IGF-1, which behaves as an essential effector of adipose cell differentiation *(26,27)*. Consistent with this observation, 850 n*M* insulin can be replaced by 1–10 n*M* IGF-1 in serum-free media *(20,26)*.

4. Purified bovine or human GH, as well as recombinant human GH can be used. The stock solution is prepared as follows: 1 mg/mL solution in 500 m*M* Tris-HCl buffer, pH 8.5, is sterilized by using a low protein adsorption hydrophilic durapore filter (Millipore), diluted 1:10 into sterile water, and stored at –20°C. After thawing, an aliquot can be stored at 4°C for several weeks.

5. 50 mg/mL stock solution of bovine fetuin is prepared in 50 m*M* phosphate buffer, pH 7.0, sterilized by filtration through 0.45 μm cellulose acetate, and stored frozen as small aliquots. After thawing, aliquots are never frozen again, but can be stored at 4°C for at least a month.

6. A partially purified mitogenic kallikrein *(28)*, present in an extract prepared from frozen female rat submaxillary glands (purchased from Pel-Freez, AR), is suitable. After homogenization of minced glands in 3 vol cold water, by using a Sorval blender for 2 min, the cytoplasmic fraction is precipitated with ammonium sulfate at 60% saturation. After centrifugation, the protein pellet is solubilized in 20 m*M* Tris-HCl buffer, pH 7.5, extensively dialyzed against the same buffer, cleared by centrifugation, and lyophilized. After solubilization and adjustment to a protein concentration of 1–2 mg/mL, the solution is sterilized by filtration, and aliquots are stored frozen.

7. LIF is required to maintain pluripotent ES cells, and is omitted to induce commitment of ES cells toward the adipogenic lineage. To produce LIF, Cos7 cells are transiently transfected with a LIF-expressing construct, using standard techniques, 4 d after the medium is saved. A titration of LIF activity in the medium conditioned by transfected Cos7 cells is then required. All these steps are described in detail in **ref. 29**.

8. Adipose cell differentiation under these conditions is not an uniform process. Differentiation occurs in clusters of variable sizes. The formation of clusters of lipid-filled cells is visible within 5–10 d, and maximal differentiation occurs within the following 10 d, according to the clonal line. Other agents can be added concomitantly to insulin and $T_3$ during the first 2–3 d, to accelerate and/or amplify the differentiation process (i.e., leading to an increase in both the number and size of cell clusters). The most popular additive is a cocktail of 10–250 n*M* DEX and 100–500 μ*M* MIX *(30)*. Among other adipogenic effectors are AA *(17)*, acting via one of its major metabolites, prostacyclin or PGI2, which has been shown to be an autocrine link between preadipocytes *(31)*, and a paracrine link between adipocytes and preadipocytes, upon hormonal stimulation *(32)*; cPGI$_2$ *(18)*, a stable analog of PGI$_2$; glucocorticoids (such as DEX), which have been described as stimulating AA metabolism in Ob1771 preadipocytes *(19)*; and peroxisome proliferator activated receptor γ ligands, such as BRL 49653 *(33)*. It is interesting to note that cAMP-elevating agents (MIX or forskolin) potentiate the adipogenic effect of the previously mentioned effectors, consistent with the fact that cAMP appears as a critical intracellular signal for inducing early events of terminal differentiation of adipose cells *(34,35)*.

9. ITT medium has been the first chemically defined medium described to also allow human adipose precursor cells to differentiate *(21)*. This serum-free medium has been later improved *(25)*, as presented in Chapter 20.

10. 5F medium may be enriched with 10 μ*M* AA *(17)* or 10 n*M* DEX *(19)*, to induce terminal differentiation. Whereas differentiation proceeds in cell clusters within 12 d upon these additions, it proceeds uniformly (affecting almost all of the cells) when 5F medium is enriched with 200 n*M* cPGI$_2$ *(18)*.

11. Other serum-free media derived from the present one have been used to induce differentiation of 3T3-L1 and 3T3-F442A preadipocytes *(36–39)*.

12. Previous filtrations through 80- and/or 40-μm nylon screens may facilitate this step, but are not usually required.

13. Staining of nuclei may be valuable when using a cell-counting chamber. For this purpose, a small aliquot of the cell suspension is diluted in the same volume of a solution of 0.02% (w/v) crystal violet in PBS.

14. The cell yield for a 6-wk-old rat is around 2 g epididymal adipose tissue/animal, giving around $2 \times 10^6$ cells for plating.

15. Contaminating erythrocytes may be eliminated by careful washing of stromal-vascular cells after attachment. Nevertheless, treatment of the cell suspension, for 10 min with a lysis buffer at RT may be used, despite some cell loss. The composition of the lysis buffer is as follows: 0.154 *M* NH$_4$Cl, 10 m*M* KHCO$_3$, and 0.1 m*M* EDTA, buffered with Tris-HCl at pH 7.65.

16. A high proportion of the cell population (40–95%) will differentiate within 7 d in ITT medium, as can be assessed by the cytoplasmic lipid accumulation, which can be easily observed under a microscope. Triglycerides accumulated in the cytoplasm can be stained with oil red O as follows: Mix 6 vol of an oil red O stock saturated solution (0.5 g in 100 mL isopropanol) with 4 vol distilled water. Let stand for 15 min and filter through paper to obtain the oil red O staining solution. Cultured cells, previously washed with PBS and fixed with 10% formaldehyde for 10 min, are washed with water and covered with the oil red O staining solution during 10–20 min. Aspirate, and wash with water.

    A marker of terminal differentiation, such as glycerol-3-phosphate dehydrogenase, indicative of the number of differentiated cells, can also be assayed by using a standard spectrophotometic assay on cytoplasmic extracts, as described by Wise and Green *(40)*.

17. ITT medium can be supplemented with other adipogenic agents such as MIX and/or DEX or cPGI$_2$ *(41)*, as mentioned in **Notes 8** and **9**.

18. All cultures should be subcultured before cells have reached confluence.

19. Attachment of ES cells to the substratum is susceptible to change according to the tissue culture material. Corning or Greiner tissue culture flasks and dishes are suitable. Coating is performed by covering the surface of a 25-cm$^2$ flask by 5 mL 0.1% gelatin for 15 min at RT followed by careful aspiration of the gelatin solution.

20. It is critical to produce a single cell suspension for subcultures. This is achieved by knocking the flask several times, to ensure complete dissociation during the trypsin treatment.

21. The concentration of RA that gives the best result is $10^{-6}$ $M$ RA. However, because of the high instability of RA, the concentration of RA able to commit ES cells into the adipogenic lineage should be determined for each new preparation ($10^{-8} - 10^{-6}$ $M$ should thus be tested).

22. Bacterial-grade Petri dishes are used to prevent cell attachment to the substrate, and plastics should be tested for their capacity to maintain cells in suspension.

23. Five-d-old Ebs have a tendency to attach to the bottom of the plastic dish. This phenomenon can be reduced by frequent but gentle pipeting to keep Ebs in suspension. Ebs that are firmly attached should be eliminated, because such Ebs seem to have no adipogenic capacity.

24. Dishes can be coated with 0.1% gelatin to increase attachment of Ebs to the substratum. A higher density of Ebs per plate can lead to a decrease of the number of Eb containing adipocyte colonies.

25. Serum included in the differentiation medium is preselected to give terminal differentiation of preadipocytes from clonal cell lines (such as 3T3-F442A or Ob1771). Addition of 1 $\mu M$ thiazolidinedione BRL49653 (SmithKline Beecham) stimulates terminal differentiation of RA-treated Ebs.

26. A wide variety of differentiated derivates, such as neuron-like cells, fibroblast-like cells, and unidentified cell types, grow up over this period. Spontaneously beating cardiomyocytes should appear 1–5 d after plating of control culture (untreated with RA). At least 40% of Ebs should contain beating cardiomyocytes. In contrast, few Ebs should contain beating cells from RA-treated cultures. Large adipocyte colonies appear late in the RA-treated culture.

## References

1. Ailhaud, G. and Hauner, H. (1997) Developement of white adipose tissue, in *Handbook of Obesity* (Bray, G., Bouchard, C., and James, P. T., eds.), M. Dekker, New York, pp. 359–378.

2. Green, H. and Kehinde, O. (1975) An established preadipose cell line and its differentiation in culture. II. Factors affecting the adipose conversion. *Cell* **5**, 19–27.

3. Green, H. and Kehinde, O. (1976) Spontaneous heritable changes leading to increased adipose conversion in 3T3 cells. *Cell* **7**, 105–113.

4. Négrel, R., Grimaldi, P., and Ailhaud, G. (1978) Establishment of preadipocyte clonal line from epididymal fat pad of ob/ob mouse that responds to insulin and to lipolytic hormones. *Proc. Natl. Acad. Sci. USA* **75**, 6054–6058.

5. Négrel, R., Grimaldi, P., Forest, C., and Ailhaud, G. (1985) Establishment and characterization of fibroblast-like cell lines derived from adipocytes with the capacity to redifferentiate into adipocyte-like cells. *Methods Enzymol.* **109**, 377–385.

6. Spiegelman, B. M., Frank, M., and Green, H. (1983) Molecular cloning of mRNA from 3T3 adipocytes. Regulation of mRNA content for glycerophosphate dehydrogenase and other differentiation-dependent proteins during adipocyte development. *J. Biol. Chem.* **25**, 10,083–10,089.

7. Rosen, B. S., Cook, K. S., Yaglom, J., Groves, D. L., Volanakis, J. E., Damm, D., White, T., and Spiegelman, B. M. (1989) Adipsin and complement factor D activity: an immune-related defect in obesity. *Science* **244**, 1483–1487.

8. Dani, C., Doglio, A., Amri, E., Bardon, S., Fort, P., Bertrand, B., Grimaldi, P., and Ailhaud, G. (1989) Cloning and regulation of a mRNA specifically expressed in the preadipose state. *J. Biol. Chem.* **264,** 10,119–10,125.

9. Amri, E. Z., Bonino, F., Ailhaud, G., Abumrad, N., and Grimaldi, P. A. (1995) Cloning of a protein that mediates transcriptional effects of fatty acids in preadipocytes. Homology to peroxisome proliferator-activated receptors. *J. Biol. Chem.* **270,** 2367–2371.

10. Abumrad, N., Raafat-El-Maghrabi, M., Amri, E., Lopez, E., and Grimaldi, P. A. (1993) Cloning of a rat adipocyte membrane protein implicated in binding or transport of long-chain fatty acids that is induced during preadipocyte differentiation. *J. Biol. Chem.* **268,** 17,665–17,668.

11. Grégoire, F. M., Smas, C. M., and Sul, H. S. (1998) Understanding adipocyte differentiation. *Physiol. Rev.* **78,** 783–809.

12. Négrel, R. (1999) Paracrine/autocrine signals and adipogenesis, in *Progress in Obesity Research: 8* (Guy-Grand, B. and Ailhaud, G., eds.), John Libbey, London, UK, pp. 55–63.

13. Amri, E., Dani, C., Doglio, A., Grimaldi, P., and Ailhaud, G. (1986) Coupling of growth arrest and expression of early markers during adipose conversion of preadipocyte cell lines. *Biochem. Biophys. Res. Commun.* **137,** 903–910.

14. Mountford, P., Zevnik, B., Duwel, A., Nichols, J., Li, M., Dani, C., Robertson, M., Chambers, I., and Smith, A. (1994) Dicistronic targeting constructs: reporters and modifiers of mammalian gene expression. *Proc. Natl. Acad. Sci. USA* **91,** 4303–4307.

15. Hooper, M., Hardy, K., Handyside, A., Hunter, S., and Monk, M. (1987) HPRT-deficient (Lesch-Nyhan) mouse embryos derived from germline colonization by cultured cells. *Nature* **326,** 292–295.

16. Gaillard, D., Négrel, R., Serrero-Davé, G., Cermolacce, C., and Ailhaud, G. (1984) Growth of preadipocyte cell lines and cell strains from rodents in serum-free hormone-supplemented medium. *In Vitro* **20,** 79–88.

17. Gaillard, D., Négrel, R., Lagarde, M., and Ailhaud, G. (1989) Requirement and role of arachidonic acid in the differentiation of preadipose cells. *Biochem. J.* **257,** 389–397.

18. Gaillard, D., Négrel, R., and Ailhaud, G. (1989) Prostacyclin as a potent effector of adipose cell differentiation. *Biochem. J.* **257,** 399–405.

19. Gaillard, D., Wabitsch, M., Pipy, B., and Négrel, R. (1991) Control of terminal differentiation of adipose precursor cells by glucocorticoids. *J. Lipid Res.* **32,** 569–579.

20. Deslex, S., Négrel, R., and Ailhaud, G. (1987) Development of a chemically defined serum-free medium for complete differentiation of rat adipose precursor cells. *Exp. Cell Res.* 168, 15–30.

21. Deslex, S., Négrel, R., Vannier, C., Etienne, J., and Ailhaud, G. (1986) Differentiation of human adipocyte precursors in a chemically defined serum-free medium. *Int. J. Obes.* **10,** 19–27.

22. Evans, M. J. and Kaufman, M. H. (1981) Establishment in culture of pluripotential cells from mouse embryos. *Nature* **292,** 154–156.

23. Martin, G. R. (1981) Isolation of a pluripotent cell line from early mouse embryos cultured in medium conditioned by teratocarcinoma stem cells. *Proc. Natl. Acad. Sci. USA* **78,** 7634–7638.

24. Dani, C., Smith, A. G., Dessolin, S., Leroy, P., Staccini, L., Villageois, P., Darimont, C., and Ailhaud, G. (1997) Differentiation of embryonic stem cells into adipocytes *in vitro*. *J. Cell. Sci.* **110**, 1279–1285.

25. Hauner, H., Entenmann, G., Wabitsch, M., Gaillard, D., Ailhaud, G., Négrel, R., and Pfeiffer, E. (1989) Promoting effect of glucocorticoids on the differentiation of human adipocyte precursor cells cultured in a chemically defined serum. *J. Clin. Invest.* **84**, 1663–1670.

26. Catalioto, R. M., Gaillard, D., Ailhaud, G., and Négrel, R. (1992) Terminal differentiation of mouse preadipocyte cells: the mitogenic-adipogenic role of growth hormone is mediated by the protein kinase C signalling pathway. *Growth Factors* **6**, 255–264.

27. Smith, P. J., Wise, L. S., Berkowitz, R., Wan, C., and Rubin, C. S. (1988) Insulin-like growth factor-I is an essential regulator of the differentiation of 3T3-L1 adipocytes. *J. Biol. Chem.* **263**, 9402–9408.

28. Catalioto, R. M., Négrel, R., Gaillard, D., and Aihaud, G. (1987) Growth-promoting activity in serum-free medium of kallikreinlike arginylesteropeptidases from rat submaxillary gland. *J. Cell. Physiol.* **130**, 352–360.

29. Smith, A. G. (1991) Culture and differentiation of embryonic stem cells. *J. Tiss. Cult. Meth.* **13**, 89–94.

30. Rubin C. S., Hirsch, A., Fung, C., and Rosen, O. M. (1978) Development of hormone receptors and hormonal responsiveness *in vitro*: insulin receptors and hormonal sensitivity in the preadipocyte and adipocyte forms of 3T3-L1 cells. *J. Biol. Chem.* **253**, 7570–7578.

31. Catalioto, R. M., Gaillard, D., Maclouf, J., Ailhaud, G., and Négrel, R. (1991) Autocrine control of adipose cell differentiation by prostacyclin and $PGF_{2a}$. *Biochim. Biophys. Acta* **1091**, 364–369.

32. Darimont, C., Vassaux, G., Ailhaud, G., and Négrel, R. (1994) Differentiation of preadipose cells: paracrine role of prostacyclin upon stimulation of adipose cells by angiotensin II. *Endocrinology* **135**, 2030–2036.

33. Lehmann, J. M., Moore, L. B., Smith-Oliver, T. A., Wilkison, W. O., Willson, T. M., and Kliewer, S. A. (1995) An antidiabetic thiazolidinedione is a high affinity ligand for peroxisome proliferator-activated receptor gamma (PPAR gamma). *J. Biol. Chem.* **270**, 12,953–12,956.

34 . Vassaux, G., Gaillard, D., Darimont, C., Ailhaud, G., and Négrel, R. (1992) Differential response of preadipocytes and adipocytes to PGI2 and PGE2: physiological implications. *Endocrinology* **131**, 2393–2398.

35. Mandrup, S. and Lane, M. D. (1997) Regulating adipogenesis. *J. Biol. Chem.* **272**, 5367–5370.

36. Hauner, H. (1990) Complete adipose differentiation of 3T3-L1 cells in a chemically defined medium: comparison to serum-containing culture conditions. *Endocrinology* **127**, 865–872.

37. Guller, S., Corin, R. E., Mynarcik, D. C., London, B. M., and Sonenberg, M. (1988) Role of insulin in growth hormone-stimulated 3T3 cell adipogenesis. *Endocrinology* **122**, 2084–2089.

38. Schmidt, W., Poll-Jordan, G., and Löffler, G. (1990) Adipose conversion of 3T3-L1 cells in a serum-free culture system depends on epidermal growth factor, insulin-like growth factor I, corticosterone, and cyclic AMP. *J. Biol. Chem.* **265,** 15,489–15,495.

39. Bachmeier, M. and Löffler, G. (1995) Influence of growth factors on growth and differentiation of 3T3-L1 preadipocytes in serum-free conditions. *Eur. J. Cell. Biol.* **68,** 323–329.

40. Wise, L. S. and Green, H. (1979) Participation of one isozyme of cytosolic glycerophosphate dehydrogenase in the adipose conversion of 3T3 cells. *J. Biol. Chem.* **254,** 273–275.

41. Vassaux, G., Négrel, R., Ailhaud, G., and Gaillard, D. (1994) Proliferation and differentiation of rat adipose precursor cells in chemically defined medium: differential action of anti-adipogenic cells. *J. Cell. Physiol.* **161,** 249–256.

# 20

## Cultures of Human Adipose Precursor Cells

### Hans Hauner, Thomas Skurk, and Martin Wabitsch

### 1. Introduction

Recently, there has been rapidly growing interest in cell culture models that allow the study of the adipose differentiation process in vitro, as well as the long-term regulation of fat cell metabolism in human adipose tissue (AT) material. Although valuable clonal preadipocyte cell lines of rodent origin have been available for more than 20 yr, it became more and more obvious from subsequent studies that substantial differences exist in the developmental stage and regulation of differentiation, as well as in specific adipocyte functions, between rodent and human AT (*1–3*). This knowledge makes it necessary to use human cell culture models, if specific questions concerning human AT metabolism and the mechanisms that may lead to either hypertrophic or hyperplastic growth in humans are investigated. Another aspect that is currently attracting considerable attention is the secretory function of adipocyte precursors and fat cells. Numerous factors are released from these cells, and maintain an intense crosstalk with distant organs, or act at the local level (*4*).

Therefore, techniques that allow stable cultivation of human adipocyte precursor cells may provide a valuable tool for extending understanding of the mechanisms that lead to obesity and the known alterations in fat cell metabolism under defined conditions. The possibility to culture human adipocyte precursor cells and in vitro differentiated adipocytes may also open new perspectives for developing targeted therapeutic interventions to prevent further AT growth (*5*).

Important progress in the culture of human adipocyte precursor cells was achieved with replacing the former serum-containing media by chemically defined serum-free media and by better characterization of the hormonal requirements. It was revealed in these studies (*1,6*) that the presence of serum almost completely prevents adipose differentiation in the human system, in

From: *Methods in Molecular Biology, vol. 155: Adipose Tissue Protocols*
Edited by: G. Ailhaud © Humana Press Inc., Totowa, NJ

contrast to the experience in cell lines or in cultured rodent adipose precursor cells *(6)*. This chapter describes in detail an improved technique that allows the study of human adipocyte precursor cells in primary culture.

## 2. Materials

Enzymes, media, tissue culture plasticware, and supplements may be purchased from any company supplying such products. Only a reagent-grade for tissue culture should be used. All material, including solutions that are used for cell culture must be handled under sterile conditions. Prior to use, all solutions are filtrated through filters with a pore size of 0.2-µm, to exclude bacterial contamination (e.g., by using Millipore material). The following solutions are required for the isolation, culture, and differentiation of the stromal cell fraction from human AT.

1. Basal medium: Dulbecco's modified Eagle's medium (DMEM)/Ham's F12 medium (50:50, v:v), supplemented with 15 m$M$ HEPES, 15 m$M$ NaHCO$_3$, 33 µ$M$ biotin, 17 µ$M$ D-pantothenate, at pH 7.4, and is used for tissue transportation from the surgery room to the laboratory.

2. Collagenase solution: Crude collagenase (e.g., Worthington CLS type 1, specific activity 172 U/mg) at a final concentration of approx 200 U/mL, pH 7.4, is dissolved in phosphate-buffered saline (PBS): 10 m$M$ KH$_2$PO$_4$, 10 m$M$ Na$_2$HPO$_4$, 2.7 m$M$ KCl, 0.137 $M$ NaCl, pH 7.4, supplemented with 2% bovine serum albumin (fraction V, according to Cohn). The solution should be freshly prepared for every preparation. However, the collagenase solution can be also stored at –20 C for a few days, with only a minor loss of enzyme activity. For use, the frozen solution is thawed at room temperature and is slowly prewarmed to 37°C. However, repeated thaw–freeze cycles must be avoided.

3. Erythrocyte lysing buffer: 155 m$M$ NH$_4$Cl, 5.7 m$M$ K$_2$HPO$_4$, 0.1 m$M$ EDTA at pH 7.3. This buffer serves to remove red blood cells, which represent the major contaminating cellular component, and may reduce or prevent adhesion of the stromal cells.

4. Inoculation medium: Basal medium supplemented with gentamicin (50 µg/mL) and 10% fetal calf serum (FCS). The FCS is used to improve cell attachment, and to promote spreading.

5. Adipogenic medium: Basal medium supplemented with 10 µg/mL human transferrin (stock solution 1 mg/mL H$_2$O), 50 µg/mL gentamicin (stock solution 10 mg/mL), 1 n$M$ triiodo-$_L$-thyronine (1 m$M$ solution is alcalized with 1 $M$ NaOH, and diluted to 2 µ$M$ stock solution in EtOH), 66 n$M$ human insulin (stock solution 22 µ$M$ in 10 m$M$ HCl), 100 n$M$ hydrocortisone (stock solution 0.1 m$M$ in 50% EtOH).

6. Adipogenic factors: To promote adipose differentiation, the following reagents may be added to the adipogenic medium: 0.2–0.5 m$M$ isobutyl-methylxanthine (IBMX), a nonselective phosphodiesterase inhibitor (stock solution 20 m$M$, alcalized with Na$_2$CO$_3$) for the initial 3 d or 1 µg/mL troglitazone, a thiazolidinedione (stock solution 1 mg/mL) for the initial 3 d. Thiazolidinediones are

activators of the nuclear receptor peroxisome proliferator activated receptor γ (PPARγ), which is a well-characterized master regulator of adipose differentiation *(7)*. Members of this class include troglitazone, pioglitazone, ciglitazone, and rosiglitazone among others. The authors frequently use troglitazone, which is dissolved in dimethyl sulfoxide (DMSO). The stock solution must be stored protected from light.

7. Oil red O staining: Oil red O is a specific neutral lipid marker. 0.5 g of this dye is dissolved in 100 mL 99% isopropanol giving a 0.5% solution. Six mL of this stock solution is mixed with 4 mL $H_2O$, resulting in a 0.3% solution ready for use.

## 3. Methods
### 3.1. Tissue Collection

1. AT specimens are usually obtained from elective surgery. Strictly sterile conditions are required, and it is recommended to collect the tissue in the surgery room. Reconstructive procedures, such as mastectomy or surgical removal of abdominal fat, are particularly suitable for tissue sampling. In studies of visceral AT, tissue collection from the intra-abdominal fat depot during abdominal surgery is needed. Because the intra-abdominal AT depots are more densely vascularized, careful electrocauterization, to avoid bleeding, is recommended. It is also possible to take tissue samples during laparoscopic interventions. However, in this case, the amount of material is usually limited. The differentiation rate depends on the age of the donor, so it is recommended to use tissue samples from donors at ages between 20 and 50 yr. The differentiation capacity is significantly higher in cultures from younger subjects, compared to older people, but the highest rates are observed in samples from children. To consider possible effects of obesity and confounding by concomitant diseases, careful documentation of basal patient characteristics is advisable.

   Another potential source of AT for primary cultures of human adipocyte precursor cells is the use of biopsy material. Between 1 and 2 g AT cylinders can be obtained by repeated aspiration using a 16-gauge steel needle *(5)*. The needle is fixed on a 20-mL sterile syringe filled with 5–10 mL 0.154 *M* NaCl and 1 mL 0.2% EDTA. However, this technique has its limitations, i.e., the amount of tissue is small, there is a high contamination by blood cells, which also reduces the yield of adipose precursor cells, and only the subcutaneous depots are accessible (*see* **Note 1**).

   For ethical reasons, informed consent must be obtained from the patient prior to tissue sampling.

2. The samples are crudely prepared in the surgical room, with surgical scissors, to remove skin and other non-AT material, such as gland tissue. The AT pieces are immediately transported to the laboratory in basal medium. Because the time of tissue collection is difficult to schedule, and sometimes may be late in the day, the samples may be stored overnight in a refrigerator at +4°C. Comparative studies (unpublished observation) have shown that there is no significant difference in adipose differentiation capacity of stromal cells isolated either immediately or after overnight storage in a refrigerator.

## 3.2. Cell Isolation

1. After transportation to the laboratory, the crudely prepared fat pads are carefully liberated from remaining connective tissue and blood vessels. This step is followed by collagenase digestion. Different types of collagenases are offered by several companies, and batches may differ substantially in their quality and activity. This problem renders it difficult, if not impossible, to standardize the conditions for collagenase digestion. However, variation in collagenase activity can be compensated by the adaption of concentration and duration of incubation. For the standard digestion procedure, a crude collagenase preparation at a concentration of 200 U/mL and 3 mL/g tissue is used. The incubation time is approx 60–90 min in a shaking water bath at 37°C.

2. To collect the disaggregated cells, samples will be centrifuged at $200g_{max}$ for 10 min. The supernatant with the adipocytes is discarded, or can be used for other purposes.

3. The pellet is resuspended in erythrocyte-lysing buffer, and incubated for not more than 10 min. This suspension is filtered through a 150 µm nylon mesh, and centrifuged once again. The resulting pellet is resuspended in an appropriate volume of basal medium. The cells are optionally filtered through a nylon mesh with a pore size of 70 µm (*see* **Note 2**).

## 3.3. Determination of Total Cell Number

1. For determination of the total cell number, a 50-µL aliquot of the cell fraction, obtained after resuspension in basal medium, is taken, and diluted with 100 µL medium and 50 µL trypan blue (0.4% in $H_2O$). 10 µL of this solution is transferred to a Neubauer chamber and counted under the microscope. This procedure is repeated and the mean value used to calculate total stromal cell number. The yield of isolated stromal cells/g wet AT is in the range of 100,000–350,000. In samples from severely obese subjects, the yield of stromal cells is usually lower than in samples from lean persons.

2. After cell counting, the still-concentrated stromal cell solution is diluted with inoculation medium to a final concentration of 150,000 cells/mL for subsequent seeding of the cells in a 4.5-cm² well. Thus, 1 mL corresponds to a seeding density of approx 33,000 cells/cm².

## 3.4. Cell Inoculation and Attachment

1. Usually, cells are resuspended in the inoculation medium containing 10% FCS, because serum contains many components that not only promote cell adhesion, but also support cell spreading and proliferation. This is the most frequently used and least expensive procedure. The advisable inoculation density is 30,000–50,000 cells/cm² to achieve optimal differentiation. For cell culture, 6- or 12-well plates, representing an area of 10 and 4.5 cm²/well, respectively, are used. Although not all cells become attached to the surface of the sterile plastic dishes, the remaining cell number is sufficient to obtain a confluent cell monolayer within a few days, when cells have spread out. For optimal cell attachment, it is recommended to keep the cells for 16–24 h in the serum-containing medium. Longer

exposure of the cells to serum is associated with an increased mitogenic activity, detectable by an increase in total cell number *(3)*. This phenomenon goes along with a progressive loss of differentiation capacity. When cells are incubated in serum-containing medium for the usual 16-d period in the presence of adipogenic factors, the differentiation rate is reduced by more than 90%, and, at the same time, total cell number increases by at least three- to fourfold (*see* **Note 3**).

2. It is also possible to inoculate the stromal cells in unsupplemented (without FCS) adipogenic medium. However, under this condition, cell adhesion is low and does not exceed 30–40%, on average. A small advantage is that the differentiation is higher (**Table 1**). Finally, cells can be inoculated in serum-free basal medium in culture dishes precoated with extracellular matrix proteins, such as fibronectin or laminin (commercially available). Frozen fibronectin prepared from human plasma, is thawed at 37°C and diluted to a concentration of 0.02 mg/mL, using aqua bidest. The surface of the culture dishes is precoated with a thin layer of this solution over night in an incubator at 37°C. Before being used, dishes are rinsed with basal medium. Precoating with human fibronectin significantly improves the attachment rate, but cannot fully replace FCS. Precoating of the dishes with fibronectin does not affect adipose differentiation in human tissue (**Table 1**).

## 3.5. Stimulation of Proliferation by FGF

The low yield of stromal cells from AT prompted attempts to stimulate stromal cell proliferation without losing the capacity for adipose differentiation. Such studies showed that fibroblast growth factor (FGF) is able to promote cell division without inhibiting adipose conversion. Cells are inoculated at a density of $10,000/cm^2$ in serum-containing inoculation medium. After 16 h, cells are repeatedly washed with PBS, then fed with basal medium supplemented with 10 µg/mL transferrin, 66 n*M* insulin, 10 n*M* cortisol (reduced concentration), and a final concentration of 1 n*M* recombinant hu-bFGF (stock solution of 1 µ*M* FGF in PBS). The medium is changed every other day until confluency is reached, usually within 5–6 d; during this time, cells divide 2–3×. The FGF stock solution is stored in aliquots at –20°C for no longer than 6 mo. Repeated thaw–freeze cycles must be avoided.

## 3.6. Adipose Differentiation in Primary Culture (see Notes 4–6)

1. After cell adhesion, cultures are washed 2–3× with PBS to remove nonattached cells, which include most contaminating blood cells, cell detritus, and serum.
2. Cells are then incubated in a serum-free, hormone-supplemented medium, to induce adipose conversion. This adipogenic medium includes 66 n*M* insulin and 1 n*M* triiodothyronine, 100 n*M* hydrocortisone, and, for the first 3 d, 0.5 m*M* IBMX to induce the rearrangement of the gene expression pattern, which will result in the expression of genes required for adipose differentiation. To facilitate lipogenesis from glucose, 33 µ*M* biotin and 17 µ*M* pantothenate are continuously present in the culture medium. Under these conditions, up to 90% of the attached cells undergo differentiation, which can be easily followed by inverse light microscopy. The average differentiation rate is 40–50%.

**Table 1**
**Effect of Inoculation Conditions on Cell Attachment
and Differentiation Capacity**[a]

| Inoculation condition | Attachment rate (%) | G3PDH (rel %) |
|---|---|---|
| 10% FCS for 24 h | 60–70 | 100 |
| 10% FCS for 16 d | 60–70 | 7 |
| Fibronectin (precoating) | 50–60 | 114 |
| Completely serum-free | 30–40 | 142 |

[a]Differentiation was assessed by determination of glycerol-3-phosphate dehydrogenase (G3PDH) activity on d 16 of culture, 100% is equivalent to a mean specific activity of 692 ± 137 mU/mg protein.

3. The adipose differentiation process can be also induced by addition of thiazolidinediones, instead of IBMX. This new class of antidiabetic drugs activates the nuclear receptor, PPARγ, and, thereby, triggers the expression of genes characteristic for fat cell development and function. Troglitazone is added, at a final concentration of 1 μg/mL, for the initial 3 d after changing to the serum-free culture medium. Continuous presence of troglitazone is not required. The average differentiation rate is in the range of 60%.

The adipogenic medium is renewed 3×/wk. Visible lipid accumulation starts within 6–8 d under these conditions, initially around the nucleus. Within 16 d, the differentiating cells are completely filled with lipid droplets, and have changed their morphology to a spherical shape. Measurement of lipogenic enzyme activities indicates characteristics of differentiated fat cells, although cells still have an multilocular appearance (**Fig. 1**).

## 3.7. Oil Red O Staining

Cells cultured in a well or dish are fixed with a 10% formaldehyde solution for 2 h, then cells are washed with PBS, and subsequently incubated with 0.3% oil red O solution (400 μL/4.5-cm$^2$ well). After 1 h, the staining solution is removed by aspiration, then cells are exposed to 60% isopropanol (500 μL/4.5-cm$^2$ well) for a few minutes. Finally, cells are washed twice with PBS. Stained cultures can be stored in a refrigerator for weeks. Cells containing Oil red O dye can be easily detected, using an inverse light microscope at an appropriate magnification.

## 4. Notes

The establishment and in vitro differentiation of human adipose precursor cells can be complicated by a number of problems, some of which are addressed under the following points:

1. To keep contamination with other cell types low, and to achieve a good yield of precursor cells, it is essential to remove any contaminating tissue, particularly

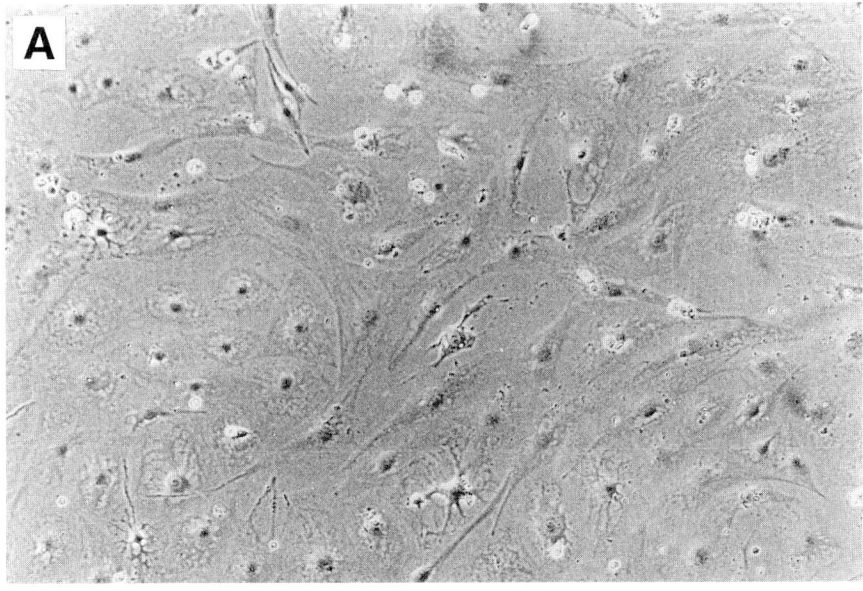

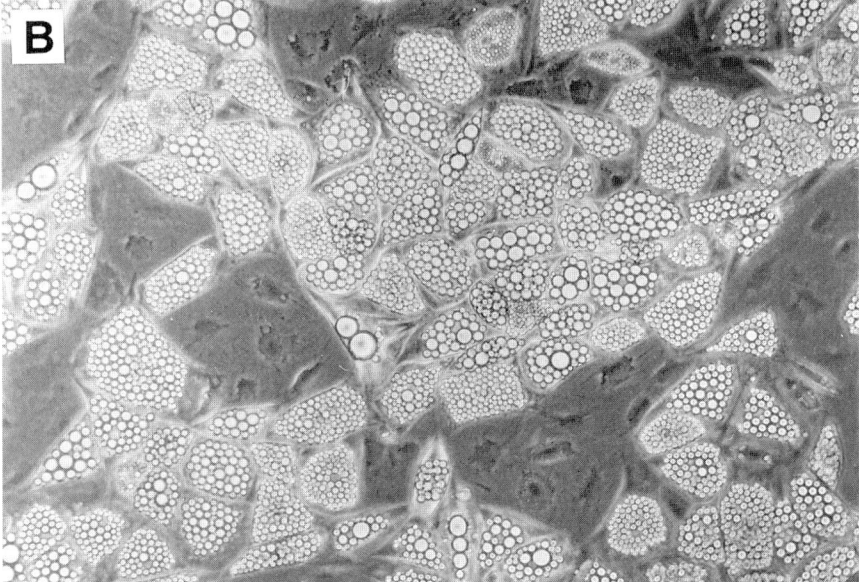

Fig. 1. Micrographs of cultured preadipose and adipose cells. (**A**) Human preadipose cells after 3 d incubation. Cells were seeded in a density of 30,000/cm² in inoculation medium for 1 d, and adipogenic medium supplemented with 1 µg/mL troglitazone. (**B**) Human preadipose cells from subcutaneous tissue in primary culture on d 16. Cells were cultured with adipogenic medium and, for the first 3 d, 1 µg/mL troglitazone. Note that human adipocytes are still multilocular at this stage.

blood vessels and fibrous material, prior to collagenase digestion. Moreover, the tissue samples should be thoroughly washed with PBS to get rid of blood that may reduce cell attachment and differentiation capacity. The AT should be finely minced to obtain an optimal and uniform enzymatic disaggregation of the tissue. More than 90% of the initial tissue should be digested.

2. AT samples from the intra-abdominal depots are densely vascularized. For this reason, there is usually a variable but significant contamination of the stromal cell fraction by endothelial cells (EC). In contrast to the stromal cell fraction from subcutaneous AT, which is essentially free of ECs, this cell type may cause several problems. The proportion of ECs should be reduced to less than 5% of total cell number. Therefore, after a first filtration through a 150-µm mesh in erythrocyte-lysing buffer, a second filtration is recommended, using a nylon mesh with a pore size of 25-µm to reduce this contamination *(2)*. ECs reaggregate rapidly, and are, for this reason, quantitatively retained by this filter size, but some contamination by ECs cannot be fully prevented. If precursor cells from various depots are compared, all samples must undergo the same filtration procedure.

3. Poor attachment and spreading of cells can be frequently encountered, if samples are extensively exposed to collagenase. Duration of digestion is critical, and can damage cells seriously. Therefore, too long enzymatic disaggregation of the samples should be avoided.

4. The rate of differentiation can vary considerably. Several aspects may be considered, but it is not always possible to prevent a low differentiation rate. Most important is that samples from young donors are used, that the tissue is transported in a suitable isotonic solution, and that collagenase digestion is adequate in terms of concentration and duration. A low differentiation rate, however, may be also an inherent characteristic of the donor. Moreover, a low inoculation density can result in suboptimal differentiation. Adipogenic factors should be freshly prepared, and stored appropriately.

5. Some compounds are dissolved in EtOH or DMSO. A combination of such additives can lead to subtoxic or toxic concentrations of the respective solvent. Detached and/or shrivelled cells can result from such effects. It is important that a final concentration of 0.1% DMSO and 0.1% EtOH is not exceeded.

6. With increasing lipid accumulation and rounding up, the developing adipocytes get more easily detached from the plastic surface. It is recommended to aspirate and replace the medium very gently, and to limit the culture period. We usually define d 16 as date of terminal differentiation, and use glycerol-3-phosphate dehydrogenase (GPDH)-activity as differentiation marker. By this time, cells have a multilocular morphology, but are completely filled with lipid droplets, and express almost maximal levels of characteristic genes. Then, cells can be cultured for another 10–14 d at maximum, to study AT metabolism. During this extended period, the adipocytes exhibit only one, or a few, large lipid droplets, and are very vulnerable to vibration or other mechanical damage *(8)*.

## References

1. Ailhaud, G. and Hauner, H. (1998) Development of white adipose tissue, in *Handbook of Obesity* (Bray, G. A., Bouchard, C., and James, W. P. M., eds.), Marcel Dekker, New York, pp. 359–378.
2. Björntorp, P., Karlsson, M., Pertoft, H., Pettersson, P., Sjöström, L., and Smith, U. (1978) Isolation and characterization of cells from rat adipose tissue developing into adipocytes. *J. Lipid Res.* **19,** 316–324.
3. Entenmann, G. and Hauner, H. (1996) Relationship between replication and differentiation in cultured human adipocyte precursor cells. *Am. J. Physiol.* **270,** C1011–C1016.
4. Flier, F. S. and Maratos-Flier, E. (1998) Obesity and the hypothalamus: novel peptides for new pathways. *Cell* **92,** 437–440.
5. Hauner, H. and Entenmann, G. (1991) Regional variation of adipose differentiation in cultured stromal-vascular cells from the abdominal and femoral adipose tissue of obese women. *Int. J. Obesity* **15,** 121–126.
6. Hauner, H., Entenmann, G., Wabitsch, M., Gaillard, D., Negrel, R., Ailhaud, G., and Pfeiffer, E. F. (1989) Promoting effects of glucocorticoids on the differentiation of human adipocyte precursor cells cultured in a chemically defined medium. *J. Clin. Invest.* **84,** 1663–1670.
7. Spiegelman, B. M. (1998) PPAR-gamma: adipogenic regulator and thiazolidinedione. *Diabetes* **47,** 507–514.
8. Hauner, H., Röhrig, K., Spelleken, M., Lin, L. S., and Eikel, J. (1998) Development of insulin-responsive glucose uptake and GLUT4 expression in differentiating human adipocyte precursor cells. *Int. J. Obesity* **22,** 448–453.

# 21

# Measurements of Peptide and Nonpeptide Secretory Products from Adipocytes

**Marie-Christine Alessi, Jérôme Aubert, and Raymond Négrel**

## 1. Introduction

In addition to their metabolic function of storing triglycerides and releasing free fatty acids upon hormonal stimulation, adipocytes are now recognized as behaving as secretory cells. During the past 10 yr, an increasing number of peptidic and nonpeptidic compounds have been demonstrated to be secreted by adipocytes (*1*). Most of these compounds are thought to exert primarily, and in an autocrine/paracrine manner, local effects that modulate adipose cell hypertrophy and/or hyperplasia of adipose tissue (AT). This is essentially the case for short-living molecules, such as the prostanoid, prostacyclin ($PGI_2$), which is discussed in this chapter. Some of these secreted factors, like leptin (*see* Chapter 26), allow communication with other peripheral tissues and the central nervous system, giving to the adipocyte the status of an endocrine cell. In addition, some of these factors may contribute to the development of morbid complications of obesity, i.e., cardiovascular diseases, hypertension, insulin-resistance, diabetes, and cancer (*2*).

In this context, this chapter focuses, among proteins as secretory products, on the measurements of angiotensinogen (AGT) and plasminogen activator inhibitor 1 (PAI-1). Among nonpeptidic secretory products, measurements of the two prostanoids that are the main metabolites of arachidonic acid (AA) in rodent and human AT, i.e., prostaglandin $E_2$ ($PGE_2$) and 6-keto-$PGF_{1\alpha}$ (the stable degradation product of $PGI_2$), are discussed.

### 1.1. Angiotensinogen

Angiotensinogen (AGT) is the unique substrate of renin, giving rise to angiotensin I (Ang I) and precursor of angiotensin II (Ang II) in the renin-

From: *Methods in Molecular Biology, vol. 155: Adipose Tissue Protocols*
Edited by: G. Ailhaud © Humana Press Inc., Totowa, NJ

angiotensin system (RAS), which is recognized as playing a major role in systemic blood pressure and renal electrolyte homeostasis. Local RASs giving rise to Ang II, and which are distinct from the plasma RAS, have been described in various organs *(3)*. Findings consistent with the occurrence of a local functional RAS in AT have been reported *(4–7)*. Therefore, the fact that AT appears as a major source of AGT, after the liver, is of interest, because epidemiological studies *(8)* have shown a strong correlation between accumulation of visceral fat and high blood pressure. Nevertheless the contribution of AGT, secreted by AT to this pathophysiological situation is not well understood. In addition, components of the RAS, such as angiotengin I generating activity, Ang II and Ang II receptors, have been implicated in AT development *(6,9–11)*, and particularly Ang II, which is able to recruit rodent adipose precursor cells to differentiate into new adipocytes via the induced release of adipogenic $PGI_2$ *(9;* and *see* **Subheading 1.3.**).

### 1.2. Plasminogen Activator Inhibitor 1

Plasminogen activator inhibitor 1 (PAI-1) is the primary physiological inhibitor of plasminogen activation through inhibition of plasminogen activators (PAs), such as tissue-type plasminogen activator (t-PA) or urokinase type PA (u-PA). PAI-1 is a single-chain protein of 50 kDa that consists of 379 aa. It is a member of the serine protease inhibitor superfamily. The reactive center of the inhibitor (Arg 346–Met 347) is contained within the exposed strained loop region of the COOH terminus of the molecule, and serves as a pseudosubstrate for the target serine protease. Inhibition of PAs by PAI-1 occurs in a rapid and stoechiometric manner, resulting in the formation of a covalent bond between the two molecules *(12)*. In animal models, increased PAI-1 expression promotes a fibrin deposit. Transgenic mice overexpressing the human *PAI-1* gene develop fibrin deposit *(13)*; mice with single deficiency of PAI-1 promotes lysis of plasma clots at a significantly higher rate than wild-type mice *(14)*. In human, plasma PAI-1 levels have been found to be elevated in myocardial infarction and deep venous thrombosis. An increased plasma PAI-1 level is considered to be a biological risk factor for the development of myocardial infarction *(15)*. This relationship disappears after adjustment for parameters of insulin resistance, such as body mass index and plasma triglycerides *(16)*. The link between insulin resistance and PAI-1 has been described mostly in cross-sectional studies as well as in intervention studies *(17)* with the aim of reducing insulin resistance. The possibility that AT by itself may directly contribute to the elevated expression of PAI-1 has gained recently some support. Plasma PAI-1 levels in obese mice were fivefold higher than in their lean counterparts, associated with an overexpression of the *PAI-1* gene in AT *(18)*. PAI-1 secretion by human AT has been documented *(19)* and shown, like in rat AT *(20)*, that it is more pronounced in visceral than in subcutaneous fat.

PAI-1 levels are also related to both lipid content and the volume of fat cells *(21)*. Fully differentiated 3T3-L1 adipocytes in culture produce significant levels of PAI-1 mRNA and protein *(22)*.

## 1.3. PGE$_2$ and PGI$_2$

PGE$_2$, which is a main metabolite of arachidonic acid (AA) in several cell types, is well known to antagonize both cAMP production and fatty acid release induced by lipolytic agents in rodent or human adipose cells *(23,24)*. This antilipolytic effect is caused by the negative coupling of the PGE$_2$ cell surface receptor (EP$_3$ receptor, *see* **Note 1** and **ref. 25**) toward the cAMP-generating enzyme, adenyl cyclase. Interestingly, this receptor is absent in preadipocytes, and is emerging during their terminal differentiation into adipocytes *(24,26; see* **Note 1**).

Prostacyclin, or PGI$_2$, was first described as a prostanoid produced by endothelial cells of blood vessels, and is known to play a major role as a relaxing agent and a potent antagonist of platelet aggregation *(27)*. PGI$_2$ is now recognized as a metabolite of AA in several cell types, including preadipose and adipose cells *(28–30)*. PGI$_2$ is unstable in aqueous media at neutral pH, and converts rapidly to the stable inactive prostanoid, 6-keto-PGF$_{1\alpha}$ (*see* **Note 2**). This stable degradation product is a valuable indicator of PGI$_2$ produced and released by the cells. Regarding AT, PGI$_2$ , which, along with PGE$_2$ , is a major metabolite of AA in AT, appears as a critical prostanoid in adipogenesis *(31)*. Under serum-free culture conditions, PGI$_2$ has been demonstrated *(32–35)* to mediate the adipogenic effect of exogenous AA in an autocrine manner, leading preadipocytes to differentiate into adipocytes. PGI$_2$, which can be released from adipocytes upon exposure to Ang II, has also been characterized as a paracrine adipogenic chemical relay able to recruit preadipocytes to differentiate into new adipocytes, not only in vitro *(9)*, but also ex vivo and in vivo (Darimont, C., Saint-Marc, P., Ailhaud, G., and Négrel, R., unpublished results). These autocrine/paracrine adipogenic properties of PGI$_2$ have been recently emphasized by the discovery that a stable analog of this prostanoid, carbaprostacyclin (cPGI$_2$), is able to regulate the expression of several genes by a mechanism distinct from the signaling of its cell surface IP receptor *(36; see* **Note 1**), the most likely via a nuclear receptor of the peroxisome proliferator activated receptors (PPAR) family, consistent with the fact that cPGI2 is a ligand of PPARs *(37)* and two of the isoforms of PPARs, namely PPAR$_\delta$ and PPAR$_\gamma$, are both known to play a crucial role as transcription factors in the sequential induction of adipose cell terminal differentiation *(38,39)*.

## 2. Materials
## 2.1. Measurement of AGT

1. Assay buffer: 50 m$M$ Tris-HCl buffer, pH 7.5, containing 0.3% bovine serum albumin (BSA).

2. IT medium: Dulbecco's modified Eagle medium (DMEM) supplemented with 5 µg/mL bovine insulin (850 n$M$) and 10 µg/mL human transferrin.

3. IT medium supplemented with 20 n$M$ Na selenite, 100 m$M$ ascorbic acid, and 200 p$M$ triiodothyronine (T$_3$).

4. A solution of renin in assay buffer, previously adjusted to a suitable concentration (*see* **Note 3; *40***), in order that addition of 5 µL will lead to complete conversion of 2 n$M$ AGT to Ang I.

5. Lyophilized antibodies (ABs) against Ang I (ICN or Sigma), reconstituted in previously heated (*see* **Note 4**) 0.15 $M$ Na phosphate buffer, pH 7.5, containing 0.03 $M$ EDTA, 0.1% human serum albumin (HSA), and 0.1% Na azide.

6. Ang I standards (Sigma or ICN): a 0.1 m$M$ stock solution, prepared in assay buffer, and stored in small aliquots at –20°C is diluted after thawing, just before use, in assay buffer, to prepare 2 n$M$ and 0.2 n$M$ solution. These latter will be used for the preparation of standards (1–200 fmol) in the radioimmunoassay (RIA) to establish the reference curve.

7. (3-[$^{125}$I]-iodotyrosyl$^4$) Ang I (5-isoleucine), reconstituted either in water or in assay buffer, according to the manufacturer instructions, and stored frozen as small aliquots in polypropylene vials, is used in 1-d experiments only, after thawing and subsequent dilution with assay buffer to a volumic radioactivity of 500,000 dpm/mL.

8. A 100X mixture of protease inhibitors in assay buffer, i.e., 500 m$M$ EDTA, 100 m$M$ phenylmethylsulfonylfluoride (PMSF), 100 m$M$ 8-hydroxyquinoline sulfate, and 1 m$M$ captopril, stored at 4°C in assay buffer to be diluted 100-fold into collected samples, in order to inhibit angiotensinases, as well as Ang I converting enzyme (ACE).

9. A freshly prepared suspension of dextran-coated charcoal containing 1 g activated Norit EXW (Sigma) and 0.01 g dextran (average mol wt: 60,000–90,000) in 100 mL assay buffer. This suspension must be gently stirred for at least 20 min at 4°C before use.

## 2.2. Measurement of PAI-1

### 2.2.1. Measurement of PAI-1 Antigen

1. Commercially available enzyme-linked immunosorbent assay (ELISA) kits using specific Abs directed against various forms of human PAI-1 (as presented in **Table 1**), or against rat PAI-1 (as discussed in **Subheadings 3.2.** and **3.2.1.** and **Notes 5–7**; see also **ref. *41***). The addresses of the distributing companies are detailed in **Note 7**.

2. Preparation 92/654 (the more recent standard) from the National Institute of Biological Standards and Controls (Hertfordshire, UK) containing approx 200 ng PAI-1/vial, to use as a standard guide (*see* **Note 5**).

### 2.2.2. Measurement of PAI-1 Activity

#### 2.2.2.1. COMMERCIALLY AVAILABLE KITS

Commercially available kits based on the inhibition of exogenous PAs, as presented in **Table 2** (*see* also **Note 7**).

**Table 1**
**Available ELISA for Measurement of Human PAI-1 Ag**

| Kit | Antibody | Specificity |
|---|---|---|
| Asserachrom PAI-1 (Diagnostica Stago) | Monoclonal | Active, latent, complexed |
| Coaliza PAI-1 (Chromogenix) | Monoclonal | Active, latent, complexed |
| Imubind PAI-1 plasma (American Diagnostica) | Monoclonal | Active, latent ≥ complexed |
| Imubind PAI-1 tissue (American Diagnostica) | Monoclonal | Active, latent ≥ complexed |
| Imulyse PAI-1 (Biopool) | Monoclonal | Active, latent ≥ complexed |
| Tintelize PAI-1 (Biopool) | Monoclonal/ polyclonal | Active, complexed ≥ latent |
| PAI-1 sntigen (Technoclone) | Monoclonal | Active, latent, complexed |
| PAI-1 ELISA kit (Monozyme) | Monoclonal | latent ≥ complexed > active |

[a] *See* **Notes 5–7**.

**Table 2**
**Commercially Available Kits for Measurement of PAI-1 Activity**

| Kit | Type of assay | PA used |
|---|---|---|
| Stachrom PAI (Stago) | Activity assay | u-PA |
| Coatest PAI (Chromogenix) | Activity assay | t-PA |
| Berichrom PAI (Behring) | Activity assay | u-PA |
| Spectrolyse PL (Biopool) | Activity assay | t-PA |
| Chromolize PAI-1 (Biopool) | Immunoactivity assay | t-PA |
| Spectrolyse/fibrin (Biopool) | Activity assay | t-PA |

2.2.2.2. REVERSE FIBRIN AUTORADIOGRAPHY

1. 50 m$M$ imidazole buffer, pH 7.35, containing 140 m$M$ NaCl.
2. Agarose (low melting point).
3. A freshly prepared 1% (w/v) human or bovine fibrinogen solution in imidazole buffer (*see* **Note 8**).
4. 10% (w/v) and 2.5% Triton X-100 solutions.
5. A thrombin solution in imidazole buffer, adjusted to 3 U/mL.
6. A 50–80 mU/mL u-PA solution prepared in imidazole buffer (*see* **Note 9**).

## 2.3. Measurement of Prostanoids

### 2.3.1. Using AA Prelabeled Cultured Cells

1. [$^3$H] AA (150–200 Ci/mmol), adjusted with unlabeled AA (Cayman, Ann Abrboe, MI; or Sigma) to a concentration of 100 μ$M$ and a specific radioactivity of 0.5–1 Ci/mmol in EtOH (*see* **Note 10**).

2. A stock solution of 5 m*M* butylated hydroxytoluene (BHT) in EtOH.
3. AA and PG standards (PGE$_2$, PGF$_{2\alpha}$, 6-keto-PGF$_{1\alpha}$) stored at –20°C as concentrated (m*M*) stock ethanolic solutions.
4. Chemicals and solvents of high chemical purity: acetic, chlorhydric, and formic acid; chloroform, cyclohexane, diethylether, EtOH, ethylacetate, hexane, iso-octane, and methanol.
5. Octadecylsilyl silica cartridges (Sep-Pak C$_{18}$ from Waters, Milford, MA) and glass, polypropylenem or polyethylene tubes resistant to organic solvents.
6. Thin layer chromatography (TLC) tanks and silicagel-coated plates (*see* **Note 11**; *42*).

### 2.3.2. Using Immunoassays

1. 5 m*M* BHT as in **Subheading 2.3.1., item 2**.
2. Solvents and/or cartridges for extraction and concentration of prostanoids, as in **Subheading 2.3.1., items 4** and **5**.
3. Commercially available kits from Cayman or R&D Systems (Abingdon, UK).

## 3. Methods
### 3.1. Measurement of AGT

The amount of AGT in collected samples (*see* **Note 12**) is determined after its complete conversion to Ang I in the presence of exogenous renin (*see* **Note 3**), and in the presence of inhibitors of angiotensinases and ACE. Generated Ang I is quantified using a specific competitive RIA, essentially as described by Sealey and Laragh for the assay of plasma renin (*43*).

### 3.1.1. Sample Collection

To facilitate the determination of AGT contents in cultured cell extracts (intracellular AGT) or in media (secreted AGT) conditionned by AT explants (*see* Chapter 17) or adipose cells in culture (*see* Chapter 19), it is essential to avoid the presence of serum. For this purpose, serum-free media have been described (*44*; and *see* **Note 12**, as well as the two next paragraphs). Samples are collected and rapidly mixed into vials containing small appropriate volumes of the antiprotease cocktail described above (*see* **Subheading 2.1., item 6**) and 10% BSA, in order that both were diluted 100-fold. These samples are immediately frozen.

1. Cultured adipose cells: Preadipocytes cultured under appropriate conditions to differentiate into adipocytes (*see* Chapter 19) are extensively washed (twice rapidly, and once for 1 h) with IT medium (*see* **Subheading 2.1., item 2**), supplemented or not with other effector(s) or hormone(s), and maintained for various periods of time at 37°C in a humidified atmosphere of 5% CO2. The culture medium (secreted AGT) is collected in the presence of BSA and antiproteolytic agents as described in the previous paragraph. Cells can be scraped, and homogenized by sonication in assay buffer containing BSA and antiproteolytic agents, as above, to assay for intracellular AGT.

2. AT explants: chopped pieces of dissected adipose depots are incubated for various periods of time in IT medium supplemented with selenite, ascorbate, and $T_3$ (*see* **Subheading 2.1., item 3**). Aliquots of the culture medium are collected after various periods of time in the presence of BSA and antiproteolytic agents, as above.

### 3.1.2. Conversion of AGT to Ang I

Frozen cellular homogenates or conditioned media are slowly thawed in an ice bath. Aliquots (250 µL) are transferred to polystyrene or polypropylene tubes, and preincubated for 5 min at 37°C, before being treated with an adequate amount of exogenous renin (*see* **Note 3**). Samples can be frozen at this stage before to be assayed in the RIA. Subsequent thawing must be performed at temperatures not exceeding 4°C, and samples are stored in an ice bath during the preparation of the RIA.

### 3.1.3. Assay of Ang I

Add to each tube (prepared in duplicate or triplicate):

1. 50 µL of the labeled Ang I tracer.
2. Various amounts of the standard Ang I solutions, to give final concentrations starting from 1 fmol and increasing up to 200 fmol (*see* **Subheading 2.1., item 4**) or volumes up to 200 mL of samples to be tested.
3. An adequate volume of assay buffer, to complete to 250 mL.
4. 100 µL Abs solution (except in the blank: 100 µL buffer isused for the dilution of the Abs (*see* **Note 4**).
5. Do not forget to prepare blank tubes, omitting both Abs and Ang I (or sample), and maximal binding tubes containing Abs, but not containing Ang I (or sample).
6. Mix and let stand the covered tubes at 4°C for 24 h.
7. Add 1 mL stirred dextran-coated charcoal suspension, and mix, to avoid settling of charcoal.
8. Centrifuge for 5 min at 10,000$g$ in the cold.
9. Withdraw 1.2 mL (taking care to avoid pipeting of charcoal) of the supernatant containing the Ang I bound to Abs.
10. Count in a scintillation spectrometer,
11. Calculate % bound, compared to maximal binding,
12. Plot % bound as a function of Ang I/assay tube (reference curve),
13. Extrapolate the content of each sample, using the reference curve.

### 3.2. Measurement of PAI-1

Determination of PAI-1 can be performed by two different approaches, based either on the immunological detection of various conformations of the human protein, or on the measurement of PAI-1 activity, using functional assays. Cautions with respect to storage and management of the samples, according to their origin, must be stressed (these are outlined in **Note 13**; *see* *also* **refs. *45–51***). Moreover, monoclonal antibodies (mAbs) raised against human PAI-1 are not suitable for other species. In that respect, mAbs, raised against mouse and rat PAI-1, have been obtained *(52,53)*, and an ELISA is commercially available (Immuclone Rat PAI-1 from American Diagnostica, Greenwich, CT) (*see* **Note 7**).

### 3.2.1. Measurement of PAI-1 Antigen

Because of the conformational changes of PAI-1 (*see* **Notes 5** and **6**), which might to considered in the interpretation of results, the specificity of each commercially available ELISA for the measurement of human PAI-1 antigen (Ag) is presented in **Table 1**. A detailed protocol is furnished with each assay kit.

### 3.2.2. First Generation PAI Activity Assays

These assays are based on the inhibition of exogenously added PAs, primarily t-PA. The amount of added t-PA must exceed that of PAI-1 (*see* **Note 14**). Residual t-PA activity is then measured in microtiter plates by an indirect chromogenic substrate assay. These assays are usually characterized by a poor reproducibility for samples of low PAI activity values. Furthermore, they are not able to distinguish between PAI-1 and PAI-2 (*see* **Note 15**), but are suitable for species other than human (*see* **Table 2**).

### 3.2.3. Immunofunctional Assays

Such assays for human PAI-1 are now commercially available. Kits are listed in **Table 2**. One of them quantifies PAI-1–t-PA complex Ag before and after exogeneous addition of t-PA. A more recent assay is based on the recognition of t-PA-coated microtiter plates by active PAI-1. t-PA-linked PAI-1 is then revealed by a mAb labeled with peroxydase. These immunoassays have the advantages of being specific for PAI-1, more accurate, and reproducible, even in the low range of PAI-1 values. In addition, the specificity of the results can be validated (*see* **Note 16**). The protocols to follow are described by the manufacturers.

### 3.2.4. Reverse Fibrin Autography

To detect and characterize plasminogen activator activity in complex biological fluids (*see* **Note 13**), Granelli-Piperno and Reich *(54)* developed the fibrin agarose zymography technique, based on an electrophoretic separation of denatured proteins by sodium dodecyl sulfate polyacrylamide gel electrophoresis (SDS-PAGE), followed by their renaturation in the presence of Triton X-100 and diffusion into a fibrin-agarose underlay. Not only free t-PA and u-PA, but also plasma kallikrein and complexes of PAs with inhibitors can be revealed by this technique *(55)*. A variant of this technique, reverse fibrin autography, allows detection ofthe presence of PAI-1 (*56,57*; and *see* **Note 17**). This technique consists of including PAs (u-PA or t-PA) into the fibrin gel. Zones of lysis resistance correspond to the presence of PAIs.

### 3.2.4.1. PREPARATION OF THE FIBRIN–AGAROSE UNDERLAY (FOR ONE GEL)

Mix, in the following order:

1. 10 mL solution of 2% agarose (may be prepared beforehand and stored at 4°C, melt in a boiling water bath just prior to use, and placed in a water bath at 54°C).
2. 15 mL 1% fibrinogen in imidazole buffer free of undissolved material (*see* **Note 8**).
3. 60 µL 10% Triton X-100.
4. 10 µL u-PA solution (*see* **Note 9**, especially for reverse fibrin).
5. Finally, 1 mL 3 U/mL thrombin in imidazole buffer.

### 3.2.4.2. GEL MANAGEMENT

1. In a vertical stand (same type as used for preparing PAG), two glass plates are mounted, separated by 0.75-mm spacers.
2. In a 50-ml Falcon tube, mix the different solutions in the specified order presented above, i.e., agarose, fibrinogen, Triton X-100, u-PA (for reverse autography), and, finally, thrombin.
3. Inject the mixture between the two glass plates, using a 50-mL syringe. This step should be executed rapidly. The mixture should be injected between the glass plates before the agarose has solidified and the fibrinogen has clotted. Leave the gel for 30 min at room temperature (RT).
4. One glass plate is gently slid off the fibrin agarose.
5. To kill bacteria that contaminate the fibrin–agarose underlay and may induce spots of premature lysis, the plates are exposed for 90 s under an UV lamp (*see* **Note 18**).
6. SDS-PAGE is carried out according to the method of Laemmli et al. *(57)* in a slab gel system. Let the front dye completely run out of the gel. The PAG is then transferred to a tray containing 2 L 2.5% Triton X-100, and left for 1 h under continuous gentle shaking.
7. Excess of Triton X-100 is washed away by two 5-min washes in 0.5 L water.

### 3.2.4.3. FINAL STEPS

1. The gel is then placed on the fibrin agarose underlay, in a moistened chamber.
2. At different standardized times, take a Polaroid picture of the gel placed on a black background, to visualize zones of lysis resistance.
3. Quantification of the size and intensity of spots of lysis resistance is achieved by comparison to those obtained with increasing amounts of standard PAI-1 on the photographs taken as a reference.

## 3.3. Measurement of PGE$_2$ and 6-keto-PGF$_{1\alpha}$

Two main procedures can be used to study AA metabolism and prostanoid production in cultured adipose cells. The first is based on prelabeling of complex cellular lipids with AA, followed by the analysis of AA metabolites released into the culture medium by TLC, as the simplest technique (*see* **Note 11**). This procedure is rather qualitative, but is nevertheless comparative, on the basis of the total and respective amount of radioactivity associated to each prostanoid *(28,59)*. The second method, using selective immunoassays by means of specific Abs raised against various prostanoids, including PGE$_2$ and 6-keto-PGF$_{1\alpha}$, is quantitative, and presents the advantage of being suitable not only for adipose cells in culture *(9,28,29,35)*, but also for other biological materials, such as AT explants or isolated adipocytes (*28*; and *see* Chapter 17), as well as the interstitial fluid of AT collected by *in situ* microdialysis (*60*; and *see* Chapter 25).

### 3.3.1. Adipose Cells in Culture Prelabelled with AA

1. Preadipose or adipose cells cultured under serum-supplemented or serum-free conditions, as described in Chapter 19, are fed with fresh culture medium enriched with tritiated AA (1 $\mu M$; 0.5–1 Ci/mmol; 0.5–1 mCi/cm$^2$), added from a 100X ethanolic stock solution and maintained in this culture medium for 24–48 h (*see* **Note 19**).

2. These prelabeled cells are washed with prewarmed culture medium, and maintained for various periods of time (usually 24 h) in a culture medium enriched or not with various effectors or hormones, as desired (*see* **ref. 59** for an example).

3. The culture medium, as well as a rapid wash of the cells, are pooled and collected in the presence of 50 $\mu M$ BHT, added as antioxydant. The sample can be either frozen or immediately processed, as described in the three next paragraphs.

4. Labeled, unmetabolized AA and its metabolites released into the culture medium can be recovered, and thus concentrated, either by solvent extraction or by chromatography on C$_{18}$ silica cartridges.

   a. Solvent extraction: After acidification to pH 3.0 with HCl, 1 vol of sample is extracted twice by vigorous agitation in the presence of 3 vol ethylacetate–cyclohexane, 1:1 (v/v), using tubes resistant to organic solvents. After separation of the aqueous and organic phases by a quick low-speed centrifugation, the upper organic phases of the two extractions are pooled and dried, either under nitrogen or in a SpeedVac evaporator. The dry residue can be stored at –20°C before analysis.

   b. Chromatography on Sep-Pak C$_{18}$ cartridges: This procedure of rapid extraction and purification of AA metabolites from biological samples has been described in details by Powell *(62)*. Briefly, the successive steps are as follows: wetting the cartridge by passing 20 mL EtOH and removal of the excess ethanol with 20 mL water; loading samples previously adjusted to pH 3.0 and 15% EtOH; washing with 20 mL aqueous EtOH to elute polar lipids; and eluting AA metabolites with 5 mL methanol, and drying as above.

5. Thin layer chromatography analysis: Such an analysis can be easily performed, using silica gel-coated-plastic plates (*see* **Note 11**) with a good efficacy, and at low cost, in any laboratory, compared to high-performance liquid chromatogra-

phy, which requires special and expensive equipment. With respect to preadipose and adipose cells, and in order to separate 6-keto-PGF$_{1\alpha}$ from PGE2 and PGF$_{2\alpha}$ (*see* **Note 20**), the developing solvent of choice is the organic upper phase of a mixture of ethylacetate–2,2,4-trimethylpentane (iso-octane)–acetic acid–water 110:50:20:100 (v/v/v/v).

After application of a small volume (5–20 µL) of known radioactivity content of each dried sample to analyze, dissolved in methanol–ethylacetate, 1:1 (v/v), and development of the plate in a TLC tank previously saturated with the vapor of the developing solvent (*see* **Note 21**), a migration of the solvent to a height of 19 cm allows a good separation of 6-keto-PGF$_{1\alpha}$, PGF$_{12\alpha}$, PGE$_2$ and unmetabolized AA. Their relative migration to the solvent front (Rf) are respectively 0.18, 0.25, 0.36, and 0,85. Unlabeled (or tritiated) standards are run in parallel.

After careful drying of the plate in an heated atmosphere under a well-ventilated chemical hood, detection of the separated compounds as yellow-brown spots can be easily performed by placing the plate in a TLC tank containing a few iodine crystals, for 1–5 min (*see* **Note 22**). The plate is cut into segments of identical area (1–1.5 cm$^2$), and the radioactivity of each zone measured by liquid scintillation counting (*see* **Note 23**).

### 3.3.2. Immunological Assay of PGs

Such determinations were initially performed using both commercial Abs and radioactive tracers in competitive RIAs *(28,63)*. These assays have been improved by means of developing immunotechnologies, and several kits have been commercialized. Among them, sensitive assays for PGE$_2$ and 6-keto-PGF$_{1\alpha}$ are presently distributed by Cayman and R&D Systems (*see* **Subheading 2.3.2., item 3** for the address, and **Notes 24** and **25**)

## 4. Notes

1. Several subtypes of cell surface receptors for prostanoids have been characterized and cloned *(25)*. According to the official nomenclature, prostanoid receptors are named according to the chemical nature of the natural prostanoid ligand (first letter, such as E for PGE or I for PGI, and second letter P for prostanoid), followed by a number characteristic of the subtype, and eventually by a Greek letter identifying isoforms. Accordingly, the receptor of PGI$_2$ is called IP receptor and that of PGE$_2$, which exists under various isoforms, is the EP$_3$ subtype in AT.

2. The half-life of PGI$_2$ is 2–3 min at neutral pH in vitro, and 10 min in plasma.

3. The specificity of renin for AGT, with respect to species, is a crucial parameter. To measure rat AGT, purified porcine renin (Sigma or ICN) can be used. In this case, incubation of samples with 0.002 U porcine renin for 1 h is adequate to generate Ang I. Human renin is able to cleave AGT from other species, but human AGT is not cleaved by other renins. To the authors' knowledge, human renin is

still not commercially available. It can be obtained through the courtesy of Roche (Basel, Switzerland). Cleavage of mouse AGT strictly requires the mouse enzyme. Mouse renin, partially purified to the acid precipitation step from male submaxillary glands, according to Jacobsen and Poulsen *(40)*, can be routinely used after extensive dialysis against 50 m*M* Tris-HCl, pH 7.5. The enzymatic activity is assayed with serial dilutions of this preparation, using the presently described RIA of Ang I to select an adequate amount of excess renin, in order that 5 μL renin solution allows complete conversion of mouse AGT within 5 min at 37°C. Porcine AGT can be used as substrate for this selection step; porcine AGT can also help as an internal standard, added into samples, to ensure that conversion of AGT and inhibition of Ang I degradation are correctly achieved.

4. This buffer must be heated for 30 min at 56°C and cooled to RT prior to reconstituting the Abs solution to inactivate any contaminating proteolytic enzyme.

5. The PAI-1 molecule exhibits a unique conformational flexibility. Three different conformations have been identified: an active form, a latent form, and a substrate form. In addition, PAI-1 may occur in an inactive form, because of complex with activators or to cleavage by the targeted protease. All these conformational changes must be considered in the interpretation of PAI-1 Ag measurements. Indeed, according to the Abs used, one form could contribute more than the others to the measured signal. Moreover, to date, no official calibrator for PAI-1 Ag is available, except the standard 92/654 mentioned in **Subheading 2.2.1., item 2.**

6. A multicenter study on the evaluation of seven commercially available kits revealed differences between four- and 10-fold in various plasmas from normal subjects *(41)*. But, after normalization of the data relative to an amino-acid calibrated PAI-1 sample, no significant difference could be observed for 6/7 kits. These results are consistent with the fact that the major cause of differences results from the calibration curve. They also suggest that, despite the use of various mAbs, the potential differential reactivity of the different PAI-1 conformations is of minor importance.

7. Addresses for the various companies distributing ELISA for the measurement of PAI-1 Ag (**Table 1** and **Subheading 3.2.**) are as follows. Diagnostica Stago: 9 rue des Frères Chausson, 92600 Asnières sur Seine, France; Chromogenix AB, Taljegardsgatan 3, S-45153 Molndal, Sweden; American Diagnostica, Inc., 222 Railroad avenue, PO Box 1165, Greenwich, CT, 06836-1165, USA; Biopool, 6025 Nicole Street, Ventura, CA, 93003, USA; Technoclone, Mulnerasse 23, Postfach 158, A1092 Vienna, Austria; Monozyme, Agernallé 3, DK-2970 Horsholm, Denmark.

8. Undissolved material should be removed by centrifugation for 20 min at RT at 20,000*g*, or by sterile filtration.

9. The concentration of u-PA to be used is tricky, and should be determined experimentally. The amount added should be such that, after an overnight incubation of the underlay, the fibrin has just lysed. A concentration of u-PA between 50 and 80 mU/mL appears to be suitable.

10. All ethanolic stock solutions are stored at –20°C.

11. The general procedure of TLC has been described in detail in several books and reviews *(42)*. Among the variety of precoated TLC silica gel plates that is avail-

able, 0.2 mm × 20 × 20 cm silica gel 60 plastic plates from Merck can be routinely used. Compared to glass-coated plates, plastic ones present the advantage of being easily cut for either recovery of the separated compounds after elution into methanol or for rapid scintillation counting (elution of the compounds adsorbed in silica with methanol [1 mL] prior to addition of the scintillation cocktail is recommended).

12. The presence of serum in samples intended to AGT measurement is to be avoided in order to minimize intrinsic components of the renin–Ang system, such as AGT, Ang I, renin, or other proteases, as well as other undesirable ill-defined serum components.

13. The samples should be frozen when prolonged storage is intended: The recommended temperature is below –30°C or lower. During thawing, prolonged exposure at 37°C must be avoided, mostly for the activity assays. Samples may be kept in a crushed ice-water mixture for several hours, until analysis. PAI-1 quantification depends on the analysed fluids.

   a. For plasma and platelets, some factors contributing to intra biological indi-vidual variabilities have been well documented. The main points are: Inter-pretation of the results should be controled for age and gender; the impact of circadian variation could be eliminated by having all samples drawn at the same time of day *(45)*; a 20-min rest period before the venepuncture should be respected, to abort stress and exercise effect; since variation in inflamma-tory status and glucidolipidic metabolism are predominantly implicated in plasma fibrinolytic variations, deviations in these factors need to be recorded and fibrinolytic results interpreted accordingly; and a moderate dose of alcohol absorbed with the evening meal greatly affects PAI activity, which remains significantly different from the control 9 h later *(46)*.

   Furthermore, blood collection and handling procedures, for assessment of t-PA and PAI-1, have been well documented *(47)*. Some points deserve special attention. Because platelet PAI-1 corresponds to 90% of total circulat-ing PAI-1, it is particularly important to perfectly deplete plasma from plate-lets before freezing *(48)*. This is achieved using centrifugation at 3000*g* or more for 20–30 min *(49)*. Only the middle layer of the plasma should be pipeted off, avoiding platelets and white cell contaminants. Citrate is the usual anticoagulant for t-PA or PAI-1 evaluation. Specific anticoagulant-limiting platelet activation, such as CTAD (Becton Dickinson, Rutherford, NJ) could be recommended, especially if sample processing is delayed. Nevertheless, the use of this anticoagulant does not exclude a perfect depletion of plasma from platelets. Incubation of plasma at 37°C, which is known to decrease PAI-1 activity, should be avoided. For PAI-1 activity determination in plasma, it has been proposed to collect blood on an acidic medium, in order to prevent t-PA–PAI-1 complex formation. Such a medium is commercially available (Stabilyte evacuated tubes or Stabilyte monovette, Biopool, *see* **Note 7**). Under these conditions, PAI activity in samples is stable up to 24 h. These samples can be stored at –20°C for several months, and thawed up to three times without loss of PAI-1 activity.

b. For conditioned medium of cultured cells, it should be taken into account that, after conformational modifications, PAI-1 is rapidly inactivated at 37°C (half-life, 2 h). Thus, conditioned media from most cultured cells, as well as from AT explants, contain mostly an inactive PAI in a latent form *(50)*. Indeed, PAI-1 has the unusual property of renaturing into an active form by treatment with denaturing agents such as SDS or guanidine hydrochloride. Nevertheless, the possible inactivation of the molecule by PAs, which may be simultaneously produced by the cells, leads to favoring PAI-1 Ag determination, rather than PAI-1 activity. Furthermore, PAI-1 synthesis is inducible and particularly by cytokines (tumor necrosis factor α) and growth factors (transforming growth factor β) *(51)*. Moreover, PAI-1 expression is strongly related to the state of proliferation of the cells: PAI-1 expression increases during the proliferation phase, but it sharply decreases at confluence. Therefore, a careful analysis of these parameters is necessary to properly interpret any variation of PAI-1 synthesis.

14. Usually 1 vol of t-PA (40 U/mL) is incubated for 15 min at RT with 1 vol of sample, acidified, then diluted to terminate the inhibition reaction.

15. PAI-2 is not present in plasma from healthy individuals, but appears in circulation during pregnancy and leukemia M4 M5. It is produced by monocytes and some cell lines.

16. Since PAI-1 is a thermolabile protein, it is possible to demonstrate a decrease in the measured inhibitory potential of the sample after an incubation of several hours at 37°C. Another procedure is to quench this inhibitory potential, using specific Abs.

17. Whereas PAI-1 activity can best be detected after SDS-PAGE, that of PAI-2 can only be detected after electrophoresis in the absence of SDS *(57)*.

18. Take care because overexposure to UV light leads to inactivation of plasminogen.

19. Under these conditions, $80 \pm 5\%$ of the radioactivity is incorporated into the cells, 95% into polar lipids. This can be easily checked, after extraction of the cellular lipids, according to Bligh and Dyer *(61)*, by TLC analysis, using hexane–diethylether–formic acid 80:20:1 (v/v) as the developing solvent.

20. Another solvent system usually used for TLC analysis of prostanoids (chloroform–methanol–acetic acid–water 90:8:1:0.8 (v/v/v/v), does not allow separation of 6-keto-PGF$_{1\alpha}$ from PGE$_2$.

21. Saturation of the atmosphere of the TLC tank with vapors of the developing solvent improves the chromatography. This can be easily achieved by placing filter paper sheets, bathed in the solvent, all around the tank for at least 1 h prior to chromatography.

22. All operations with solvents and iodine are to be performed under a well-ventilated chemical hood.

23. Elution of the adsorbed compounds into the silica gel with 1 mL methanol should improve scintillation counting.

24. Both the solvent and the Sep-Pak extraction procedures may be used in order to concentrate the prostanoids present in the surrounding medium of AT explants or isolated adipocytes before quantification by immunoassay.

25. PGF$_{2\alpha}$, a minor metabolite of AA in AT, can also be assayed using specific immunoassay kits.

## References

1. Négrel, R. (1999) Paracrine/autocrine signals and adipogenesis, in *Progress in Obesity Research* (Guy-Grand, B. and Ailhaud, G., eds.), John Libbey, London, UK, pp. 55–63.
2. Arner, P. (1999) Physiopathology of visceral adipose tissue, in *Progress in Obesity Research* (Guy-Grand, B. and Ailhaud, G., eds.), John Libbey, London, UK, pp. 567–572.
3. Lee., M. A., Bohm, M., Paul, M., and Ganten, D. (1993) Tissue renin-angiotensinogen systems. Their role in cardiovascular disease. *Circulation* **87,** 7–13.
4. Saye, J. A., Ragsdale, N. V., Carey, R. M., and Peach, M. J. (1993) Localization of angiotensinogen peptide-forming enzymes of 3T3-F442A adipocytes. *Am. J. Physiol.* **264,** C1570–C1576.
5. Crandall, D. L., Herzlinger, H. E., Saunders, B. D., and Kral, J. G. (1994) Development aspects of the adipose tissue renin-angiotensin system: therapeutic implications. *Drug Dev. Res.* **32,** 117–125.
6. Harp, J. B. and DiGirolamo, M. (1995) Components of the renin-angiotensin system in adipose tissue: changes with maturation and adipose mass enlargement. *J. Gerontol. Biol. Sci. Med. Sci.* **50,** B270–B276.
7. Schling, P., Mallow, H., Trindl, A., and Löffler, G. (1999) Evidence for a local renin angiotensin system in primary cultured human preadipocytes. *Int. J. Obes.* **23,** 336–341.
8. Rocchini, A. P. (1991) Insulin resistance and blood pressure regulation in obese and nonobese subjects. *Hypertension* **17,** 837–842.
9. Darimont, C., Vassaux, G., Ailhaud, G., and Négrel, R. (1994) Differentiation of preadipose cells: paracrine role of prostacyclin upon stimulation of adipose cells by angiotensin II. *Endocrinology* **135,** 2030–2036.
10. Zorad, S., Fickova, M., Zelezna, B., Macho, J., and Kral, J. G. (1995) The role of angiotensin II and its receptors in regulation of adipose tissue metabolism and cellularity. *Gen. Physiol. Biophys.* **14,** 383–391.
11. Jones, B. H., Standridge, M. K., and Moustaid, N. (1997) Angiotensin II increases lipogenesis in 3T3-L1 and human adipose cells. *Endocrinology* **138,** 1512–1519.
12. Van Meijer, M. and Pannekoek, H. (1995) Structure of plasminogen activator inhibitor 1 (PAI-1) and its function in fibrinolysis: an update. *Fibrinolysis* **9,** 263–276.
13. Erickson, L. A., Fici, G. J., Lund, J. E., Boyle, T. P., Polites, H. G., and Marotti, K. R. (1990) Development of venous occlusions in mice transgenic for the PAI-1 gene. *Nature* **346,** 74–76.
14. Carmeliet, P., Stassen, J. M., Schoonjans, L., Ream, B., Van den Oord, J. J., De Mol, M., Mulligan, R. C., and Collen, D. (1993) Plasminogen activator inhibitor 1 gene deficient mice. II. Effects on hemostasis thrombosis and thrombolysis. *J. Clin. Invest.* **92,** 2756–2760.
15. Hamsten, A., De Faire, U., Walldius, G., Dahlen, G., Szamosi, A., Landou, C., Blombäck, M., and Wiman, B. (1987) Plasminogen activator inhibitor in plasma: risk factor for recurrent myocardial infarction. *Lancet* **II,** 3.

16. Juhan-Vague, I., Pyke, S., Alessi, M. C., and Thompsom, S. (1996) Fibrinolytic factor and the risk of myocardial infarction of sudden death in patients with angina pectoris: results of ECAT study. *Circulation* **94**, 2057–2063.

17. Juhan-Vague, I., Alessi, M. C., and Morange, P. (1999) Obesity and insulin resistance, in *Contemporary Endocrinology Insulin Resistance* (Reaven, G. and Laws, A., eds.), Humana Press Inc., Totowa, NJ, pp. 317–332.

18. Samad, F., Yamamoto, K., Pandey, M., and Loskutoff, D. (1997) Elevated expression of transforming growth factor-b in adipose tissue from obese mice. *Mol. Med.* **3**, 36–47.

19. Morange, P. E., Aubert, J., Peiretti, F., Lijnen, H. R., Vague, P., Verdier, M., Negrel, R., Juhan-Vague, I., and Alessi, M. C. (1999) Glucocorticoids and insulin promote plasminogen activator inhibitor 1 production by human adipose tissue. *Diabetes* **48**, 890–895.

20. Shimomura, I., Funahashi,T., Takahashi, M., Maeda, K., Kotani, K., Nakamura, T., Yamashita, S., Miura, M., Fukuda, Y., Takemujra, K., Tokunaga, K., and Matsuzawa, Y. (1996) Enhanced expression of PAI-1 in visceral fat possible contributor to vascular disease in obesity. *Nat. Med.* **2**, 800–803.

21. Eriksson, P., Reynisdottir, S., Lönnqvist, F., Stemme, V., and Arner, P. (1998) Adipose tissue secretiion of plasminogen activator inhibitor-1 in non obese individuals. *Diabetologia* **41**, 65–71.

22. Lundgren, C. H., Brown, S. L., Nordt, T. K., Sobel, B. E., and Fujii, S. (1996) Elaboration of type 1 Plasminogen Activator inhibitor from adipocytes: a potential pathogenetic link between obesity and cardiovascular disease. *Circulation* **93**, 106–110.

23. Richelsen, B. (1991) Prostaglandins in adipose tissue. *Danish Med. Bull.* **38**, 228–244.

24. Vassaux, G., Gaillard, D., Darimont, C., Ailhaud, G., and Négrel, R. (1992) Differential response of preadipocytes and adipocytes to PGI2 and PGE2: physiological implications. *Endocrinology* **131**, 2393–2398.

25. Kiriyama, M., Ushikubi, F., Kobayashi, T., Hirata, M. Sugimoto, Y., and Narumiya, S. (1997) Ligand binding specificities of the eight types and subtypes of the mouse prostanoid receptors expressed in Chines hamster ovary cells. *Br. J. Pharmacol.* **122**, 217–224.

26. Börglum, J. D., Pedersen, S. B., Ailhaud, G., Négrel, R., and Richelsen, B. (1999) Differential expression of prostaglandin receptor mRNAs during adipose cell differentiation. *Prostaglandins Other Lipid Mediators* **57**, 305–317.

27. Weksler, B. B., Marcus, A. J., and Jaffe, E. A. (1977) Synthesis of prostaglandin I2 (prostacyclin) by cultured human and bovine endothelial cells. *Proc. Natl. Acad. Sci. USA* **74**, 3922–3926.

28. Négrel, R. and Ailhaud, G. (1981) Metabolism of arachidonic acid and prostaglandin synthesis in the preadipocyte clonal line Ob 17. *Biochem. Biophys. Res. Commun.* **98**, 768–777.

29. Hyman, B. T., Stoll, L. L., and Spector, A. A. (1982) Prostaglandin production by 3T3-L1 cells in culture. *Biochim. Biophys. Acta* **713**, 375–385.

30. Axelrod, L., Minnich, A. K., and Ryan, C. A. (1985) Stimulation of prostacyclin production in isolated rat adipocytes by angiotensin II, vasopressin and bradikinin: evidence for two separate mechanisms of prostaglandin synthesis. *Endocrinology* **116,** 2548–2553.
31. Négrel, R. (1999) Prostacyclin as a critical prostanoid in adipogenesis. *Prostaglandins Leukotrienes Essential Fatty Acids* **60,** 383–386.
32. Gaillard, D., Négrel, R., Lagarde, M., and Ailhaud, G. (1989) Requirement and role of arachidonic acid in the differentiation of preadipose cells. *Biochem. J.* **257,** 389–397.
33. Négrel, R., Gaillard, G., and Ailhaud, G. (1989) Prostacyclin as a potent effector of adipose cell differentiation. *Biochem. J.* **257,** 399–405.
34. Vassaux, G., Gaillard, D., Ailhaud, G., and Négrel, R. (1992) prostacyclin is a specific effector of adipose cell differentiation: its dual role as a cAMP- and Calcium- elevating agent. *J. Biol. Chem.* **267,** 11,092–11,097.
35. Catalioto, R. M., Gaillard, D., Maclouf, J. Ailhaud, G., and Négrel, R. (1991) Autocrine control of adipose cell differentiation by prostacyclin and PGF$_{2\alpha}$. *Biochim. Biophys. Acta* **1091,** 364–369.
36. Aubert, J., Ailhaud, G., and Négrel, R. (1996) Evidence for a novel regulatory pathway activated by (carba)prostacyclin in preadipose and adipose cells. *FEBS Lett.* **397,** 117–121.
37. Forman, B. M., Chen, J., and Evans, R. M. (1997) Hypolipidemic drugs, polyunsaturated fatty acids and eicosanoids are ligands for peroxisome proliferator-activated receptors. *Proc. Natl. Acad. Sci. USA* **94,** 4312–4317.
38. Bastié, C., Holts, D., Gaillard, D. Jehl-Pietri, C., and Grimaldi, P. A. (1999) Expression of peroxisome proliferator-activated receptor PPARd promotes induction of PPARg and adipocyte differentiation in 3T3C2 fibroblasts. *J. Biol. Chem.* **274,** 21,920–21,925.
39. Brun, R. P., Tontonoz, P., Forman, B. M., Ellis, R., Chen,J., Evans, R. M., and Spiegelman, B. M. (1996) Differential activation of adipogenesis by multiple PPAR isoforms.*Genes Dev.* **10,** 974–984.
40. Jacobsen, J. and Poulsen, K. (1985) Characterization of angiotensin I, II and III from mouse as position-5 isoleucine angiotensins? An HPLC study. *J. Hypertension* **3,** 155–157.
41. Declerck, P. J., Moreau, H., Jespersen, J., Gram, J., and Kluft, C. (1993) Multicenter evaluation of commercially available methods for the immunological determination of plasminogen activator inhibitor 1 (PAI-1) *Thromb. Haemost.* **70,** 858–863.
42. Kates, M., ed. (1986) *Laboratory Techniques in Biochemistry and <olecular Biology. Techniques in Lipidology: Isolation, Analysis and Identification of Lipids.* Elsevier, Amsterdam.
43. Sealey, J. E. and Laragh, J. H. (1975) Radioimmunoassay of plasma renin activity. *Semin. Nucl. Med.* **5,** 189–202.
44. Aubert, J., Darimont, C., Safonova, I., Ailhaud, G., and Négrel, R. (1997) Regulation by glucocorticoids of angiotensin gene expression and secretion in adipose cells. *Biochem. J.* **328,** 701–706.

45. Juhan-Vague, I., Alessi, M. C., Raccah, D., Aillaud, M. F, Billerey M., Ansaldi J., Philip-Joet C., and Vague P. (1992) Daytime fluctuation of plasminogen activator inhibitor-1 (PAI-1) in populations with high PAI-1 levels. *Thromb. Haemost.*, **67**, 76–82.

46. Hendriks, H. F., Veenstra, J., Velthuis-te Xierik, E. J. M., Schaafsma, G., and Kluft, C. (1994) Effect of moderate dose of alcohol with evening meal on fibrinolytic factors. *Br. Med. J.* **308**, 1003–1006.

47. Kluft, C. and Verheijen, J. H. (1990) Leiden fibrinolysis working party: Blood collection and handling procedures for assessment of tissue-type plasminogen activator (t-PA) and plasminogen activator inhibitor-1 (PAI-1) *Fibrinolysis* **4**, 155–160.

48. Booth, N. A., Simpson, A. J., Croll, A., Bennett, B., and Mac Gregor, I. R. (1988) Plasminogen activator inhibitor (PAI-1) in plasma and platelets. *Br. J. Haematol.* **70**, 327–333.

49. Sidelmann, J. (1994) The influence of centrifugation load on platelet number and PAI-1 antigen concentration in human plasma. *Fibrinolysis* **8**, 148–150.

50. Kooistra, T., Sprengers, E. D., and van Hinsbergh, V. W. M. (1986) Rapid inactivation of plasminogen activator inhibitor upon secretion from cultured human endothelial cells. *Biochem. J.* **239**, 497–503.

51. Sawdey, M. and Loskutoff, D. J. (1991) Regulation of murine type 1 plasminogen activator inhibitor gene expression *in vivo*. Tissue specificity and induction by lipopolysaccharide, tumor necrosis factor-a, and transforming growth factor-b. *J. Clin. Invest.* **88**, 1346–1353.

52. Declerck, P. J., Verstreken, M., and Collen, D. (1995) Immunoassay of murine t-PA, u-PA and PAI-1 using monoclonal antibodies raised in gene-inactivated mice. *Thromb. Haemost.* **74**, 1305–1309.

53. Ngo, T. H., Verheyen, S., Knockaert, I., and Declerck, P. J. (1998) Monoclonal antibody-based immunoassays for the specific quantitation of rat PAI-1 antigen and activity in biological samples. *Thromb. Haemost.* **79**, 808–812.

54. Granelli-Piperno, A. and Reich, E. (1978) A study of proteases and protease-inhibitor complexes in biological fluids. *J. Exp. Med.* **148**, 223–234.

55. Hauert, J., Nicoloso, G., Schleuning, W. D., Bachmann F., and Schapira, M. (1989) Plasminogen activators in dextran-sulfate activated euglobulin fractions: a molecular analysis of factor XII and prekallikrein dependent fibrinolysis. *Blood* **73**, 994–999.

56. Erickson, L. A., Lawrence, D. A., and Loskutoff, D. J. (1984) Reverse fibrin autography: a method to detect and partially characterize protease inhibitors after sodium dodecyl sulfate polyacrylamide gel electrophoresis. *Anal. Biochem.* **137**, 454–463.

57. Cajot, J. F., Kruithof, E. K. O., Schleuning, W. D., Sordat, B., and Bachmann, F. (1986) Plasminogen activator, plasminogen activator inhibitors and procoagulant analyzed in twenty human tumor cell lines. *Int. J. Cancer.* **38**, 719–727.

58. Laemlli, U. K. (1970) Cleavage of structural proteins during the assembly of the head of bacteriophage T4. *Nature* **227**, 680–685.

59. Gaillard, D., Wabitsch, M., Pipy, B., and Négrel, R. (1991) Control of terminal differentiation of adipose precursor cells by glucocorticoids. *J. Lipid Res.* **32,** 569–579.
60. Darimont, C., Vassaux, G., Gaillard, D., Ailhaud, G., and Négrel, R. (1994) *In situ* microdialysis of prostaglandins in adipose tissue: stimulation of prostacyclin release by angiotensin II. *Int. J. Obes.* **18,** 783–788.
61. Bligh, E. G. and Dyer, W. J. (1959) A rapid method of total lipid extraction and purification. *Can. J. Biochem. Physiol.* **37,** 911–917.
62. Powell, W. S. (1982) Rapid extraction of arachidonic acid metabolites from biological samples using octadecylsilyl silica. *Methods Enzymol.* **86,** 467–477.
63. Négrel, R., Grimaldi, P., and Ailhaud, G. (1981) Differentiation of Ob17 preadipocytes to adipocytes. Effects of prostaglandin$F_{2a}$ and relationship to prostaglandin synthesis. *Biochim. Biophys. Acta.* **666,** 15–24.

# 22

## Assessment of White Adipose Tissue Metabolism by Measurement of Arteriovenous Differences

### Keith N. Frayn and Simon W. Coppack

## 1. Introduction

### 1.1. Principle of Arteriovenous Difference Measurement

Adipose tissue (AT) is more than a collection of adipocytes. It is a highly organized tissue in which different cell types interact, and in which a complex mix of hormones and substrates arriving in the plasma, together with neural input and the rate of blood flow (BF), all regulate metabolic activity. These multiple, interacting influences cannot be reproduced in vitro, and, hence, if we wish to understand the integration of AT metabolism in the whole body, it is essential to perform studies of AT metabolism in vivo. There are a number of ways in which this can be done *(1)*. One of the most specific and informative, in a quantitative sense, is the measurement of arteriovenous (A-V) differences across the tissue.

The principle of this technique is simple: Differences in the composition of blood sampled from the arterial supply to the tissue, and from the venous drainage from the tissue, reflect the net metabolic activity of the tissue. In a straightforward example, if the glucose concentration in venous blood is less than that in arterial blood, then the tissue is consuming glucose in a net sense. The net uptake of glucose can be measured quantitatively by estimating the A-V difference for glucose and the rate of BF. The net rate of uptake is then given by (*see* **Notes 1** and **2**):

$$\text{Net uptake} = \text{A-V difference} \times \text{BF}$$

In the case of AT we may also be interested in the release of substances such as nonesterified fatty acids (NEFA), glycerol, and other adipocyte products, such as leptin. Then the venous concentration will be greater than the arterial, and the net rate of release is given by:

$$\text{Net release} = \text{V-A difference} \times \text{BF},$$

From: *Methods in Molecular Biology, vol. 155: Adipose Tissue Protocols*
Edited by: G. Ailhaud © Humana Press Inc., Totowa, NJ

where V-A difference is the V-A concentration difference.

If a tissue were both utilizing a substrate from blood, and releasing it into the blood, then only the net difference between the rates of those processes would be reflected in a measurement of A-V concentration difference. The addition of a tracer can help to measure absolute rates of substrate uptake or release. One commonly used example is the measurement of NEFA uptake by skeletal muscle. Most muscle beds also contain AT, and net A-V difference measurements may show a net release of NEFA from such a tissue bed. If an isotopically labeled fatty acid (FA) is infused intravenously, it will be found that labeled FAs are extracted from blood across the tissue. This is assumed to represent absolute utilization. Once absolute utilization has been measured, it can be added to net release, to give absolute release. Details of the combination of A-V difference and tracer techniques are not covered here; the reader is referred to the book by Wolfe *(2)* for further information.

The principles of calculation of net substrate uptake or release, and of measurement of absolute rates of uptake or release with tracers, require the maintenance of steady-state conditions (*see* **Note 3**).

## 1.2. Choosing a Suitable Site for Blood Sampling

Although the principle is simple, the practice is not easy. The starting point is to identify a vessel that carries the specific venous drainage from AT. Although such vessels must exist in principle, they are probably too small to be identified. Vessels large enough to be identified and to have blood sampled from them probably always carry some drainage from other tissues, such as skin. However, there are some sites where the contribution of AT seems to predominate, probably because other tissues, such as skin, may be less active metabolically, and also because AT may predominate in mass terms. In humans, this applies to the veins draining from the subcutaneous (sc) anterior abdominal AT into the inferior epigastric vein *(3)*, and probably to no other readily accessible vessels (**Fig. 1**). There are vessels that fulfill this criterion in other species (**Table 1**), but this chapter is confined to humans.

Blood sampling from the necessary vein in humans is best achieved by introduction of a flexible, indwelling catheter. Although, in principle, it may be possible to take single samples by venipuncture, in practice the technique requires that a flexible catheter is threaded down a small vein, until it is in a vessel of reasonable size for drawing samples. In addition, the presence of an indwelling catheter means that samples can be taken over a reasonable period, for instance, before and for some hours after eating a test meal, or before and during insulin infusion *(11,12)*.

As well as identifying and sampling from a suitable vein, it is necessary to obtain arterial blood for comparison. The assumption always made is that arte-

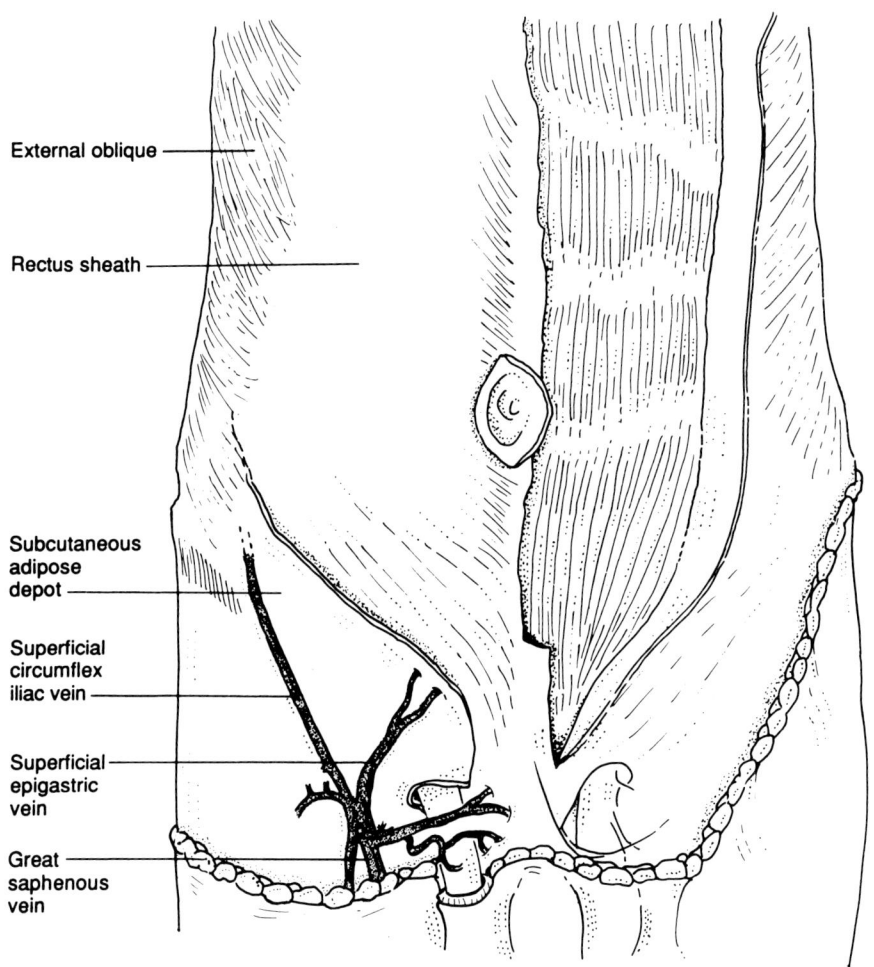

External oblique

Rectus sheath

Subcutaneous adipose depot

Superficial circumflex iliac vein

Superficial epigastric vein

Great saphenous vein

Fig 1. Anatomy of the venous drainage from the sc AT of the anterior abdominal wall in humans. Reproduced with permission from **ref. 8**.

rial blood is the same all over the body, and so access specifically to the artery supplying the adipose depot to be studied is not necessary. In humans, arterial blood is usually taken from the radial or brachial artery, although the femoral artery may be used, if this were to be cannulated for other reasons.

## 1.3. Derived Measurements of AT Metabolism

Measurement of A-V or V-A differences will give estimates of net uptake or release of substances by the adipose depot studied. These may be informative in their own right, but often further information can be gained by looking at the

**Table 1**
**Measurement of While AT Metabolism by A-V Differences in Different Species**

| Species | Key ref(s) | Notes |
|---|---|---|
| Dog | (4,5) | |
| Fat-tailed sheep | (6) | The Syrian fat-tailed sheep has a large tail composed almost entirely of AT |
| Humans | (3) | The depot sampled is the sc anterior abdominal. |
| Rat | (7) | The venous drainage from the inguinal fat pad is sampled |

relationships between these measurements *(13)*. For instance, lactate release may be expressed relative to glucose uptake, to give an estimate of the proportion of glucose uptake that is released as lactate. The ratio of glycerol to NEFA release is another example. From this ratio, an estimate of FA esterification in the adipose depot can be made. Such ratios have the advantage over raw A-V or V-A differences, because they are independent of BF *(13)*. Hence, useful quantitative metabolic information may be derived without measurement of BF.

If some assumptions are made, then further estimates of AT metabolic function can be made. The removal of plasma triacylglycerol (TAG) during passage through the tissue seems primarily to reflect the action of lipoprotein lipase (LPL) in AT capillaries, although there is some uptake of intact particles from the larger TAG-rich lipoprotein fractions *(9)*. Therefore, net TAG removal from plasma may be taken as a measure of the physiological rate of LPL action in AT. This increases after a meal *(14)*, as would be expected from the known activation of adipose tissue LPL at this time *(15)*.

The hydrolytic action of LPL on TAG in vitro leads to the accumulation of 2-monoacylglycerol (MAG) *(16)*, but, in vivo, no net release of MAG from adipose tissue can be detected, even during high rates of TAG hydrolysis *(10)*. The assumption can then be made that glycerol is released mole for mole with hydrolysis of TAG by LPL. Total net glycerol release can be measured. If it is assumed that this arises from the combined actions of LPL in AT capillaries, and hormone-sensitive lipase (HSL) within adipocytes, then subtraction of the rate of action of LPL from total glycerol release will give an estimate of the rate of action in vivo of HSL *(14)* (*see* **Fig. 2** and **Notes 1–4**).

## 2. Materials

### 2.1. Instrumentation

Little instrumentation is needed for a basic study. However, it is helpful to use a clinical infusion pump to perfuse the catheter with saline (0.9% NaCl)

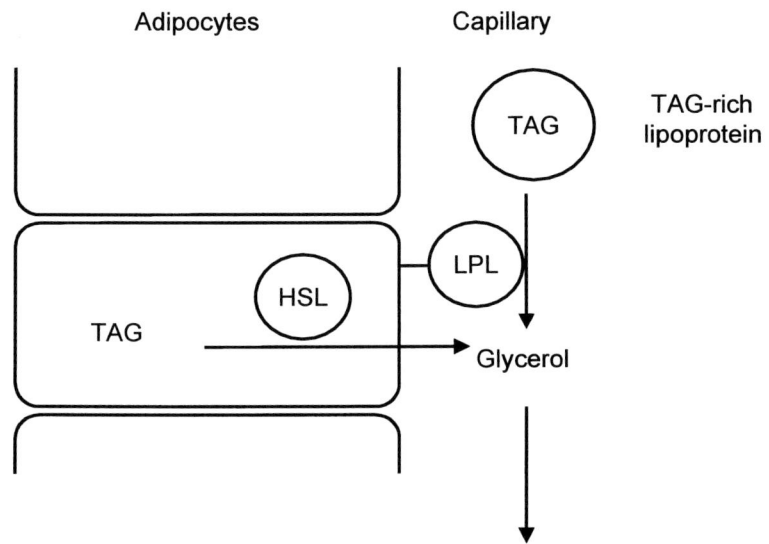

Fig. 2. Assumptions involved in the calculation of the rates of action in vivo of LPL and hormone-sensitive lipase (HSL) in white adipose tissue. LPL acts on circulating triacylglycerol- (TAG)-rich lipoproteins; HSL acts upon TAG stored in adipocytes. Both are assumed to release glycerol mole for mole with TAG hydrolysis. It is assumed that there is no other removal of TAG from plasma (known not to be entirely true *[9]*). Partial hydrolysis of TAG (generation of mono- or diacylglcyerols) is assumed to be insignificant, for which there is good evidence in plasma *(10)*. Then:

$$\text{Rate of LPL action} = (\text{TAG}_a - \text{TAG}_v) \times (1 - H) \times \text{BF}$$

where $\text{TAG}_a$ is the concentration of TAG measured in arterial plasma, $\text{TAG}_v$ that in AT venous plasma; BF is AT BF; $H$ is hematocrit expressed as a fraction (*see* **Note 4**).

$$\text{Net glycerol release} = (\text{Glycerol}_v - \text{Glycerol}_a) \times \text{BF}$$

where glycerol concentrations in artery (a) and vein (v) are in whole blood.

$$\text{Rate of HSL action} = \text{Net glycerol release} - \text{Rate of LPL action}.$$

between blood samples, at a rate of approx 35 mL/h. This rate is not critical, and almost any clinical infusion pump will suffice. An electric warming pad of the sort available from a high-street pharmacy is useful. A microhematocrit centrifuge is required for some measurements (*see* **Notes 2** and **4**).

Measurement of AT BF by the most common method, the washout of [133]Xe, necessitates an external radioactivity monitoring device. We have used a lightweight detector based on a CsI crystal (Mediscint, Oakfield Instruments, Eynsham, UK) *(17)*. *See* Chapter 23 for further details on the measurement of AT BF.

### 2.2. Catheters and Other Clinical Disposables

Ohmeda (Swindon, UK) Secalon Hydrocath 22-gage × 10 cm, or similar catheter with introducer and guidewire; sterile gloves; sterile saline; local anesthetic (e.g., lignocaine 1%; [lidocaine in US]); sterile forceps (plastic are satisfactory).

## 3. Methods

### 3.1. Obtaining Informed Consent

The procedure must be approved by the local research ethics committee. Obtain informed consent from the subject/patient in advance. One requirement of research ethics committees is that the procedure is fully explained to the subject. In making this explanation, it is reasonable to say that the procedure should not hurt (other than the injection of a small amount of local anesthetic), but that it may require more than one attempt.

### 3.2. Identification of Site

1. Relax the subject in a warm room (30°C if possible), and cover the abdomen with a warming pad.
2. Inspect the abdomen visually (subdued light is best) for veins draining the sc abdominal region. Identify veins that are straight rather than tortuous, medial rather than lateral, and low on the abdomen rather than supraumbilical.
3. Mark the course of the veins with a water-soluble pen.

### 3.3. Catheterization

1. Have an assistant standing ready, with the guidewire supplied with the catheter, held in sterile gloves.
2. Introduce a small bleb of intradermal local anesthetic at the proximal end of a straight segment of vein.
3. Using the introducing needle provided with the catheter, advance very slowly through the skin (toward the groin), and move the needle slowly forward, watching carefully for flashback of blood into the needle hub. Flashback will be slow, hence the need for very slow movement of the needle.
4. When flashback is obtained, the assistant should thread the guidewire through the needle (straight end first) until >10 cm is inside. If resistance is felt, catheterization may not have been successful, and another attempt may be necessary.
5. If the guidewire threads in without obvious resistance, then the assistant should use sterile forceps to hold the guidewire in place, while the introducing needle is withdrawn over the guidewire. Take care when pulling back over the J-end of the guidewire, that the guidewire is not pulled back.
6. Thread the catheter over the guidewire until it is about to enter the skin; about 1 cm of guidewire should protrude from the catheter hub. Lubricate entry into the skin with drops of sterile saline. While holding the exposed end of the guidewire,

thread the catheter through the skin, until as much of its length as will easily go has disappeared under the skin. The guidewire can be withdrawn.

7. Flush the catheter with a small amount (1 mL) of sterile saline, and connect the infusion pump (35 mL/h). Tape the catheter firmly in place. Do not attempt to withdraw blood samples for 10–15 min while potential venous spasm subsides. During this time, return the room temperature to that desired for the study to be performed (e.g., 23°C).

8. Possible complications of siting the venous cannula include bruising and superficial infection. However, these are unusual (<3%) with good technique, and such a cannula can be left *in situ* without problem for 24 h.

## 3.4. Taking Blood Samples

1. Disconnect the infusion pump.
2. Connect a small syringe (1 mL is preferred) to the catheter, and gently withdraw the plunger until approximately 100–200 µL mixed blood and saline has entered the syringe, to clear the catheter of saline (dead volume approx 100 µL). Remove and discard this syringe.
3. Connect the syringe in which the sample is to be collected. Small syringes (up to 5 mL) are preferred to larger ones, because the tendency is less to pull too hard and collapse the vein. Withdraw blood samples slowly, over a period of a few minutes, if necessary (*see* **Note 5**).
4. After collecting the desired volume of sample (*see* **Note 6**), flush the catheter gently with 5 mL saline, and reconnect the saline infusion.
5. Measure hematocrit, if required (*see* **Notes 2 and 4**).

## 3.5. Collecting Arterial Blood

Methods for arterial cannulation are outside the scope of this review. There is a small but significant morbidity associated with arterial puncture *(18)*, and this should only be attempted by a skilled operator (*see* **Note 7**).

## 3.6. Analytical Techniques

Details of analytical technique are outside the scope of this review, but there are two major points to be borne in mind.

1. Some substances may be labile. This applies, for instance, to plasma TAG, which hydrolyzes on standing (presumably through the action of LPL in plasma), so that the plasma TAG concentration will decrease, but the plasma NEFA concentration will increase. For this reason, draw blood samples into chilled syringes, keep on ice, and separate plasma as rapidly as possible (*see* **Note 5**).
2. When measuring A-V differences, analytical precision may be crucial. If both arterial and venous concentrations are measured with a standard deviation (SD) of 100 µmol/L (as an example), then the SD of the A-V difference will be $\sqrt{2} \times$ 100 µmol/L or 141 µmol/L. Because the A-V difference may be small, this SD may render its precise measurement impossible. Potential solutions are to take multiple samples, or to make replicate analyses to improve precision.

## 4. Notes

1. Units of measurement. If concentrations are measured in μmol/L, and BF in mL blood 100 g tissue/min, then the product of A-V difference and BF gives the net rate of uptake or release in nmol/100 g tissue/min. Removal of substances may be expressed simply as fractional extraction (= A-V difference/arterial concentration), as absolute extraction (= A-V difference × BF) or as clearance (= fractional extraction × BF). If units for A-V difference and BF are as above, units will be: fractional extraction, dimensionless (often expressed as percentage); absolute extraction, nmol/100 g tissue/min; clearance, mL/100 g tissue/min.

2. For substrates that are essentially confined to the plasma compartment (e.g., NEFA, lipids), absolute exchange rates are obtained by multiplication of the plasma A-V difference by the rate of plasma flow. The latter is calculated from the BF by multiplication by (1 – hematocrit). For substances carried partly in plasma and partly in red blood cells, it is best to measure total concentrations in whole blood, e.g., after deproteinization with perchloric acid; misleading results from A-V difference studies have been reported when this is not done *(19)*. For AT glucose uptake, there may be special considerations. Glucose uptake is small in comparison with arterial concentration (typical fractional extraction is 3–4%; *[20]*). Accurate measurement of the A-V difference demands high analytical precision. We have found that glucose A-V differences measured on whole blood are less precise than those measured in plasma (probably because additional dilution factors are involved), and measuring glucose in plasma is preferred. Conversion to whole-blood concentrations is then necessary to estimate net uptake, and should be done using the formula proposed by Dillon *(21)*: $B = P (1 – 0.3 H)$, where $B$ is glucose concentration in whole blood, $P$ concentration in plasma, and $H$ is hematocrit expressed as a fraction.

3. The requirement for steady-state conditions has been discussed in the literature *(22,23)*. It is most easily seen if the arterial concentration of a substance is rising. Then, if simultaneous samples are taken from artery and vein, the venous sample will match the arterial concentration at some earlier time, depending on the time for transit of blood through the tissue. In addition, small molecules will equilibrate throughout the interstitial space, so, as the arterial concentration rises, there will be apparent uptake of substance, which does not represent true metabolic utilization. If the arterial concentration falls, the reverse will be true. One way of mostly overcoming nonsteady-state problems is to analyze the areas-under-curves for arterial and venous concentrations over reasonably long periods of time, rather than concentrations at individual time-points. For metabolic responses following a meal, for instance, this might be the whole postprandial period (e.g., 6 h), during which concentrations of many substances will rise, then fall again.

4. If concentrations of substances are to be compared, only some of which are confined to the plasma compartment (e.g., NEFA, TAG, and glycerol, when glycerol is distributed throughout whole blood), then it is necessary to convert all concentrations to those in whole blood. For NEFA and TAG, this is done as follows: $B = P (1 – H)$, where $B$ is glucose concentration in whole blood, $P$

concentration in plasma, and $H$ is hematocrit expressed as a fraction. Measurement of hematocrit in AT venous blood is not required, because it is not measurably different from that in arterial blood *(9)*.

5. Blood flow from the catheter may be slow. Samples of up to 10 mL may be collected from a freely flowing vein, but may require several minutes for withdrawal. A smaller vein, or one that has a tendency to spasm, may yield smaller samples. If the substances to be measured in the blood are labile, consider using a series of smaller syringes, and keep them on ice when not connected to the catheter.

6. If the sample volume desired cannot be collected, the following tips may help. Use finger pressure on the skin near the estimated catheter tip position to attempt to change the geometry of the vein–catheter relationship: This may allow blood to flow again. Ask the subject to move the leg (e.g., flex the hip), or even to cough, for the same reasons. If this fails, but the catheter still appears to be in the vein, the vein may be in spasm, and it is best to wait 60 min or so, with the warming pad in place over the abdomen, and the subject relaxed, to see if the situation has improved. As a last resort, a new catheterization may have to be made.

7. The potential problems associated with arterial cannulation have led some to use arterialized venous blood from a vein draining a warmed hand, as an alternative. Concentrations of many substances are indistinguishable in arterial and arterialized venous blood *(24)*. However, this is not true for $CO_2$ *(25)*, and may not be for catecholamines *(26)*; it should be verified for each substance to be measured in this way. The degree of arterialization may vary according the position of the catheter, the method of warming the hand, the state of the subject, and the ambient temperature, and it should be checked by measurement of blood $O_2$ saturation. The best method in our experience is to cannulate the cephalic vein in a retrograde direction, so that the tip of the cannula lies above or distal to the wrist joint, then place the hand in a box with still air warmed to 60°C. Keep the cannula patent between samples, by slow infusion (50 mL/h) of saline.

## Acknowledgment

We thank the colleagues who have worked with us in developing and exploiting this technique, particularly Sandy Humphreys for her skilled and dedicated contribution.

## References

1. Frayn, K. N., Fielding, B. A., and Summers, L. K. M. (1997) Investigation of human adipose tissue metabolism *in vivo*. *J. Endocrinol.* **155,** 187–189.
2. Wolfe, R. R. (1984) *Tracers in Metabolic Research: Radioisotope and Stable Isotope/Mass Spectrometry Methods.* Alan R. Liss, New York.
3. Frayn, K. N., Coppack, S. W., Humphreys, S. M., and Whyte, P. L. (1989) Metabolic characteristics of human adipose tissue *in vivo*. *Clin. Sci.* **76,** 509–516.
4. Bülow, J. (1982) Subcutaneous adipose tissue blood flow and triacylglycerol-mobilization during prolonged exercise in dogs. *Pflügers Archiv: Eur. J. Physiol.* **392,** 230–234.

5. Holloway, B. R., Stribling, D., Freeman, S., and Jamieson, L. (1985) Thermogenic role of adipose tissue in the dog. *Int. J. Obesity* **9,** 123–432.

6. Gooden, J. M., Campbell, S. L., and van der Walt, J. G. (1986) Measurement of blood flow and lipolysis in the hindquarter tissues of the fat-tailed sheep in vivo. *Quart. J. Exp. Physiol.* **71,** 537–547.

7. Kowalski, T. J., Wu, G., and Watford, M. (1997) Rat adipose tissue amino acid metabolism in vivo as assessed by microdialysis and arteriovenous techniques. *Amer J. Physiol.* **273,** E613–E622.

8. Frayn, K. N. (1992) Studies of human adipose tissue *in vivo*, in *Energy metabolism: tissue determinants and cellular corollaries* (Kinney J. M. and Tucker, H. N., eds.), Raven Press, Philadelphia, pp. 267–295.

9. Karpe, F., Humphreys, S. M., Samra, J. S., Summers, L. K. M., and Frayn, K. N. (1997) Clearance of lipoprotein remnant particles in adipose tissue and muscle in humans. *J. Lipid Res.* **38,** 2335–2343.

10. Fielding, B. A., Humphreys, S. M., Shadid, S., and Frayn, K. N. (1995) Arteriovenous differences across human adipose tissue for mono-, di- and triacylglycerols before and after a high-fat meal. *Endocrinol. Metab.* **2,** 13–17.

11. Coppack, S. W., Frayn, K. N., Humphreys, S. M., Dhar, H., and Hockaday, T. D. R. (1989) Effects of insulin on human adipose tissue metabolism in vivo. *Clin. Sci.* **77,** 663–670.

12. Frayn, K. N., Shadid, S., Hamlani, R., Humphreys, S. M., Clark, M. L., Fielding, B. A., Boland, O., and Coppack, S. W. (1994) Regulation of fatty acid movement in human adipose tissue in the postabsorptive-to-postprandial transition. *Am. J. Physiol.* **266,** E308–E317.

13. Frayn, K. N., Lund, P., and Walker, M. (1993) Interpretation of oxygen and carbon dioxide exchange across tissue beds in vivo. *Clin. Sci.* **85,** 373–384.

14. Coppack, S. W., Evans, R. D., Fisher, R. M., Frayn, K. N., Gibbons, G. F., Humphreys, S. M., et al. (1992) Adipose tissue metabolism in obesity: lipase action in vivo before and after a mixed meal. *Metabolism* **41,** 264–272.

15. Ong, J. M., and Kern, P. A. (1989) Effect of feeding and obesity on lipoprotein lipase activity, immunoreactive protein, and messenger RNA levels in human adipose tissue. *J. Clin. Invest.* **84,** 305–311.

16. Nilsson-Ehle, P., Egelrud, T., Belfrage, P., Olivecrona, T., and Borgström, B. (1973) Positional specificity of purified milk lipoprotein lipase. *J. Biol. Chem.* **248,** 6734–6737.

17. Samra, J. S., Frayn, K. N., Giddings, J. A., Clark, M. L., and Macdonald, I. A. (1995) Modification and validation of a commercially available portable detector for measurement of adipose tissue blood flow. *Clin. Physiol.* **15,** 241–248.

18. Groome, J., Vohra, R., Cuschieri, R. J., and Gilmour, D. G. (1989) Vascular injury after arterial catheterization. *Postgrad. Med. J.* **65,** 86–88.

19. Aoki, T. T., Brennan, M. F., Müller, W. A., Moore, F. D., and Cahill, G. F. (1972) Effect of insulin on muscle glutamate uptake. Whole blood versus plasma glutamate analysis. *J. Clin. Invest.* **51,** 2889–2894.

20. Coppack, S. W., Frayn, K. N., Humphreys, S. M., Whyte, P. L., and Hockaday, T. D. R. (1990) Arteriovenous differences across human adipose and forearm tissues after overnight fast. *Metabolism* **39,** 384–390.
21. Dillon, R. (1965) Importance of hematocrit in interpretation of blood sugar. *Diabetes* **14,** 672–678.
22. Zierler, K. L. (1961) Theory of the use of arteriovenous concentration differences for measuring metabolism in steady and non-steady states. *J. Clin. Invest.* **40,** 2111–2125.
23. Elia, M., Folmer, P., Schlatmann, A., Goren, A., and Austin, S. (1988) Carbohydrate, fat, and protein metabolism in muscle and in the whole body after mixed meal ingestion. *Metabolism* **37,** 542–551.
24. Frayn, K. N. and Macdonald, I. A. (1992) Methodological considerations in arterialization of venous blood. *Clin. Chem.* **38,** 316–317.
25. Forster, H. V., Dempsey, J. A., Thomson, J., Vidruk, E., and DoPico, G. A. (1972) Estimation of arterial $PO_2$, $PCO_2$, pH, and lactate from arterialized venous blood. *J. Appl. Physiol.* **32,** 134–137.
26. McLoughlin, P., Popham, P., Linton, R. A., Bruce, R. C., and Band, D. M. (1992) Use of arterialized venous blood sampling during incremental exercise tests. *J. Appl. Physiol.* **73,** 937–940.

# 23

# Measurement of Adipose Tissue Blood Flow

## Jens Bülow

## 1. Introduction

Adipose tissue (AT) is a well-vascularized tissue with a capillary density corresponding to that of skeletal muscle. The blood supply to the tissue is tightly coordinated with the tissue metabolism. AT blood flow (BF) is increased during exercise and starvation, i.e., in situations in which fat is mobilized from the tissue. The BF is also increased during fat deposition, especially after food intake, which coinsides with the idea that the increase in BFw in these situations is necessary either to carry nonesterified fatty acid mobilized during lipolysis away from the tissue or to supply lipoprotein-trigylcerol (TAG), especially very low density lipoprotein-TAG, to the tissue during lipid depositions.

Changes in AT BF may be brought about by a number of different effectors. Some of these are listed in **Table 1**. There are substantial differences between the BF levels in the various AT depots in the body, the visceral depots omental, mesenterial, perirenal, and epicardial having the highest flow levels. Flow values in the various AT depots in the resting, fasting dog originally described in **ref. _1_** is given in **Table 2**.

This chapter describes the various methods that have been applied to measure AT BF in humans and in experimental animals: tracer washout methods, microdialysis methods, and the microsphere method.

### 1.1. Tracer Washout Techniques

The theoretical background of these techniques has been described in detail in **ref. _2_**. In brief, the kinetic model used is a saturation–desaturation, with residue detection model implying that the parameter that will be measured is the initial slope of the desaturation curve, when the constant tracer infusion instantaneously is stopped. Measurement of this initial slope of the washout

From: _Methods in Molecular Biology, vol. 155: Adipose Tissue Protocols_
Edited by: G. Ailhaud © Humana Press Inc., Totowa, NJ

**Table 1**
**Effect of Various Physiological Stimuli on the Perfusion of AT**

| | |
|---|---|
| Starvation | Vasodilatation |
| Exercise | Vasodilatation |
| Feeding | Vasodilatation |
| Hypoglycemia | Vasodilatation |
| Orthostatic maneuvers | Vasoconstriction |
| Heat stress | Vasodilatation |
| Mental stress | Vasodilatation/constriction |
| Sympathetic stimulation | Vasodilatation/constriction |
| Insulin (physiological level) | No effect |
| Insulin (pharmacological level) | Vasodilatation ? |
| Adenosine | Vasodilatation |
| High NEFA:albumin ratio | Vasoconstriction |

**Table 2**
**Perfusion Coefficients (mL/100 g/min) in Various AT Depots in Resting, Fasting Dog[a]**

| | |
|---|---|
| Inguinal, sc | $6.2 \pm 1.3$ |
| Preperitoneal | $13.3 \pm 4.1$ |
| Omental | $7.6 \pm 1.9$ |
| Mesenteric | $9.5 \pm 2.0$ |
| Perirenal | $10.2 \pm 2.6$ |
| Parametrial | $6.9 \pm 2.0$ |
| Pericardial | $10.8 \pm 1.8$ |
| Popliteal | $2.8 \pm 0.6$ |

[a]Adapted with permission from **ref. 1**.

curve is, however, only possible when the tissue behaves as a monocompartmental system with homogenously distributed BF and homogenous tissue composition. These prerequisites are normally fulfilled or nearly fulfilled in AT, implying that the washout curves are monoexponential. The perfusion coefficient in the tissue region studied can then be calculated as the product of the exponential rate constant (k) and the tissue/blood distribution coefficient ($\lambda$) of the tracer applied as $f = k \times \lambda \times 100$ where f is the perfusion coefficient in mL/(100 g/min), $k$ is the exponential washout rate-constant in $min^{-1}$ and $\lambda$ is given in mL/mL.

## 1.2. Microdialysis (see Notes 1–3)

Microdialysis has been applied for measurements of relative changes in AT BF in humans *(3)*. The measurements are based on the principle that, when a substance, which is not metabolized locally in the tissue, and which does not affect the local BF, is added to the microdialysis perfusate, the rate of diffusion out of the probe will depend on the concentration gradient between the perfusate and the interstitial space. The concentration in the interstitial space will depend on the rate with which the substance is supplied from the microdialysis probe, and the rate with which it is removed by the local BF. Thus, the change in concentration from perfusate to dialysate will be a function of the local BF. Traditionally, the changes in outflow:inflow ratio, i.e., the ratio between the concentration in the dialysate and in the perfusate, has been used as an index of changes in local BF. This concept has been validated against the [133]Xe-washout technique *(3)*. The major drawback of the method is, that it only provides qualitative information about the changes in BF, i.e., information about whether BF has increased, decreased, or has not changed during the experiment. On the other hand, it is well suited to study effects of local physiological or pharmacological perturbations. Such local perturbations can be obtained by infusion of vasoactive substances or metabolically active substances in the microdialysis fiber.

### 1.3. Microsphere Method (see Notes 4–9)

The microsphere method is only suited for measurements of BF in experimental animals, because the method demands that the animal is sacrificed at the end of the experiment *(4–6)*. The advantage of the method is that perfusion coefficients can be measured simultaneously in many different tissue regions and types. A disadvantage is that it only provides a snapshot of the BF distribution. Microspheres can be obtained with different diameters in the range from approx 7 µm, which may pass capillaries, to 50 µm, which will be caught in arteriovenous anastomoses. For BF measurements, a 15-µm diameter is most commonly applied. In studies of shunting fractions, different diameters can be applied.

The microsphere technique is based on the bolus fractionation principle, which states that a bolus of microspheres injected centrally in the arterial system will be distributed throughout the body in accordance with the distribution of the cardiac output, provided that the spheres are well mixed in the total cross-sectional area at the injection site, and that they are trapped in the capillaries *(5)*.

Microspheres are available labeled with 12 different radioactive isotopes. Thus, it is possible to perform up to 12 flow measurements in each experimental animal.

## 2. Materials
### 2.1. Tracers for Washout Techniques

Several substances have been used. [133]Xe is the most widely applied since it was introduced in 1966 *(7)* for this purpose. Other tracers that have been used are [127]Xe, [81]Kr [99m]Tc, and [131]I- antipyrin. Of these the inert gases (Xe and Kr) are the most lipid-soluble, i.e., they have a high $\lambda$. A high $\lambda$ will result in a low *k*, thus allowing extended measurements on the same isotope depot.

## 2.2. Detector Systems for Washout Techniques

Several scintillation detector systems have been used. Most commonly NaI(Tl) crystal detectors have been applied. Frequently, the detector is placed at some distance from the subject (and depot) in a special holder. In this case, the detector is traditionally referred to as a stationary detector, in contrast to portable detectors, which are in principle equivalent to the stationary systems, but are much smaller, the crystals are either CdTe(Cl) or CsI(Na) crystals, but also NaI(Tl) detectors may be portable, when the crystal is small, i.e., less than 1/4 in. in diameter. **Figure 1** shows examples of a stationary and a portable detector.

A third possibility is to apply a γ-camera. However, this is not very practical due to the dimensions of the system and the high cost, compared to the smaller detector systems. On the other hand the γ-camera has some features that make it attractive under certain circumstances. First, it has the advantage that it is possible to image the isotope depot throughout the study, which makes it much easier, in the postprocessing of the washout curves, to deal with changes in the geometry between the detector and the isotope depot.

The geometry between detector and isotope depot (the counting geometry) is of crucial importance for both types of detector systems. With respect to the portable detectors, the smaller crystals imply that the detector field of view is narrow, compared to the stationary detectors. This has the implication that the portable detectors may be very sensitive to changes in the geometry between the detector and the isotope depot caused by diffusion of the depots toward or away from the detector in the vertical direction. This may change the counting efficiency dramatically, if the isocount profiles of the detector are close to each other, which may be the case, especially when the detector is collimated. The problem is most serious when the distance between the detector and the depot is short, especially when the detector is attached directly to the skin immediately above the depot. This may be solved by lifting the detector some mm away from the skin surface *(8,9)*. Another geometrical problem which may arise is that the isotope depot not only may move due to diffusion, but also because of countercurrent exchange between the vein draining the depot and the concomitant artery. This process implies that the depot changes shape with time, which means that the counting efficiency also may change because of lateral movement of the depot (*10*; e.g., *see* **Fig. 2**).

With respect to the stationary detector systems, the geometry problems are solely caused by changes in the distance between the isotope depot and the detector. Since the γ-rays are emitted in a spherical fashion from the depot, this implies that the counting efficiency will change with the square of the change in distance. To solve this problem, the distance between the depot and detector can be lengthened, so that the effect of the geometry changes occurring during movements of the body is minimized. This normally demands a distance of

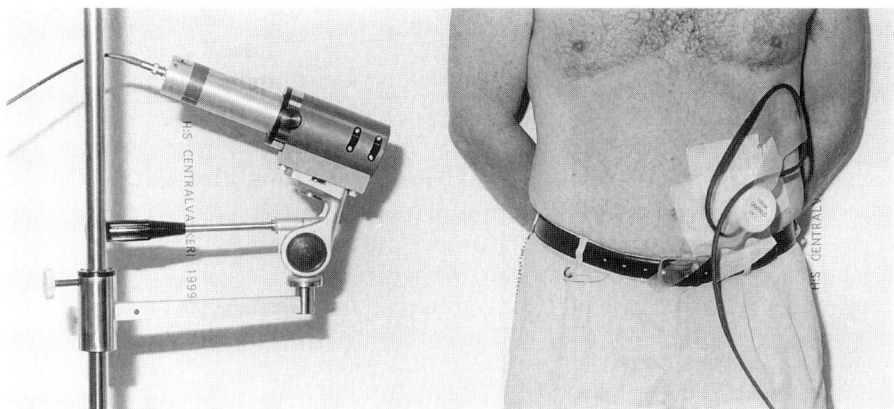

Fig. 1. Examples of a stationary detector (*left*) and a portable detector (*right*).

Fig. 2. The setup used for determination of maximal count rate.

0.7–1 m from detector to depot. The drawback of this is the necessity of using higher doses of radioactivity, in order to obtain the same count rate. Another solution is to use an external source with constant activity as reference source. This technique is described in detail in **ref. *11***. The advantage of the technique is that it also corrects for systematic changes in the counting. The major disadvantage is that it demands at least a two-channel discriminator system.

## 3. Methods (*see* Notes 10–15)

### 3.1. Registration of Count Rate

1. The registration of the radioactive decay is done by γ-spectrometers, which are either mono- or multichannel systems.

2. When using these systems, it is important to know the maximal count rate of the system.
3. Because it takes a fixed time to detect a pulse (a time period in which other pulses cannot be detected), it is crucial to know the effect of this period, known as "system dead time."
4. The effective dead time determines the maximal number of counts that can be acquired per second without significant loss. The dead time is normally specified by the manufacturers, and information about the maximal count rate is given in the system specification. However, this count rate is frequently measured by use of a tone generator, which feeds pulses equidistantly into the system. This will result in a much higher maximal count rate than when the pulses are fed into the system in a statistically randomized fashion, as is the case with radioactive decay. In this case, the maximal count rate will often be lower than given in the system specifications, and will depend on the width of the window. A narrow window will reduce the maximal count rate in many systems. It is therefore necessary to determine the maximal count rate in each system in the laboratory. In many systems, it will be in the range from 1000–5000 counts/s.

### 3.2. Study Protocols

1. Depending on the length of the study, a lipophilic tracer ($^{133}$Xe) or a hydrophilic tracer ($^{99m}$Tc) may be used.
2. In studies lasting more than 1 h, it is necessary to use $^{133}$Xe; $^{99m}$Tc may be used in experiments lasting less than 60 min, because of the washout rate constant for $^{133}$Xe is 10- to 15-fold larger than that for $^{99m}$Tc, because the difference between the respective tissue/blood distribution coefficients.
3. On the other hand, when using $^{99m}$Tc (or $^{131}$I-antipyrin), it is necessary to correct for the recirculation of the tracer since the arterial tracer concentration is not zero, as it is for the gaseous tracers, e.g., Xe. This correction is most easily performed by measurement of the count rate in a tissue area symmetrical to the isotope depot, then subtracting this from the count rate in the depot.
4. When using a tracer with high $\lambda$, it is crucial to measure changes in BF in the same isotope depot throughout the whole experiment, because the solubility coefficients may be somewhat different in different depots. Therefore, experiments should be designed so that there are sufficient amounts of activity in the depot for the whole experiment, in order to avoid reinjection of tracer.
5. When the tracer has been injected in the tissue, it is necessary to wait for the hyperemia after the injection trauma settles. This normally lasts about 30 min, but in some subjects it can last more than 60 min. This problem may be overcome by using so-called "atraumatic labeling," a technique originally described for skin BF measurement *(12)*. It may be used also in subcutaneous (sc) AT, but it implies a number of difficulties when using portable detectors because of the diffusion of the depot.

## 4. Notes

1. Microdialysis probes are commercially available or may be manufactured in the laboratory. Commercially available probes most commonly used are the double-

lumen probes from CMA Microdialysis (Stockholm, Sweden) with membrane length 10, 20, or 30 mm. Since the highest sensitivity for detection of BF changes is obtained when the outflow:inflow ratio is low, longer membranes are preferable. Probes manufactured in the laboratory can be made of different dialysis tubes. An easy way to obtain tubes with proper pore size, membrane stability, and reasonable dimensions is from artificial dialysis kidneys, as, e.g., GFE12 or GFS16 (Gambro, Lund, Sweden). One fiber from such a kidney is cut to the desired membrane length, and both ends are glued to a nylon tubing of 0.25 mm inner diameter and 0.4 mm outer diameter. Cyanoacrylate is a glue with appropriate viscosity characteristics and in vivo stability. The advantage of the commercially available probes is that they are easily available, the insertion trauma may be smaller than the trauma induced when homemade probes are inserted. The latter procedure demands that a G18 cannula is passed in and out of the region in which the probe will be positioned. This procedure may give rise to unintended bleeding. The disadvantage of the commercially available probes is that they are expensive. The price of a probe is about 100-fold higher than the price of a homemade probe.

2. For BF measurements, Ringer's acetate, containing 4 m$M$ glucose, 1 m$M$ lactate, and either 10 m$M$ EtOH, or 3.9 kBq/mL $^{14}$C-EtOH, or 3.7 kBq/mL $^3$H$_2$O. Glucose and lactate are added to the perfusate, in order to prevent drainage of these substances from the interstitial space, with possible changes in local tissue metabolism as a result. With this perfusate composition, the perfusion rate of the probe should be 2 µL/min when the membrane length is 10 mm. However, it is crucial to select the perfusion rate properly. It must be selected taking both the membrane length and the local perfusion rate into account, in order to optimize the outflow:inflow ratio to be most sensitive to BF changes. A ratio below 0.5 should be aimed for.

3. Dialysate should be collected in capped glass tubes (e.g., 300 µL), in order to prevent evaporation during collection. Immediately after sampling, especially when EtOH is used, the dialysates should be analyzed either by scintillation counting or chemically. Samples containing EtOH can not be frozen.

4. In order to obtain complete mixing of the microspheres in the blood stream the bolus must be injected in the left ventricle, or better, in the left atrium. Preferentially, a catheter with pig tail end curve should be used for the injection.

5. In order to be able to quantitate the regional perfusion coefficients, an external reference organ must be established. A catheter for this purpose is inserted into the abdominal aorta, and blood is drawn at a constant rate from this during the passage of the bolus.

6. At the end of the experiment, biopsies are taken from the organs of interest, and the radioactivity is measured in a well-counter. The biopsy material must be of a size containing a proper number of spheres, in order to minimize the effects of random distribution of the spheres in the tissue. If the biopsy contains about 624 spheres and the reference blood sample contains about 1000 spheres, the coefficient of variation on consecutive flow determinations during otherwise steady

state conditions will be 10% (*4*). Thus, in the planning of the experiment, it is necessary to have an estimate of the flow level in the tissue and the amount of tissue that can be collected for counting, in order to calculate the minimum number of spheres needed.

7. Since the perfusion rate of adipose tissue is in the range of 2–10 mL/100 g/min during rest in the fasting state, increasing to about 8–30 mL/100 g/min, biopsies of 10 g are often needed, when a bolus of $0.5 \times 10^6$/kg body wt is given. This dose is well tolerated when given as 15 µ spheres; spheres of larger diameter may give rise to transient circulatory disturbances (*5*). When counting biopsies of 10-g size, it is necessary to split the biopsy into smaller pieces, in order to avoid loss of counting efficiency caused by so-called "edge phenomena." Whether such edge phenomena become significant depends primarily on the thickness of the scintillation crystal and the energy of the isotope. However, as a rule, it is advisable not to fill the well to more than half its depth. An experimental way to determine the effective upper height allowed for each isotope is to put a drop of the isotope solution in a counting vial, then add solute to the vial until the counting efficiency begins to decrease.

8. Regional perfusion coefficients are calculated from the activity in the biopsy, the weight of the biopsy, and the activity in the reference blood sample as $f = (C_{tissue} \times v/C_{blood} \times w) \times 100$, where $C_{tissue}$ and $C_{blood}$ are the number of counts in the tissue and 1 mL blood, respectively, $w$ is the weight of the biopsy, and $v$ is the rate at which the blood is drawn.

9. When several BF determinations are performed, it is necessary to count a number of isotopes simultaneously. In order to separate the count rates of the different isotopes, a set of linear equations with $n$ unknowns ($n$ = number of isotopes) must be solved. To do this implies that the crosstalk of each isotope in all channels must be determined. This is done by counting a pure sample of the isotope in all channels. The solution of the equations may be obtained by commercially available software. A description of a program for this purpose is given in **ref. 6**.

10. Set the total equipment to the desired high voltage, gain, and window width. Consult the user handbook from the manufacturer. In modern equipment, these settings are rather stable, but it is wise to check the settings at least every second month, in order to assure a symmetrical setting around the principal photo peak.

11. Determine the maximal count-rate of the system before, e.g., 5% of the counts are lost in the actual setting. This is most easily done by placing a (point) source at the end of a ruler, e.g., 2 m away from the detector. The strength of the source is chosen so that the count rate in this geometry is well below the maximal count rate, e.g., 400 cps. The source is now gradually moved toward the detector, and the count rate in the new geometry is registered. The theoretical count rate in this geometry, is calculated from the count rate in the previous geometry according to the distance rule, which says that the count rate varies with the square of the change in counting geometry. When the observed count

rate deviates from the theoretical count rate, the upper count rate has been found. An example of the procedure is given in **Table 3** and **Fig. 3**. In that case the maximal count rate was found to be 2300 cps.

12. When using the traumatic labeling technique, the isotope is injected into the tissue dissolved in an isotonic solution of sodium chloride, using a subdermal cannula (e.g., 0.4 mm in outer diameter ~G25). Optimmally, the injection should be performed in the middle of the tissue layer. In order to minimize the injection trauma, the injection should be performed with gentle movements, and with a minimum of sideway movements of the tip during the injection. If the cannula is retracted immediately after the injection, this may result in loss of isotope through the injection channel, especially when the procedure gives rise to bleeding. There-fore, it is wise to keep the cannula *in situ* for at least 15–20 s after the injection, to minimize this loss. The volume injected should not exceed 0.1 mL, since larger volumes may give rise to initial problems with very low, or no washout caused by compression of the microcirculation, resulting in the depot not being washed out for an undefined period of time. The amount of radioactivity injected should be chosen carefully, in order to minimize the radiation to which the subject is exposed. In case of stationary detectors, the dose of radioactivity needed usually is between 3.7 MBq (100 µCi ) and 18.5 MBq (500 µCi). For portable systems, the amount needed is approx 1–3 kBq (0.3–1 µCi). The initial amount of isotope is determined by the counting efficiency of the system, the length of the study in relation to the expected biological washout rates, and the length of the periods in which counts are collected, i.e., the time resolu-tion in the registration. In order to minimize the random statistical noise of the registrations, it is advisable to collect more than 10,000 counts in every registration.

Atraumatic labeling is performed by taping a gas-tight membrane (a mylar membrane is well suited), to the skin above the AT region which is to be examined. A droplet of the isotope solution is then injected into the space between the membrane and the skin. The Xe dissolved in the saline will then diffuse into the tissues below the membrane. After 30 min, the membrane is removed, and the rest of the isotope on the skin surface is removed. By this procedure, both skin and AT will be labeled. However, because of the low solubility in skin, compared to AT, the contribution from skin will be negligible after about 30 min.

The perirenal AT is a depot, which, in humans, is accessible for labeling. The technique demands access to ultrasound guided puncture facilities, but, apart from this, the techniques used to study this depot do not differ basically from those used to study the sc depot *(13)*.

13. In normal-weight persons, the abdominal sc $\lambda_{t/b}$ for Xe is an average of approx 8 *(14)*. Traditionally, a value of 10 has been applied. In the perirenal depot, the average value is approx 5 *(14)*. An estimate of the actual $\lambda_{t/b}$ can be obtained from measurement of the skinfold thickness (SFT) either by a Harbenden skinfold caliper or by ultrasound imaging. The correlation between local SFT and the sc abdominal $\lambda_{t/b}$ is given by the equation $\lambda_{t/b} = 0.22.SFT + 2.99$ *(14)*.

Another way to obtain an estimate of the tissue–blood partition coefficient is to mix two tracers with very different λ values in the same depot, then register

**Table 3**
**Relation Between Source to Detector Distance (D) and Observed (R) and Theoretical (T) Count Rates**

| D cm | 160 | 140 | 120 | 100 | 80 | 70 | 65 | 60 | 55 | 50 | 45 |
|---|---|---|---|---|---|---|---|---|---|---|---|
| R cps | 350 | 450 | 600 | 900 | 1380 | 1800 | 2100 | 2400 | 2800 | 3400 | 4000 |
| T cps | | 457 | 622 | 895 | 1398 | 1825 | 2116 | 2483 | 2955 | 3575 | 4414 |

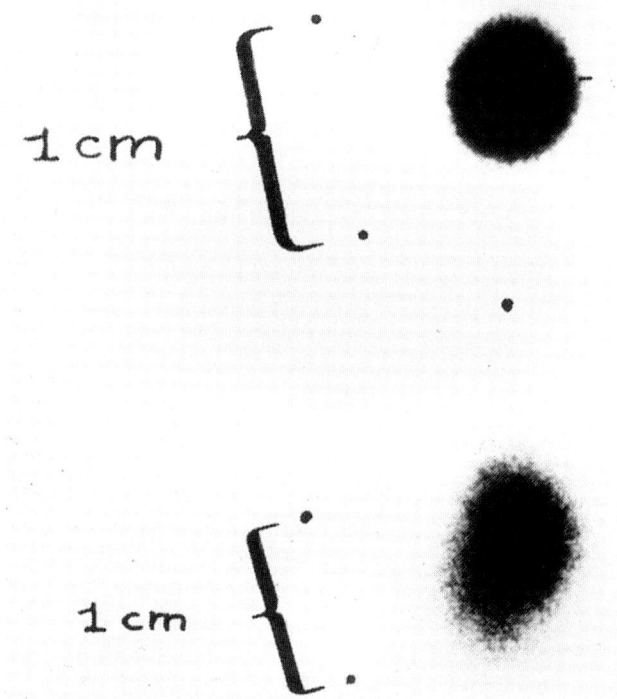

Fig. 3. Initial (*upper*) and late (*lower*) shape of a Xe depot. 0.1 mL saline solution was injected.

the washout of the two tracers simultaneously. The idea is to use a tracer with a λ virtually independent of the lipid content of the tissue, i.e., a hydrophilic tracer as, e.g., $^{99m}$Tc, which has a λ about 0.7. Another possibility is $^{131}$I-antipyrin, with a λ of about 1.1 mL/mL *(15)*. When the washout rates of the two tracers are measured simultaneously, the unknown λ can be calculated from the ratio between the washout rates and the known λ of the hydrophilic tracer.

The tissues–blood distribution coefficient may be determined directly in a biopsy from the region in which the washout has been measured, which demands a tissue mass of at least 400 mg. This can only be obtained as a surgical biopsy. Needle biopsies will normally result in biopsies of approx 100–200 mg, so, unfortunately, it is not possible to use this technique for this purpose.

The biopsy is transferred to a preweighted glass vial. After determination of the initial weight of the vial and biopsy, they are transferred to a dessicator in order to remove the tissue water. This process may take 2–3 wk. When the weight of the biopsy has stabilized, the difference between the original weight and the new weight is regarded as the result of evaporation of tissue water. The lipid in the biopsy is then extracted 3× with methanol:heptane:chloroform (1.4:30:40), dried, and weighed. When the weight is stabilized, the lipid content is calculated from the weight difference. The remaining material is regarded as being protein. The concentration of Xe in tissue is then calculated as $C_t = F_L \times \lambda_L + F_P \times \lambda_P + F_W \times \lambda_W$. The blood concentration is calculated as $C_b = Hct \times \lambda_{bc} + (1 - Hct) \times \lambda_{pl}$. $\lambda_{t/b}$ is then $C_t \times C_b^{-1}$. In these equations, $F_L$, $F_P$, and $F_W$ are the fractional contents of lipid, protein, and water in the tissue $Hct$ is the hematocrit, and $\lambda_L$, $\lambda_P$, $\lambda_W$, $\lambda_{bc}$, and $\lambda_{pl}$ are the Ostwald solubility coefficients in lipid, protein, water, blood cells, and plasma. In **Table 4** are given the $\lambda$ values measured in the different tissue phases at 37°C. These values were originally published in **ref. 16**.

14. Correction for changes in counting geometry is especially relevant during exercise experiments in which stationary detectors are used. This can be done by applying an external reference source virtually placed in the same counting geometry as the AT tracer depot. Sources of [57]Co (half-life, 270 d) or [99m]Tc (half-life, 6.04 h) may be used. The drawback of using the latter is that it needs to be manufactured for each experiment, and the short half-life needs to be taken into account when the correction factors are calculated. The procedure was originally described in **ref. 11**, and an example of the effect of the procedure is given in **Fig. 4**.

    Correction terms for the effect of diffusion on the washout rate registered with a portable detector may be obtained in separate experiments from depots placed in a tissue area without circulation. This situation is most easily established experimentally in a lower extremity by inflation of a blood pressure cuff placed proximal on the thigh, to a pressure above systolic pressure. If this is done in periods of, e.g., 30 min duration at different time-points after the depot has been injected, the influence of the movements of the depot caused by diffusion may be obtained. The overall rate constants can then be corrected for the fraction caused by diffusion, the remaining part only representing removal of tracer resulting from the tissue perfusion.

15. A local depot of [133]Xe of 3.7MBq (100 μCi) will give rise to a whole body radiation dose less than 0.5 mSv, which is only about 10% of the natural background radiation in most countries. The other isotopes applied for this type of measurement give rise to radiation doses of the same size. Thus, in general, the method may be applied without radiation safety problems, especially when the portable systems are used.

**Table 4
Ostwald Solubility
Coefficients for Xe
in mL/g at 37°C[a]**

| | |
|---|---|
| Water | 0.0827 |
| 0.9% NaCl | 0.0797 |
| Plasma | 0.0939 |
| Erytrocytes | 0.2710 |
| AT | 1.7151 |
| Olive oil | 1.8276 |

[a]Adapted with permission from **ref. 8**.

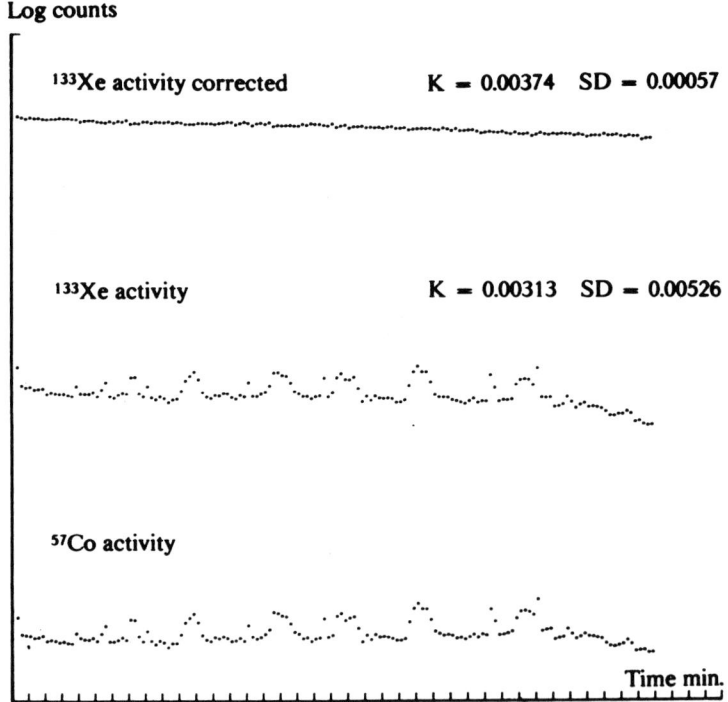

Fig. 4. Example of a correction of a Xe washout curve with an external Co source during exercise.

## References

1. Bülow, J. and Tøndevold, E. (1982) Blood flow in different adipose tissue depots during prolonged exercise in dogs. *Pflügers Arch.* **392,** 235–238.

2. Lassen, N. A., Henriken, O., and Sejrsen, P. (1983) Indicator methods for measurement of organ and tissue blood flow, in *Handbook of Physiology*, vol. 3, *The Cardiovascular System*. (Shepard, J. T. and Abboud, F. M., eds.), Physiological Society, MD, pp. 21.

3. Fellander, G., Linde, B., and Bolinder, J. (1996) Evaluation of the microdialysis ethanol technique for monitoring of subcutaneous adipose tissue blood flow in humans. *Int. J. Obes.* **20,** 220–226.

4. Dole, W. P., Jackson, D. L., Rosenblatt, J. I., and Thomson, W. L. (1982) Relative error and variability in blood flow measurements with radiolabeled microspheres. *Am. J. Physiol.* **243,** H371–H378.

5. Hales, J. R. S. (1974) radioactive microsphere techniques for studies of the circulation. *Clin. Exp. Physiol. Pharmacol.* **Suppl. 1,** 31–46.

6. Saxena, P. R., Schamhardt, H. C., Forsyth, R. P., and Hoeve, J. (1980) Computer programs for the radioactive microsphere technique. Determination of regional blood flows and other haemodynamic variables in different experimental circumstances. *Compu. Prog. Biomed.* **12,** 63–84.

7. Larsen, O. A., Lassen, N. A., and Quaade, F. (1966) Blood flow through human adipose tissue determined with radioactive Xenon. *Acta Physiol. Scand.* **66,** 337–345.

8. Bojsen, J., Staberg, B., and Kølendorf, K. (1984) Subcutaneous measurements of $^{133}$Xe disappearance with portable CdTe(Cl) detectors: Elimination of interference from combined convection and diffusion. *Clin. Physiol.* **4,** 309–320.

9. Samra, J. S., Frayn, K. N., Giddings, J. A., Clark, M. L., and Macdonald, I. A. (1995) Modification and validation of a commercially available portable detector for measurement of adipose tissue blood flow. *Clin. Physiol.* **15,** 241–248.

10. Bojsen, J., Kølendorf, K., and Staberg, B. (1983) Comparison of portable CdTe(Cl) detectors with stationary NaI(Te) detectors for subcutaneous $^{133}$Xe disappearance measurements. *Clin. Physiol.* **3,** 325–334.

11. Bülow, J. and Madsen, J. (1975) Compensation for geometric changes during monitoring of $^{133}$Xe washout from subcutaneous adipose tissue. Scand. *J. Clin. Lab. Invest.* **35,** 641–644.

12. Sejrsen, P. (1971) Measurements of cutaneous blood flow by freely diffusible radioactive isotopes. *Dan. Med. Bull.* **Suppl. 18,** 9–38.

13. Bülow, J. and Madsen, J. (1978) Human adipose tissue blood flow during prolonged exercise II. *Pflügers Arch.* **376,** 41–45.

14. Bülow, J., Jelnes, R., Astrup, A., Madsen, J., and Vilmann, P. (1987) Tissue/blood partition coefficients for xenon in various adipose tissue depots in man. *Scand. J. Clin. Lab. Invest.* **47,** 1–3.

15. Jelnes, R., Astrup, A., and Bülow, J. (1985) The double isotope technique for *in vivo* determination of the tissue-to-blood partition coefficient for Xenon in human subcutaneous adipose tissue, an evaluation. *Scand. J. Clin. Lab. Invest.* **45,** 565–568.

16. Chen, R. Y. Z., Fan, F.-C., Kim, S., Jan, K.-M., Usami, S., and Chien, S. (1980) Tissue-blood partition coefficient for xenon: Temperature and hematocrit dependence. *J. Appl. Physiol.* **49,** 178–183.

# 24

## Respiratory and Thermogenic Capacities of Cells and Mitochondria from Brown and White Adipose Tissue

### Barbara Cannon and Jan Nedergaard

## 1. Introduction

Adipose tissues (ATs) in mammals are distinguished as being brown or white. Brown adipose tissues (BAT) has as its function the production of heat, and thus has a high oxidative capacity, evidenced by the extraordinarily high density of mitochondria in the cells. White adipose tissue (WAT) is primarily an energy-storing tissue, with low oxidative capacity. However, studies of metabolic activity are relevant for both tissues.

Because thermogenesis (heat production) is the function of BAT, it would be a natural choice to measure this directly. However, this has only been done infrequently, mostly because of technical limitations (microcalorimeters are still not widespread laboratory equipment, and are expensive). However, it has been calculated that respiratory determinations are indeed satisfactory measures of heat production *(1)* (this is probably true for most mammalian organs with good blood supply). It is therefore routine to perform respiratory measurements, and to equate the result with thermogenesis (and metabolic activity in general).

Respiratory measurements can be performed on isolated mitochondria and isolated cells (and on tissue pieces). In order to obtain sound values, it is essential that the measurements are made under optimal conditions, including provision of an adequate $O_2$ supply throughout the experiment, and, also, of a substrate for respiration that is not limiting.

Because of the requirement for sufficient $O_2$, tissue pieces can often be problematic: $O_2$ supply may be limited by the diffusion of $O_2$ through the piece of tissue. Dispersed cells and mitochondria can be more easily oxygenated, but are obviously more artificial in other respects.

From: *Methods in Molecular Biology, vol. 155: Adipose Tissue Protocols*
Edited by: G. Ailhaud © Humana Press Inc., Totowa, NJ

The supply of a suitable substate for respiration is often difficult. To estimate maximal capacity, the rate of substrate supply must exceed that of the ongoing respiration. For BAT, respiration is uncoupled from phosphorylation under conditions when thermogenesis is activated, and the rate is thus limited by the capacity of the uncoupling protein (UCP1) or by the respiratory chain. In WAT, respiration is normally coupled to adenosine diphosphate (ADP) phosphorylation, and the rate is therefore determined by the rate of utilization of adenosine triphosphate (ATP). Choice of a nonoptimal substrate, the transport of which is rate-limiting, can provide spurious results, leading to erroneous conclusions.

Isolated mitochondria from BAT are relatively easy to study, because the mature cells contain such high mitochondrial density that the mitochondrial population, isolated after homogenization of whole tissue, is statistically representative for the mature adipocytes. For WAT, the mitochondrial density in the adipocytes is low; a mitochondrial preparation from total tissue may therefore not be representative for white adipocyte mitochondria, and it may therefore be necessary to isolate mitochondria from isolated cells (which, however, leads to very low yields).

Mature adipocytes can be conveniently isolated from both tissues, based on Rodbell's classic collagenase digestion procedure *(2)*, because the fat-containing cells readily float, and can thus be separated from tissue debris in aqueous media.

## 2. Materials

### 2.1. Isolation of Mitochondria

1. 0.25 $M$ sucrose (*see* **Note 1**).
2. 100 m$M$ KCl containing 20 m$M$ K-TES ($N$-tris[hydroxymethyl]methyl-2-amino-ethanesulfonic acid), pH 7.2.
3. Bovine serum albumin (BSA), (fraction V), fatty-acid (FA)-free.
4. Glass homogenizer with tight-fitting Teflon pestle.
5. Small glass hand homogenizer.
6. High-speed centrifuge with fixed-angle rotor, tube size ≈50 mL.
7. Gauze.

### 2.2. Isolation of Adipocytes

1. Krebs-Ringer phosphate buffer (KRPB) with the following composition: 148 m$M$ Na$^+$, 6.9 m$M$ K$^+$, 1.5 m$M$ Ca$^{2+}$, 1.4 m$M$ Mg$^{2+}$, 119 m$M$ Cl$^-$, 1.4 m$M$ SO$_4^{2-}$, 5.6 m$M$ H$_2$PO$_4^-$, 16.7 m$M$ HPO$_4^{2-}$, 10 m$M$ glucose, and 10 m$M$ fructose. Include 4 % crude BSA. Adjust pH with Tris-OH or HCl to 7.4.
2. Krebs-Ringer bicarbonate buffer with the following composition: 145 m$M$ Na$^+$, 6.0 m$M$ K$^+$, 2.5 m$M$ Ca$^{2+}$, 1.2 m$M$ Mg$^{2+}$, 128 m$M$ Cl$^-$, 1.2 m$M$ SO$_4^{2-}$, 25.3 m$M$ HCO$_3^-$, 1.2 m$M$ H$_2$PO$_4^-$, 10 m$M$ glucose, 10 m$M$ fructose, and 4% FA-free BSA. Bubble the buffer with 5% CO$_2$ in air at 37°C, and adjust the pH with Tris-OH or HCl to 7.4; keep the buffer at 37°C, and continuously bubble with a small stream of 5% CO$_2$ in air until use.
3. Crude and FA-free BSA (fraction V).

4. Crude collagenase (type I, clostridiopeptidase A, EC 3.4.24.3).
5. Water shaker at 37°C.
6. Silk filter cloth (Joymar Scientific, Hicksville, NY).
7. Bürker chamber for cell counting.

## 2.3. Oxygen Electrode

1. Clark type oxygen probe. Available from, e.g., Radiometer, Rank Bros. or Yellow Springs Instruments (Yellow Springs, OH). Purchased as a complete system with measuring chamber and magnetic stirrer.
2. Chart MacLab application program.
3. Chart recorder.

# 3. Methods
## 3.1. Isolation of BAT Mitochondria

For a routine preparation of BAT mitochondria, use one adult Syrian hamster *(Mesocricetus auratus)*, acclimated to 4°C for at least 4 wk (*see* **Note 2**). All procedures are carried out at 0–4°C.

1. The hamster is killed by $CO_2$ anesthesia, and decapitation. Dissect out the peri-aortic, interscapular, cervical, and axillary BAT carefully, and, combined, rinse in sucrose.
2. Mince the tissue with scissors, and homogenize the mince in about 40 mL 250 m$M$ sucrose solution (approx 5% [w/v]), in a glass homogenizer with Teflon pestle. Five to six strokes are required.
3. Filter the homogenate through two layers of gauze, and centrifuge (all centrifugation steps for 10 min) at 8500$g$.
4. Discard the hardpacked fat layer and supernatant by rapidly inverting the tube, and wipe the walls of the tube clean with a paper tissue.
5. Resuspend the pellet (containing cell debris, nuclei, and mitochondria) in a small volume of sucrose, and transfer to a clean tube. It is then diluted again to about 40 mL sucrose solution, and centrifuged at 800$g$.
6. Carefully transfer the supernatant to a clean tube. Discard the pellet (which contains debris and nuclei). Centrifuge the supernatant at 8500$g$.
7. Resuspend the resulting mitochondrial pellet in 5 mL sucrose solution with 2% FA-free BSA and centrifuge at 8500$g$.
8. Further wash the albumin-washed mitochondrial pellet by one of two procedures: For respiratory studies, resuspend the pellet in about 15 mL of the KCl-TES buffer (*see* **Note 3**). For other studies, resuspend the pellet in 15 mL sucrose solution. In both cases, centrifuge the suspensions at 8500$g$.
9. Resuspend the resulting pellet in a minimal volume of the respective medium, by hand homogenization.
10. Measure the protein concentration in the final, albumin-washed mitochondrial pellet, and dilute the suspension with KCl-TES buffer or sucrose solution to a stock concentration of 10 or 20 mg/mL, for storage on ice.

## 3.2. Isolation of Brown Adipocytes

For a routine preparation of brown adipocytes, two adult (10- to 30-wk-old) Syrian hamsters (*M. auratus*) of either sex are used. The hamsters are kept at 20–22°C, 1–3 per cage, with food and water *ad libitum* (*see* **Note 4**).

1. The animals are killed by $CO_2$, and decapitated. Dissect out the interscapular, axillary, and cervical BAT and carefully clean from contaminating tissues.
2. Place the BAT in a polyethylene vial containing 3 mL KRPB with 4% crude BSA and 0.83 mg/mL collagenase.
3. Preincubate the tissue for 5 min in a shaking (1.7 Hz) water bath at 37°C. Add 7 mL buffer, and vortex the vial for 5 s.
4. Filter the contents of the vial onto silk cloth, and discard this first filtrate. With scissors, mince the tissue pieces collected on the silk, and incubate this mince in 3 mL fresh, albumin- and collagenase-containing buffer, as above, for 25 min, with 5-s vortexing every fifth minute.
5. Add 7 mL buffer, vortex the vial for 15 s, and filter the contents as above. However, now collect the filtrate and centrifuge it (5 min, 65*g*). Discard the infranatant by suction with a Pasteur pipet with plastic tubing on the tip connected to a water suction pump (*see* **Fig. 1**); add 10 mL buffer, and allow the cells to stand at 4°C.
6. Incubate the remaining tissue pieces on the filter, as above, for 15 min, and collect the cells; the tissue pieces remaining can be incubated for 10 min, and the cells collected, in order to increase the yield.
7. Discard the infranatants in all three tubes, and combine the cells. Add 10 mL buffer, and centrifuge the cells for 2 min.
8. Discard the infranatant, and count the cells in a Bürker chamber. Store the cells on ice until use (mouse cells at room temperature), at a concentration of $1–3 \times 10^6$ cells/mL; very little deterioration of cell response is observed during a working day (*see* **Notes 5** and **6**).
9. The cells are generally used immediately (but can be stored overnight at 4°C in 10 mL KRPB, in which case the cells are washed again the next day, by centrifugation).

## 3.3. Measurement of Respiratory Rate

The rate of $O_2$ consumption of both isolated brown-fat mitochondria and cells can be readily measured polarographically with a Clark-type $O_2$ probe *(3)*. Such a probe (obtainable from Radiometer, Rank Bros., or Yellow Springs Instruments) determines $O_2$ concentration in aqueous solutions. The current produced by the electrode is proportional to the $O_2$ tension in the solution. The electrode chamber must be continuously stirred (most practically, magnetically). Preferably, the electrode chamber must also be temperature-controlled (e.g., by circulating water from a water bath). Calibrate as follows:

1. Fill the electrode chamber with $dH_2O$ at the experimental temperature, do not add any lid (to allow equilibration with atmospheric $O_2$), and allow the output to stabilize. At 37°C, this corresponds to 217 nmol $O_2$/mL (*see* **Note 7**); at 25°C

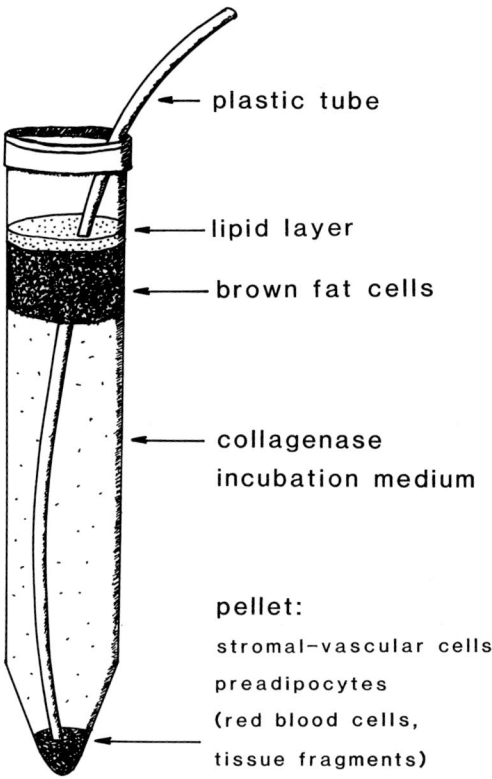

plastic tube

lipid layer

brown fat cells

collagenase
incubation medium

pellet:

stromal–vascular cells

preadipocytes

(red blood cells,

tissue fragments)

Fig. 1. The separation of mature brown adipocytes from lipid droplets and other constituents.

(the traditional, but clearly less physiological, temperature classically used for mitochondrial experiments), this corresponds to 253 nmol $O_2$/mL.

2. Add a few crystals of sodium dithionite to the solution; this reduces all $O_2$ and thus provides for determination of zero $O_2$ level.

3. Remove the calibration solution, and add the relevant buffer. Note that the output may not fully return to the level it had with $dH_2O$; this is correct, because salt-containing media dissolve less $O_2$ than $dH_2O$.

   a. For mitochondrial studies, use a medium consisting of 100 m$M$ KCl, 20 m$M$ K-TES, 4 m$M$ $KH_2PO_4$, 2 m$M$ $MgCl_2$, and 1 m$M$ EDTA, with final pH 7.2. For introduction of energy conservation and choice of substrate, *see* **Notes 8** and **9**.

   b. For cell studies, use the Krebs-Ringer bicarbonate buffer, bubbled with $CO_2$, as described above (*see also* **Note 10**).

4. Add 0.2–0.5 mg mitochondrial protein or 50,000–80,000 cells/mL buffer in the electrode chamber. Close the chamber, and allow the system to stabilize. Make further additions with a Hamilton syringe through a small hole in the cover of the chamber. Note that an addition artifact may (voluntarily) be produced by

allowing the syringe to momentarily stop the magnetic stirring. Note also that the addition of EtOH may lead to a small baseline shift.
5. The electrode output can be connected to an Apple-Macintosh computer via an analog-to-digital converter and the data acquired in the Chart MacLab application program. Alternatively, a regular chart recorder can be used.

## 4. Notes

1. Some authors use a lightly buffered sucrose solution, containing, e.g., 5 m$M$ K-TES. The authors' experience is that this has no obvious beneficial effect on the preparation.
2. Mitochondria can be isolated by the same method from the BAT of other mammals (most commonly, rats or mice). They can also be isolated from tissue taken from animals kept at warmer environmental temperatures. In this case, the tissue contains more triglyceride and fewer mitochondria per gram tissue. The triglyceride disturbs the homogenization, and a lower relative yield of mitochondria may be obtained.
3. When BAT mitochondria are isolated, they are uncoupled *(4,5)* and have a collapsed membrane potential. They demonstrate high permeability to many monovalent ions. Presumably as a consequence of this, they have lost the ability to retain osmotic support in the mitochondrial matrix, which is therefore highly condensed after preparation, and oxidation of substrates in the matrix is markedly inhibited *(6–8)*. To re-expand the matrix, the mitochondria may be incubated in an iso-osmotic medium of permeant ions (such as KCl). Matrix expansion can also be achieved with low-osmolarity sucrose (100 m$M$), although this probably gives a less-controlled expansion.
4. Brown adipocytes can also be prepared by the same method from rats *(9,10)* and mice *(11)*. It is the authors' experience that the cells from Syrian hamsters are the most robust, and are therefore very suitable for many studies.
5. The details given here for preparation of brown adipocytes are given as examples of incubation times with collagenase, collagenase concentrations (and types), centrifugation times (or flotation without centrifugation), which can be successfully applied. Different workers tend to develop personal modifications of these, particularly at times and under circumstances when, for unclear reasons, the preparations are less successful. It is difficult to find convincing evidence that these modifications are of major significance, but this idiosyncrasy also indicates that the details specified here are for guidance, and need not be adhered to exactly.
6. The preparation technique is dependent on the property of the cells to float on top of an aqueous medium. If cells are isolated from animals that are cold-exposed, the triglyceride concentration may be so low that the cells sink in the medium, and can therefore not be separated. The yield of cells will thus be low *(12,13)* and their representativeness can perhaps be discussed. Also, if cells are prepared from animals living at thermoneutrality, the cells are replete with triglycerides and their diameters are larger than those of cells from animals at room temperature. This large cell size seems to make the cells more fragile and sensitive to mechanical manipulation.

7. This value is, of course, only valid for normal atmospheric pressure. However, the effects of normal fluctuations in atmospheric pressure are generally ignored.

8. To transfer the mitochondria into a state of energy conservation, incubation should be performed in the presence of FA-free albumin (0.1– 0.5%), which removes FAs and related substances, and of purine nucleotides. The nucleotides bind to the brown-fat specific UCP1, and, in so doing, close the proton leak through this protein *(8)*. The most commonly used nucleotide is guanosine diphosphate, which is used at a concentration of 0.1–1 m*M*. Other di- and triphosphate purine nucleotides are also more or less efficient. The nucleotide binding site is accessible from the outer side of the inner mitochondrial membrane.

9. When respiratory studies are performed on isolated mitochondria, it is of great importance that a suitable substrate is utilized. The most relevant is a FA or its derivative, such as long-chain acyl-coenzyme A or acyl-carnitine. In all cases, in order to permit complete FA oxidation, malate (in m*M* concentrations) must be added to the buffer to replace citric acid cycle intermediates which have been lost during isolation *(14)*. In some species, the reuptake of malate is low and this may even limit FA combustion. When acyl-carnitine esters are used (50 μ*M*), no further additions (except malate) are required. For acyl-coenzyme A derivatives (similar concentrations), a further addition of L-carnitine (2 m*M*) guarantees that availability of this compound does not limit oxidation. If free FAs are used, further additions of ATP (100 μ*M*) and coenzyme A (5 μ*M*), in addition to carnitine, allow unlimited FA oxidation.

   Brown fat mitochondria also demonstrate a high rate of oxidation of glycerol-3-phosphate (used at m*M* concentrations), a flavoprotein-coupled substrate that is oxidized on the external face of the inner membrane, and thus does not require transport *(15)*. In many species, succinate (which is a classical substate for studies of liver mitochondria) permeates only poorly into the mitochondria and its use may therefore lead to estimates of oxidative capacity that are too low. This is also often the case for other potential substrates, chiefly the intermediates of the citric acid cycle, which have a low rate of permeation in certain species *(16)*.

10. Respiration in the brown adipocytes is most notably stimulated by the physiological agent, norepinephrine, in which case endogenous lipolysis provides the substrate, and also permits uncoupling of respiration from the constraints of a requirement for ATP utilization *(9,17)*. This uncoupling is entirely dependent upon the presence of UCP1 *(18)*. Free FAs can also be added to the cell suspension, and provide an adequate substrate *(17)*. Their combustion is also fully dependent on the presence of UCP1, which demonstrates that FAs can directly or indirectly activate UCP1 *(18)*. If other substrates are utilized, there is a transport requirement into the cells, in addition to which rates of respiration are generally low, unless respiration is artificially uncoupled with, e.g., carbonylayanide-*p*-(trifluormethoxy)phenylhydrazene (FCCP) (20 μ*M*). The maximum respiratory rates then seen are usually much lower with exogenous substrates, such as pyruvate (5 m*M*), than with the endogenously generated or exogenously added FAs.

    White adipose tissue: Mitochondria from WAT. In order to obtain mitochondria representative of the mature adipocytes in the tissue, isolated adipocytes, prepared essentially as described for the brown adipocytes, can be used as starting material. The cells are homogenized and the mitochondria isolated by routine differential centrifugation, as described. The yields are very low. The mitochondria are well-coupled, and can be stimulated to respire on citric acid cycle intermediates, in the presence of ADP *(19)*.

Adipocytes from WAT: As noted above, isolated adipocytes can be readily prepared *(2)*. Because of the appearance of the cells (one unilocular fat droplet filling most of the cell volume), intact cells are not easily distinguished from large fat droplets. Few respiratory studies have been performed on such cells. Their rate of respiration is very low, and high cells densities must be used. Basal metabolic/respiration rates can be estimated, and hormonal stimulation can be performed, but this is generally evaluated in terms of metabolic changes other than respiration. A number of microcalorimetric studies have been performed on isolated white adipocytes from humans. Basal metabolism has been determined *(20)* and comparisons made between tissue taken from, e.g., obese and lean *(21)* or hypo/euthyroid *(22)* individuals. Effects of hormone stimulation can be determined.

## References

1. Nedergaard, J., Cannon, B., and Lindberg, O. (1977) Microcalorimetry of isolated mammalian cells. *Nature (Lond)* **267,** 518–520.
2. Rodbell, M. (1964) Metabolism of isolated fat cells. 1. Effects of hormones on glucose metabolism and lipolysis. *J. Biol. Chem.* **239,** 375–380.
3. Robinson, P. K. (1994) The Clark oxygen electrode, in *Principles and Techniques of Practical Biochemistry*, (Wilson, K. and Walker, J., eds.), Cambridge University Press, Cambridge, pp. 555–562.
4. Smith, R. E., Roberts, J. C., and Hittelman, K. J. (1966) Nonphosphorylating respiration of mitochondria from brown adipose tissue of rats. *Science* **154,** 653,654.
5. Lindberg, O., DePierre, J., Rylander, E., and Afzelius, B. A. (1967) Studies of the mitochondrial energy-transfer system of brown adipose tissue. *J. Cell Biol.* **34,** 293–310.
6. Nicholls, D. G., Grav, H. J., and Lindberg, O. (1972) Mitochondria from hamster brown-adipose tissue. Regulation of respiration in vitro by variations in volume of the matrix compartment. *Eur. J. Biochem.* **31,** 526–533.
7. Nicholls, D. G., and Lindberg, O. (1973) Brown-adipose-tissue mitochondria. The influence of albumin and nucleotides on passive ion permeabilities. *Eur. J. Biochem.* **37,** 523–530.
8. Nicholls, D. G. (1974) Hamster brown-adipose-tissue mitochondria. The control of respiration and the proton electrochemical potential gradient by possible physiological effectors of the proton conductance of the inner membrane. *Eur. J. Biochem.* **49,** 573–583.
9. Fain, J. N., Reed, N., and Saperstein, R. (1967) The isolation and metabolism of brown fat cells. *J. Biol. Chem.* **242,** 1887–1894.
10. Zhao, J., Cannon, B., and Nedergaard, J. (1998) Thermogenesis is $\beta_3$- but not $\beta_1$-adrenergically mediated in rat brown fat cells, even after cold acclimation. *Am. J. Physiol.* **275,** R2002–R2011.

11. Zhao, J., Cannon, B., and Nedergaard, J. (1998) Carteolol is a weak partial agonist on $\beta_3$-adrenergic receptors in brown adipocytes. *Can. J. Physiol. Pharmacol.* **76,** 428–433.
12. Nedergaard, J. (1982) Catecholamine sensitivity in brown fat cells from cold-acclimated hamsters and rats. *Am. J. Physiol.* **242,** C250–C257.
13. Svartengren, J., Svoboda, P., and Cannon, B. (1982) Desensitisation of $\beta$-adrenergic responsiveness in-vivo. Decreased coupling between receptors and adenylate cyclase in isolated brown fat cells. *Eur. J. Biochem.* **128,** 481–488.
14. Cannon, B. (1971) Control of fatty-acid oxidation in brown-adipose-tissue mitochondria. *Eur. J. Biochem.* **23,** 125–135.
15. Bukowiecki, L., and Lindberg, O. (1974) Control of sn-glycerol 3-phosphate oxidation in brown adipose tissue mitochondria by calcium and acyl-CoA. *Biochim. Biophys. Acta* **348,** 115–125.
16. Cannon, B., Bernson, V. M. S., and Nedergaard, J. (1984) Metabolic consequences of limited substrate anion permeability in brown fat mitochondria from a hibernator, the golden hamster. *Biochim. Biophys. Acta* **766,** 483–491.
17. Prusiner, S. B., Cannon, B., and Lindberg, O. (1968) Oxidative metabolism in cells isolated from brown adipose tissue. I. Catecholamine and fatty acid stimulation of respiration. *Eur. J. Biochem.* **6,** 15–22.
18. Matthias, A., Jacobsson, A., Cannon, B., and Nedergaard, J. (1999) Bioenergetics of brown-fat mitochondria from UCP1-ablated mice: UCP1 is not involved in fatty acid-induced de-energization "uncoupling". *J. Biol. Chem.* **274,** 28,150–28,160.
19. Marshall, S. E., McCormack, J. G., and Denton, R. M. (1984) Role of $Ca^{2+}$ ions in the regulation of intramitochondrial metabolism in rat epididymal adipose tissue. Evidence against a role for $Ca^{2+}$ in the activation of pyruvate dehydrogenase by insulin. *Biochem. J.* **218,** 249–260.
20. Monti, M., Nilsson-Ehle, P., Sörbis, R., and Wadsö, I. (1980) Microcalorimetric measurement of heat production in isolated human adipocytes. *Scand. J. Clin. Lab. Invest.* **40,** 581–587.
21. Olsson, S. A., Monti, M., Sörbis, R., and Nilsson-Ehle, P. (1986) Adipocyte heat production before and after weight reduction by gastroplasty. *Int. J. Obs.* **10,** 99–105.
22. Valdemarsson, S., Fagher, B., Hedner, P., Monti, M., and Nilsson-Ehle, P. (1985) Platelet and adipocyte thermogenesis in hypothyroid patients: a microcalorimetric study. *Acta Endocrinol. (Copenh.)* **108,** 361–366.

# 25

## Measurements of White Adipose Tissue Metabolism by Microdialysis Technique

**Pierre Barbe, Christian Darimont, Perla Saint-Marc, and Jean Galitzky**

## 1. Introduction

The microdialysis method was introduced 25 yr ago to measure neurotransmitter concentrations in the brains of laboratory animals. For 10 yr, this technique has been adapted, in metabolic studies, to monitor the interstitial concentrations of small-mol-wt compounds present in the extracellular space of various tissues, specially skeletal muscle and adipose tissue (AT). This development concerned animal experiments, as well as clinical research in humans. The microdialysis probe mimics the passive function of an artificial small blood vessel implanted into the tissue (**Fig. 1**). The characteristics of AT microdialysis are as follows (*1*):

1. It collects a representative sample of all substances in the extracellular fluid.
2. It causes minimal damage to the tissue.
3. It makes possible continuous sampling for hours or days after a single penetration of the tissue.
4. It allows recovery of endogenous substances, and makes them accessible to analytical techniques.
5. It permits introduction of exogenous substances in the tissue, in order to study the resulting local biochemical effect, and to avoid general effects.
6. It allows study of the local response inside the tissue during systemic drug administration, or when a physiological test is performed (such as exercise).

Beside in vitro studies on isolated adipocytes collected by biopsies, the microdialysis technique provides an alternative approach to study the adipose cells in their actual milieu.

From: *Methods in Molecular Biology, vol. 155: Adipose Tissue Protocols*
Edited by: G. Ailhaud © Humana Press Inc., Totowa, NJ

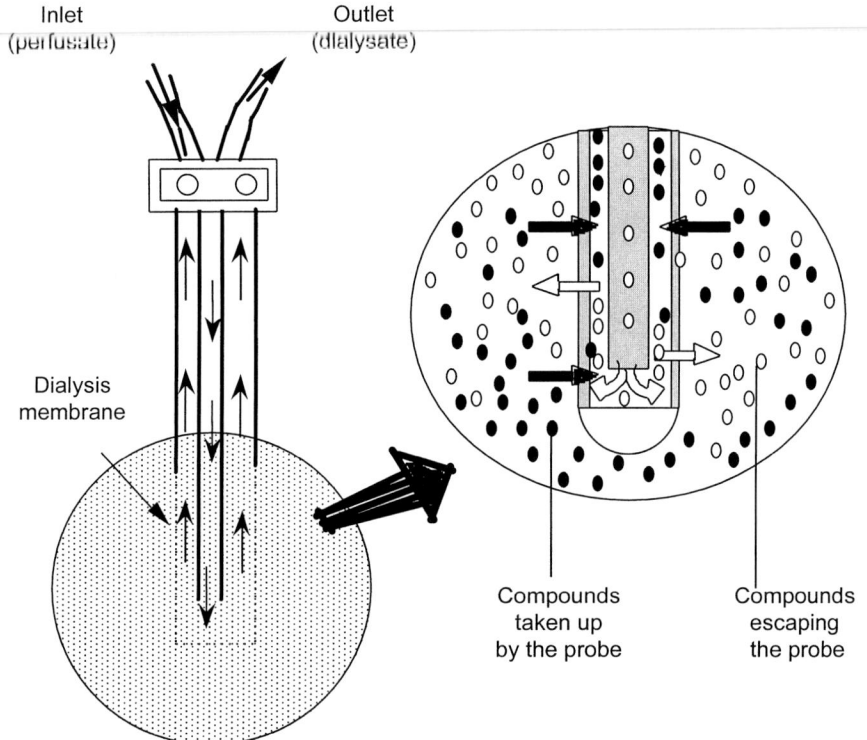

Fig. 1. Schematic representation of a concentric microdialysis probe used for the measurement of AT metabolism.

## 1.1. General Principles of Microdialysis Technique

The microdialysis technique applies the dialysis principle (from Greek: to separate), in which a membrane, permeable to small-mol-wt compounds and water, separates two fluid compartments. Provided that no differences of osmotic pressure and electrical potential exist across the membrane, molecule transport between the probe compartment and the extracellular compartment is governed only by concentration gradients.

Typically, a microdialysis probe consists of a tubular dialysis membrane connected with inlet and outlet tubes, which are continuously perfused with an isotonic solution (perfusate) at a low flow rate (0.3–5 μL/min) on the inside, the outside of the probe is in contact with the surrounding medium. Thus, a substance present in the interstitial fluid and absent from the perfusate can diffuse toward the compartment of the lowest concentration. Conversely, a pharmacological compound perfused in the probe will diffuse to the tissue. The composition of the liquid retrieved from the outlet tube (dialysate) is thus a mirror of the liquid composition surrounding the probe. The limits of the technique depend on the capacity of a compound of interest to cross the membrane, and on the sensivity of the detection assays for the measured substances and metabolites (*see* **Note 1**).

The *in situ* microdialysis technique can be used to investigate metabolism in rodents. The small size of dialysis probes makes them ideal for the study of rat AT. Concentrations of small metabolites present in the interstitial fluid of rat AT can easily be measured using this technique. Furthermore, by perfusing compounds with a low mole wt into the vicinity of cells, regulation of metabolite secretion can be performed. The accessibility of different adipose depots (visceral, subcutaneous [sc], periepididymal, and parametrial pads), in anesthetized rats allows comparative studies to be performed *(2–5)*. Furthermore, the availability of different animal models, including hyperinsulinemic or hypertensive rats, permits the study of AT metabolism in several pathological conditions *(5,6)*. The microdialysis technique has therefore proven to be a valuable and flexible tool for *in situ* measurement of AT.

This chapter describes an application of the *in situ* microdialysis technique for the periepidydimal fat pad of anesthetized male Wistar rats. Measurement of the extracellular concentration of prostaglandins (PG), and particularly prostacyclin ($PGI_2$), was chosen to illustrate the procedure.

$PGI_2$, secreted in vitro by preadipocytes and adipocytes, plays a key role in preadipocyte differentiation *(7,8)*. Indeed, this PG is a potent adipogenic agent acting via a membrane receptor, and increasing intracellular cyclic adenosine monophosphate and calcium *(9)*. Recently, $PGI_2$ was reported to activate the peroxisome proliferator activating receptor γ, which stimulates the expression of genes implicated in adipocyte differentiation *(10)*. Measurement of the extracellular concentration of $PGI_2$ in rat AT was a prerequisite for the validation of the in vivo function of this PG in AT development *(11)*.

## 1.2. The Microdialysis Probes

Different kinds of microdialysis probes have been used for AT studies. The tubular and concentric probes are more common. The tubular probes have been used especially in human experiments; they are made with a hollow dialysis membrane connected to inlet and outlet inert polyethylene tubing in a serial arrangement. This is a simple and inexpensive device often handmade by investigators. The major advantage is that the length of the membrane, which improves the recovery (*see* **Subheading 1.3.**) is not limitative, the major disadvantages are their weak reproductibility and the need to perform two cutaneous perforations (inlet and outlet). The concentric probes are made with dual concentric tubing, with a dialysis membrane at one end (a schematic view is presented in **Fig. 1**). They are manufactured and are more expansive than the tubular probes. Probes of different lengths (5–30 mm) are available. They can be used in humans, as well as in small animals. A single penetration of the tissue is needed. They have an excellent reproductibility from test to test. This review presents the authors' experience with this kind of microdialysis probes.

## 1.3. Recovery

One challenge during microdialysis experiments is to establish to what extent the measured level of a substance in the dialysate reflects its true concentration in the extracellular space surrounding the probe. Under most circumtances, the chosen perfusion rate is a compromise between the volume of the fraction required for analytical procedure and the frequency of sampling necessary to record the effect. At the usual perfusion rate, a fraction of the concentration of a substance in the extracellular space is only measured in the dialysate. The recovery of a substance is described in terms of absolute and relative recovery. The absolute recovery of a substance is the total amount of the substance recovered from the extracellular space during a definite period of time (expressed in µmol/min, for example); the relative recovery at a given flow rate is defined as the ratio between the concentration in the dialysate and the concentration in the fluid surrounding the probe (expressed in % values).

### 1.3.1. Factors of Recovery

Recovery is independant of outside substance concentration, but various factors affect the recovery during microdialysis. Some of these are linked to the technique:

1. The perfusion rate (the recovery decreases as the perfusion rate increases).
2. The surface of exchange (linked to the length of dialysis membrane, and to the design of the probe).
3. The composition of the perfusate (ideally, the same as the surrounding medium, minus the presence of the substance of interest).
4. The nature of the dialysis membrane (usually polycarbonates/polyether membranes are used, these membranes interact weakly with low-mol-wt compounds).
5. The size of a membrane's pores (a dialysis membrane with a 20,000 cutoff is limiting for high-mol-wt compounds).
6. The temperature inside and outside the probe.

Some of these are linked to the tissue:

1. The structure of the tissue: A tissue is a complex medium comprising the interstitial, intracellular, and vascular compartments; furthermore, the extracellular fluid is not a real aqueous solution, but is more related to a colloid medium, as a result of the presence of extracellular matrix; the so-called tortuosity factor accounts for these factors.

2. The speed of diffusion of the substance through the extracellular fluid: The speed of diffusion is linked to the nature of the substance, as well as to the elimination from the extracellular space by uptake into the cells and local blood flow (BF) (*see* **Note 2**).
3. The pressure and the temperature of the tissue.

### 1.3.2. Methods for In Situ Microdialysis Calibration

The relative recovery assessment of a substance is mandatory to measure its real interstitial concentration. Numerous attempts have been made to determine the extracellular concentration by extrapolation from in vitro recovery experiments, but there is general agreement that this is not reliable, because of the difference in dialysis between aqueous solutions and tissues.

Different methods have been developed to measure *in situ* recovery during in vivo experiments. The relative recovery can be calculated by varying the perfusion rate, and measuring the changes in the concentration of the substance at different perfusion rates, then extrapolating to the concentration at zero flow (perfusion rate method) *(12)*; and by perfusing, with varying concentrations of the substance, with a constant flow rate, and measuring the changes in the dialysate concentration, then calculating the concentration at which the substance in the perfusate does not change, because it has the same concentration inside the probe as in the surrounding medium (no-net flux method) *(13)*. These two methods are presented below. It is noticeable that, whatever the methods, they are time-consuming. Steady-state conditions are required to keep constant the concentration of the substance in the extracellular space, throughout the experiment (*see* **Note 3**).

A third method is to use a reference substance in the perfusate during the experiment *(14)*. Usually, the reference is a tracer dose of the substance of interest labeled by isotope (e.g., $^{14}C$-glycerol or $^{13}C$-glucose). The method is based on the hypothesis that recovery through the membrane is the same in both directions, i.e., the percentage of loss of the reference substance from the perfusate to the tissue will be the same as the percentage recovery of the substance from the tissue to the probe. If it is true in vitro, but it is not so real in vivo because of the difference in dialysis between aqueous solutions and tissues. This method needs further validation studies, before being applied to clinical research in human.

### 1.4. Analytical Procedures Used with Microdialysis

The analytical procedures used to measure the dialysate concentrations of different metabolites are critical for the application of the microdialysis technique. Very sensitive assays are needed, because of the small dialysate concentrations and the small volume of the collected fractions. Conventional biochemical analyses are occasionally valid (such as for EtOH or urea *[15]*), but it is often necessary to use radioenzymatic assays or bioluminescent enzymatic assays, such as for glucose or glycerol analysis *(16)*.

## 2. Materials

### 2.1. Microdialysis Applied to Human SC AT

This subheading describes the general procedure that the authors used to assess lipolysis in human experiments *(17–21)*.

1. Anesthetic solution: 1% lidocaine (Roger-Bellon, France).
2. Sterile Ringer's solution for clinical use (without lactate): 139.3 m$M$ sodium (Na), 2.7 m$M$ potassium (K), 0.9 m$M$ calcium, 140.5 m$M$ chloride, 2.4 m$M$ bicarbonate, 5.6 m$M$ glucose (B. Braun Medical, Boulogne, France).
3. Microdialysis probes: CMA20 with a 20,000 Dalton molecular cutoff polycarbonate membrane, 0.5 mm outer diameter and 20 mm length (Carnegie Medicine, Stockholm, Sweden).
4. Microperfusion pump: microinjection pump 22 (Harvard Apparatus, South Natick, MA).
5. Syringe: 5 mL Plastipak syringe (Becton Dickinson, Le Pont de Claix, France).
6. Needle: 23G × 1 needle (Terumo Europe, Leuven, Belgium).
7. Connecting tubing: polyethylene catheter, 0.58 mm internal diameter and 0.96 outer diameter (od) (Merck-Clevenot, Chennevières Les Louves, France). The connecting tubing is adapted to the syringe with the needle, and to the inlet of the microdialysis probe.

### 2.2. Microdialysis Applied to Rat AT

1. Anesthetic solution: 0.1 g/mL chloral hydrate diluted in 0.9% NaCl.
2. Ringer's solution: 147 m$M$ NaCl, 2 m$M$ $CaCl_2$, 4 m$M$ KCl. Filtered on 0.45 μm membrane, and stored at 4°C.
3. Microdialysis probes (CMA/20; Carnegie) with a 20,000 Dalton molecular cutoff, 0.5 mm od and 10 mm length.
4. Micro infusion pump (CMA/100, Carnegie).
5. Micro fraction collector (CMA/140; Carnegie).
6. Peristaltic pump (LKB, Sweden).
7. Heating pad (Chromex, France) connected to a temperature control unit (Horst, Germany), and a rectal thermometer (Bioblock, France).

## 3. Methods

### 3.1. General Procedure for Microdialysis of SC AT in Humans

#### 3.1.1. Preparation of Microdialysis Probe

Glycerol is added to the probe, by the manufacturer for its preservation. The probes must be prepared the day before the dialysis experiment:

1. For one probe, prepare two syringes containing 20% EtOH–sterile water solution, the second containing Ringer's solution alone (*see* **Note 4**).
2. Set the flow rate at 10 μL/min, and start the perfusion before connecting the probe, to avoid air bubbles.

3. Wash 1 h with the 20% EtOH solution at 10 µL/min, with the probe bathing in the same solution, in order to remove the glycerol present in the dialysis membrane.
4. Then wash 1 h with Ringer's solution at 10 µL/min, with the probe bathing in the same solution
5. Stop perfusion rate, and leave the probe in the sterile Ringer's solution, at ambient temperature for 1 d.

### 3.1.2. General Procedure for Implantation of Microdialysis Probe into Human Abdominal SC AT

The subject is kept in the supine position during the implantation.

1. Clean up and disinfect the implantation area of the probe (to a distance of 10 cm to the right or the left side of the navel).
2. Induce sc anesthesia by light intradermal injection of 100–200 µL lidocaine (*see* **Note 5**).
3. Make a small skin prehole (around 1 mm) with a sterile surgical blade.
4. Implant the introducer needle surrounded with a guide made with a split plastic tubing (both provided in the microdialysis probe package) into the abdominal fat.
5. Carefully remove the introducer needle, and leave the split plastic tubing inside the tissue.
6. Introduce the microdialysis probe, continuously perfused by the Ringer's at 10 µL/min, into the guide, then tear it carefully, in order to remove only the split plastic tubing, and to leave the probe inside the tissue.
7. Fix the probe to the skin with a sticking plaster (*see* **Note 6**),
8. Keep the flow rate at 10 µL/min for at least 15 min, in order to wash the dialysis probe, and to avoid possible plugging of the probe.
9. Change the flow to the perfusion rate selected (2.5 µL/min for example, or 0.8 µL/min if recovery is measured, *see* **Subheading 3.1.3.**) for 30 min to equilibrate the dialysis probe.

A schematic presentation of the microdialysis device of human abdominal sc AT is presented in **Fig. 2** (*see* **Note 7**).

### 3.1.3. Evaluation of Relative In Situ Recovery of Probe

This subheading specifically describes the perfusion rate method developped in this laboratory for in vivo experiments using CMA 20 mm probes *(18)*.

1. Discard the 25 min of equilibration period at 0.8 µL/min, and collect another 30-min dialysate sample (24 µL).
2. Wash the probe for 2 min at 10 µL/min, and decrease the flow rate to 1.5 µL/min, and equilibrate for 15 min.
3. Discard the last fraction and collect a 15-min dialysate sample (22.5 µl).
4. Wash the probe for 2 min at 10 µL/min, and decrease the flow rate to 3.5 µL/min, then equilibrate for 10 min.
5. Discard the last fraction, and collect 10-min dialysate sample (25 µL).

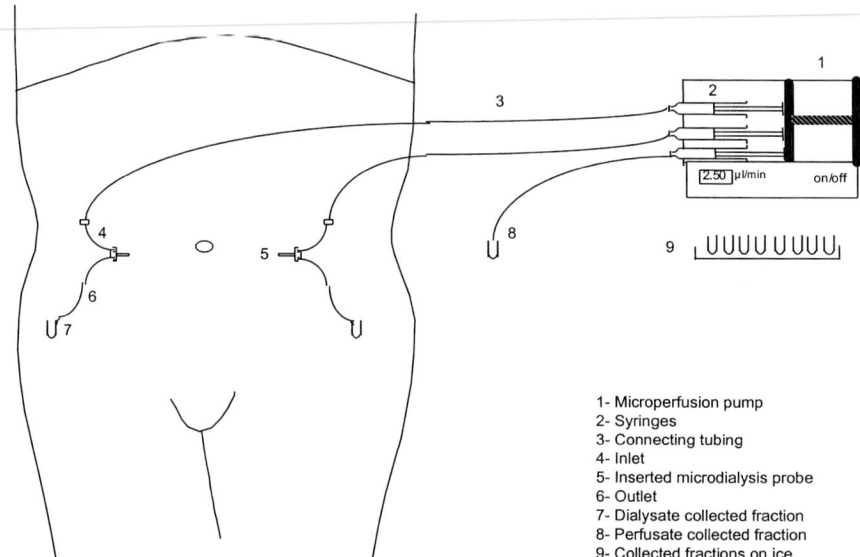

1- Microperfusion pump
2- Syringes
3- Connecting tubing
4- Inlet
5- Inserted microdialysis probe
6- Outlet
7- Dialysate collected fraction
8- Perfusate collected fraction
9- Collected fractions on ice

Fig. 2. Schematic representation of the general procedure for the measurement of adipose tissue metabolism by *in situ* microdialysis of sc abdominal AT in humans.

6. Wash the probe for 2 min at 10 μL/min, and decrease the flow rate to 2.5 μL/min, and equilibrate for 15 min.
7. Discard the last fraction, and collect for at least 2–3 10-min fractions. These fractions represent the basal period of the dialysis experiment. The perfusion rate was 2.5 μL/min in the authors' experiments, in order to collect 25-μL fractions every 10 min, which is the volume required for glycerol, glucose, and EtOH measurements.

The concentration of each metabolite of interest is determined at each perfusion flow rate. A plot between these concentrations values and the flow rate is drawn, in order to calculate the zero-flow concentration ($C_0$) (**Fig. 3**). If the relation is not linear, then transform the concentration values (1/concentration or logarithmic) and, by linear regression, calculate the $C_0$. The relative in vivo recovery at 2.5 μL/min is the ratio between the concentration measured at 2.5 μL/min and the calculated $C_0$ concentration (*see* **Note 8**).

With higher length dialysis membrane (≥30 mm), a $C_0$ evaluation can be made by measuring the dialysate concentration of the metabolite of interest at a very low flow rate (0.3–0.5 μL/min) (*see* **Note 9**) *(22)*.

### 3.1.4. Measurements of AT BF Changes During Microdialysis

In addition to local production (such for glycerol) or uptake (such for glucose), the changes of local BF can affect the interstitial concentration of a

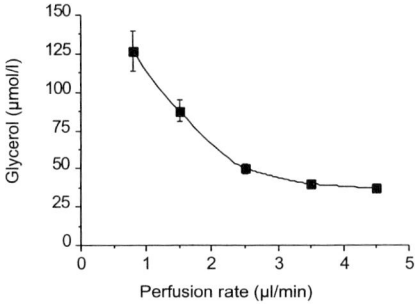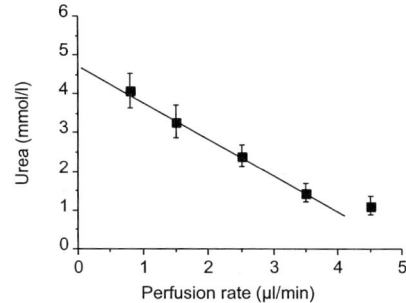

Fig. 3. Dialysate concentrations of glycerol (left part) and urea (right part) with various perfusion rates in the same experiment. Data are expressed as mean ± SEM. Individual values obtained in seven healthy subjects during microdialysis of abdominal sc AT with two probes (20 mm length) implanted in each experiment.

metabolite by changing the delivery or the removal of this compound in the extracellular space. Perfusions of vasoactive agents modified, in opposite ways, the glycerol and the glucose concentrations in sc AT (**Fig. 4**). Indeed, monitoring local BF is mandatory, to interpret the changes of interstitial concentration metabolites.

EtOH escape is the most commonly used method to measure local BF changes during microdialysis *(23)*. Practically:

1. Add 40 m*M* of EtOH to the Ringer's perfusion solution (this concentration of EtOH is devoid any metabolic or vasoacting effects).
2. Use this perfusate throughout the experiment.
3. Measure the concentration of EtOH in the perfusate and in each of the 10-min fractions of dialysate, which must be recovered at 4°C throughout the experiment and until the dosage.
4. Calculate the EtOH outflow:inflow concentration ratio. The changes of the ratio are linked to the changes in local BF (**Fig. 5**). The decrease of the ratio reflects a vasodilation; conversely, an increase of the ratio means that a vasoconstriction process has occurred.

In the range of concentrations tested (5–100 m*M*), EtOH has no apparent effect on adipocyte metabolism or BF itself *(23)*. This method appears sensitive enough to indicate small variations of local BF in the extracellular space surrounding the microdialysis probe, which is of particular interest when an agent, potentially active on adipocyte metabolism, and with known or unknown vasoactive effects, is added to the perfusate.

An endogenous compound delivered by the local BF, highly diffusible in the extracellar water space, which is not produced locally, and without interaction with the tissue metabolism, such as urea, could also be used to monitor the

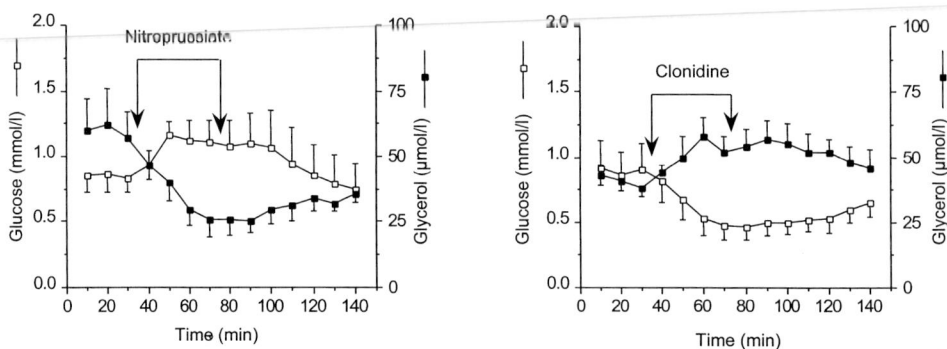

Fig. 4. Effects of *in situ* administration of sodium nitroprussiate (1.25 g/L) and clonidine ($10^{-5}$ *M*) on dialysate concentrations of glucose and glycerol in abdominal sc AT. These two vasoactive compounds were added 40 min to the perfusate. They have opposite effects: Na nitroprussiate promotes vasodilation, clonidine promotes vasoconstriction. Data are expressed as mean ± SEM. Individual values obtained in six healthy subjects during microdialysis of abdominal sc AT with two implanted 20 mm probes (one for Na nitroprussiate and one for clonidine) in each experiment. Glucose and glycerol concentrations were determined on the same fractions. The perfusion rate was 2.5 μL/min.

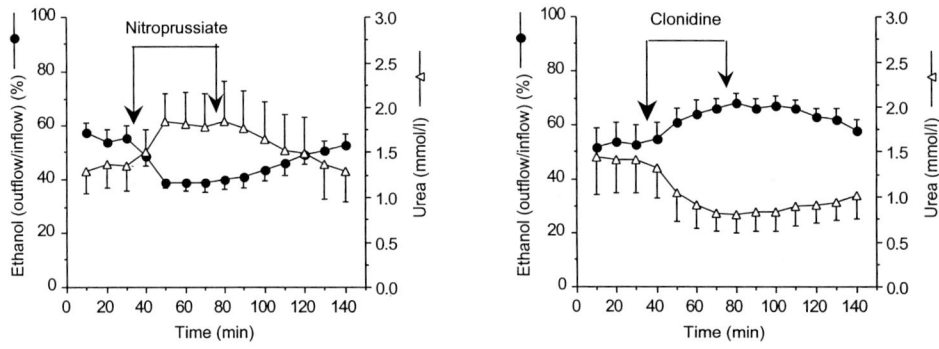

Fig. 5. Effects of *in situ* administration of sodium nitroprussiate (1.25 g/L) and clonidine ($10^{-5}$ *M*) on EtOH ratio and urea dialysate concentrations in abdominal sc AT. Data are expressed as mean ± SEM. Individual values obtained in six healthy subjects. EtOH and urea concentrations were determined on the same fractions. The perfusion rate was 2.5 μL/min (*see* **Fig. 4** for further details).

changes of local BF. The dialysate urea concentrations and the EtOH outflow:inflow ratio are modified in opposite ways during the *in situ* perfusion of vasoactive agents (**Fig. 5**). The dialysate urea concentrations reflect the changes of local BF if plasma urea levels are kept constant during the experiment.

### 3.2. General Procedure for Microdialysis of Periepididymal Fat Pad of Anesthetized Rat

#### 3.2.1. Implantation of Dialysis Probe

Male Wistar rats (250–350 g), kept at room temperature with 12 h dark–light cycle, were anesthetized with an intraperitoneal injection of 0.33 mL anesthetic solution per 100 g rat body wt (final concentration injected: 0.2 g chloral hydrate/kg).

After the rat is completely anesthetised (*see* **Notes 10** and **11**), the procedure is as follows (**Fig. 6**):

1. Place the rat on its back on a heating pad. To maintain the anesthetized animal at a constant temperature, connect the heating pad to a thermoregulator controlled by a rectal thermometer.
2. Into the peritoneal cavity, implant a needle connected to a peristaltic pump continously infusing 0.25% anesthetic solution (20 µL/min).
3. Shave the lower abdomen, and make an abdominal incision (approx 3 cm).
4. Carefully remove a periepididymal fat pad from the peritoneal cavity.
5. Implant a steel needle surrounded with split plastic tubing (both available with dialysis probes) into the periepididymal fat pad.
6. Carefully remove the steel needle, and leave the split plastic tubing inside the tissue.
7. Introduce the dialysis probe into the catheter, then tear it very carefully, in order to remove only the split plastic tubing, and to leave the dialysis probe inside the tissue.
8. Connect the inlet catheter of the probe to the microinfusion pump.
9. Wash the probe with dionized water perfused at 10 µL/min for 10 min, in order to avoid possible plugging of the dialysis membrane.
10. Equilibrate the dialysis probe by perfusing Ringer's solution for 30 min at a rate of 1 µL/min.
11. At the end of the equilibration step, connect the outlet catheter of the probe to a microfraction cellector, in order to collect samples at defined time intervals.
12. Finally, initiate perfusion with Ringer's solution alone or supplemented with agents able to increase $PGI_2$ in vitro.

Depending on the sensitivity of the kit used for detection of dialysate compounds, fractions of 10 µL or more can be collected. Because of the very short half-life of $PGI_2$, its stable degradation product, the 6-keto-$PGF_{1\alpha}$, can be detected in 20-µL fraction, using a commerciallly available kit.

#### 3.2.2. Calibration of PG Microperfusion in AT

The dialysis recovery of compounds collected in the dialysate is necessary to determine the precise concentration in the interstitial fluid. Different calibration methods have been described in the literature *(24)*. Both in vitro and in vivo calibration methods are described in the following subheading.

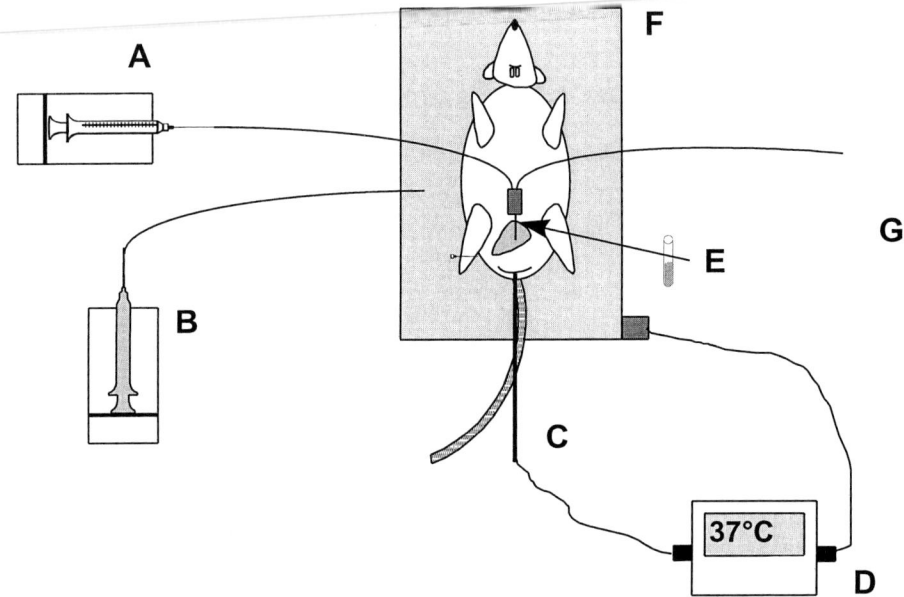

Fig. 6. Schematic representation of the general procedure for the measurement of adipose tissue metabolism by *in situ* microdialysis in anaesthetised rat. (**A**) Microinfusion pump; (**B**) peristaltic pump perfusing the anesthetic solution; (**C**) rectal thermometer; (**D**) thermoregulator; (**E**) dialysis probe; (**F**) heating pad; (**G**) microfraction collector.

### 3.2.2.1. IN VITRO RECOVERY FROM DIALYSIS MEMBRANE (STOP-FLOW METHOD)

1. Perfuse Ringer's solution at 1 µL/min through the dialysis probe immersed in a tube containing 1.5 mL labeled 6-keto-PGF$_{1\alpha}$ solution at 37°C.
2. Collect the perfusate every 15 min for 45 min, and count the radioactivity present in each fraction of 15 µL.
3. Perform **steps 1** and **2** with 2 and 5 µL/mL as dialysis rate.

**Figure 7** represents the dialysate concentration divided by the perfusate concentration (recovery) as a function of dialysis rate. The best recovery is obtained at low dialysis rate: 1 µL/min (*see* **Note 12**).

### 3.2.2.2. IN VIVO RECOVERY

The no-net-flux method is described in this subheading for the measurement of 6-keto-PGF$_{1\alpha}$ concentration in the interstitial fluid. For this method, several different concentrations of 6-keto-PGF$_{1\alpha}$ are perfused, as described below:

1. After the implantation of the probe in the AT, as described before, perfuse 0.1 n*M* 6-keto-PGF$_{1\alpha}$ at 1 µL/mL.
2. Collect three fractions of 10 µL, and freeze it immediately.

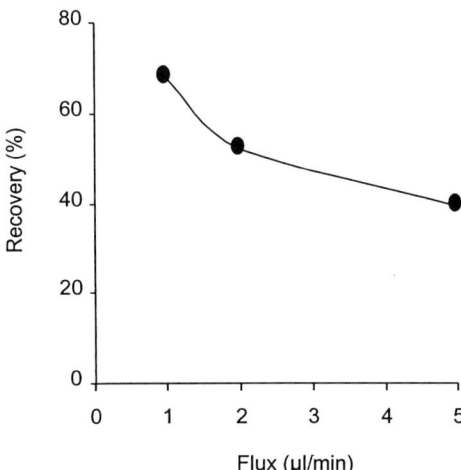

Fig. 7. Recovery of 6-keto-PGF$_{1\alpha}$ *in vitro*. The relative recovery of 6-keto-PGF$_{1\alpha}$ at various perfusion rates (1, 2, and 5 μL/min) was determined at 37°C as a percentage of 6-keto-PGF$_{1\alpha}$ (2 n*M*) present in the test solution. Individual values obtained in three separate experiments, using three different probes, are presented *(11)*.

3. Wash the probe for 20 min with Ringer's solution.
4. Perform **steps 1–3** with 0.25, 0.6, and 2 n*M* 6-keto-PGF$_{1\alpha}$.

The difference between the concentration in the dialysate minus the concentration in the perfusate as a function of perfused 6-keto-PGF$_{1\alpha}$ concentration, is represented in **Fig. 8**. A linear regression is necessary to calculate the point at which no perfusion occurred (when dialysate and perfusate concentrations are identical), which indicates the concentration of 6-keto-PGF$_{1\alpha}$ in interstitial fluid of AT (*see* **Notes 13** and **14**).

## 4. Notes

1. Determination of steady-state levels of glucose, glycerol, lactate, and adenosine in human sc AT have been made using the equilibrium technique. Steady-state levels of glycerol are 2–3× higher in sc AT than in venous blood *(25)*. Fasting interstitial glycerol levels are higher in abdominal than in femoral sc AT in lean subjects, and higher in abdominal sc AT in the obese patients than in the lean subjects *(25)*. Interstitial levels of glucose are identical with those in blood. Lactate production occurs in human AT in vivo *(26,27)*. Adenosine can reach concentrations in the extracellular space of AT sufficient for an important modulating effect on lipolysis in vivo *(28)*.
2. When the speed of diffusion through the extracellular fluid of a substance supplied by the local BF is low, the dialysis could drive to a decrease of the extracellular concentration of this compound *(29)*. The mass transfer from the vascular

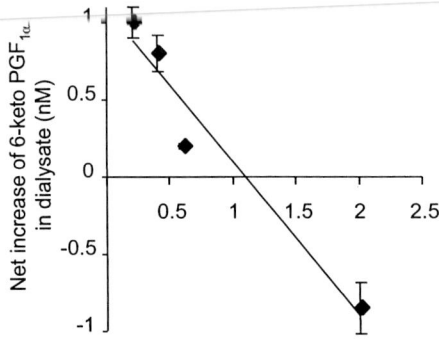

Fig. 8. Recovery of 6-keto-PGF$_{1\alpha}$ in vivo: Determination of the extracellular 6-keto-PGF$_{1\alpha}$ concentrations. Microdialysis probe implanted in the rat periepididymal AT was perfused with increasing concentrations of 6-keto-PGF$_{1\alpha}$ at a rate of 1 μL/min. The relationship between the initial concentrations of 6-keto-PGF$_{1\alpha}$ in perfusate and the net change of this PG in the dialysate is presented. According to the linear relationship, which exists between the concentration of the perfused solutes and their concentration gradients across the dialysis membrane, intercepts on the perfusate axis indicate 6-keto-PGF$_{1\alpha}$ concentrations in equilibrium between the perfusate and the interstitial fluid surrounding the tissue. Data are the mean ± SEM of four separate experiments *(11)*.

compartment to the extracellular space does not compensate the mass transfer from the extracellular space to the dialysate. This decreased extracellular concentration can interact with adipose metabolism. The relationships between extracellular glucose concentration and adipocyte lactate production is an illustration of this phenomenon *(26)*.

3. Measurement of *in situ* relative recovery is not required if you perform kinetic experiments (e.g., to assess the lipolytic effect of a new drug added to the perfusate). In vivo calibration is mandatory, to compare the data obtained from different probes, from different subjects, or from different experimental conditions *(23)*.

4. Prepare the solution in advance at room temperature, to avoid air bubbles.

5. A light intradermal injection is required in order to avoid diffusion of lidocaine throught the AT. This kind of drug could modify local metabolism and BF.

6. It is not necessary to stitch, if the subject is always at rest, but for exercise, stitches are required in order to keep the probe in the position.

7. All solutions, syringes, and connecting tubing must be prepared and placed on the microperfusion pump at the beginning of the experiment. Do not stop the perfusion more than 1 min during the experiment, when you change the perfusate in order to keep the equilibrium inside the probe. It is possible to place 2 or 4 probes (i.e., 2 on each side) in the sc abdominal wall in one subject. Keep 10 cm or more between two probes implanted in the same side of the abdominal wall, to avoid diffusion from one probe to the other.

8. The relative recovery of a substance at a given perfusion rate is different in different fat deposits. We reported different relative recoveries for glycerol between abdominal and femoral subcutaneous ATs *(21)*.

9. It is not possible to reach a 100% recovery with short-length dialysis membrane (20 mm or less) perfused at a very low flow rate (0.3–0.5 µL/min). When dialysis membranes ≥30 mm are used, the $C_0$ evaluation must be validated with one of the two reference methods (no net flux or perfusion rate), in order to define the most appropriate perfusion rate between 0.3–0.5 µL/min.

10. Implantation of the dialysis probe in anesthetized animal is recommended. Indeed, a long-term implantation of dialysis probe in ATs of awake animal leads to several problems, including possible movement of the probe out of the AT, or potential breakage when the animal is moving.

11. Implantation of two dialysis probes into different adipose pads can be performed in the same animal, for a comparative study. In order to check the possible diffusion of perfused compounds in other depots, a second probe could be inserted in the contralateral fat pad *(4)*.

12. Perfusion rate is a key parameter for the optimization of recovery, as mentioned in **Subheading 2.3.1.** Indeed, a slow-speed flow is recommended, to achieve the best diffusion of compound through the dialysis membrane.

13. Body temperature must be maintained at 37°C, in order to avoid changes in global metabolism, and also in blood pressure, which can locally affect the measurement of interstitial concentrations.

14. Local BF has been shown to affect the measurement of extracellular glycerol concentration in rat AT *(30)*. Perfusion of lipolytic agents with vasodilator properties, such as isoproterenol in the interstitial fluid of AT can delay or mask a lipolytic effect *(30)*. Vasodilator agents stimulate tissue drainage, and consequently decrease the concentration of compounds in the vicinity of the probe. In order to prevent such effects, previous vasodilation will avoid the vasodilating effect of agent perfused later. The potent vasodilator agent, hydralazine, can be perfused first, and concomitantly to the lipolytic agent, in order to measure its lipolytic effect alone.

## Acknowledgment

The authors are grateful to Dr. R. Reimer (Lausanne, Switzerland) for careful reading of the manuscript, and for helpful suggestions.

## References

1. Ungerstedt, U. (1991) Microdialysis: principles and applications for studies in animals and man. *J. Intern. Med.* **230,** 365–373.
2. Iwao, N., Oshida, Y., and Sato, Y. (1997) Regional differences in lipolysis caused by a beta-adrenergic agonist as determined by the microdialysis technique. *Acta Physiol. Scand.* **161,** 481–487.
3. Kowalski, T. J., Wu, G., and Watford, M. (1997) Rat adipose tissue amino acid metabolism *in vivo* as assessed by microdialysis and arteriovenous techniques. *Am. J. Physiol.* **273,** E631–E642.

4. Darimont, C., Delansorne, R., Paris, J., Ailhaud, G., and Négrel, R. (1997) Influence of estrogenic status on the lipolytic activity of parametrial adipose tissue *in vivo*: an *in situ* microdialysis study. *Endocrinology* **138**, 1092–1096.

5. Cimmino, M., Agosto, A., Minaire, Y., and Geloën, A. (1995) *In situ* regulation of lipolysis by insulin and norepinephrine: a microdialysis study during euglycemic-hyperinsulinemic clamp. *Metabolism* **44**, 153–1518.

6. Cabassi, A., Vinci, S., Calzolari, M., Brushi, G., and Borghetti, A. (1998) Regional sympathetic activity in pre-hypertensive phase of spontaneously hypertensive rats. *Life Sci.* **62**, 1111–1118.

7. Richelsen, B. (1991) Prostaglandins in adipose tissue. *Danish Med. Bull.* **38**, 228–244.

8. Négrel, R., Gaillard, D., and Ailhaud, G. (1989) Prostacyclin as a potent effector of adipose cell differentiation. *Biochem. J.* **257**, 399–405.

9. Vassaux, G., Gaillard, D., Ailhaud, G., and Négrel, R. (1992) Prostacyclin is a specific effector of adipose cell differentiation. Its dual role as cAMP and $Ca^{2+}$ elevating agent. *J. Biol. Chem.* **267**, 11,092–11,097.

10. Aubert, J., Ailhaud, G., and Négrel, R. (1996) Evidence for a novel regulatory pathway activated by (carba)prostacyclin in preadipose and adipose cells. *FEBS Lett.* **397**, 117–121.

11. Darimont, C., Vassaux, G., Gaillard, D., Ailhaud, G., and Négrel, R. (1994) *In situ* microdialysis of prostaglandins in adipose tissue: stimulation of prostacyclin release by angiotensin II. *Int. J. Obes.* **18**, 783–788.

12. Stähle, L., Segersvärd, S., and Ungerstedt, U. (1991) A comparison between three methods for estimation of extracellular concentrations of exogenous and endogenous compounds by microdialysis. *J. Pharmacol. Meth.* **25**, 41–52.

13. Jansson, P.-A., Fowelin, J., Smith, U., and Lönnroth, P. (1988) Characterization by microdialysis of intercellular glucose levels in subcutaneous tissue in humans. *Am. J. Physiol.* **255**, E218–E220.

14. Jansson, P.-A., Veneman, T., Nurjhan, N., and Gerich, J. (1994) An improved method to calculate adipose tissue interstitial substrate recovery for microdialysis studies. *Life Sci.* **54**, 1621–1624.

15. Bernst, E. and Gutmann, I. (1974) Determination of ethanol with alcohol deshydrogenase and NAD, in *Methods of Enzymatic Analysis* (Bergmeyer, H. U., ed.), Verlag, Weinheim, pp. 1499–1505.

16. Bradley, D. C. and Kaslow, H. R. (1989) Radiometric assays for glycerol, glucose and glycogen. *Anal. Biochem.* **180**, 11–16.

17. Galitzky, J., Lafontan, M., Nordenström, J., and Arner, P. (1993) Role of vascular alpha2-adrenoceptors in regulating lipid mobilization from human adipose tissue. *J. Clin. Invest.* **91**, 1997–2003.

18. Barbe, P., Millet, L., Galitzky, J., Lafontan, M., and Berlan, M. (1996) *In situ* assessment of the role of β1-, β2-, β3-adrenoceptors in the control of lipolysis and nutritive blood flow in human subcutaneous adipose tissue. *Br. J. Pharmacol.* **117**, 907–913.

19. Barbe, P., Stich, V., Galitzky, J., Kunesova, M., Hainer, V., Lafontan, M., and Berlan, M. (1997) *In vivo* increase of β-adrenergic lipolytic response in subcutaneous adipose tissue of obese subjects submitted to hypocaloric diet. *J. Clin. Endocrinol. Metab.* **82**, 63–69.

20. Barbe, P., Galitzky, J., Gilsezinski, I., Rivière, D., Thalamas, C., Senard, J. M., Crampes, F., Lafontan, M., and Berlan, M. (1998) Simulated microgravity increases β-adrenergic lipolysis in human adipose tissue. *J. Clin. Endocrinol. Metab.* **83,** 619–625.

21. Millet, L., Barbe, P., Lafontan, M., Berlan, M. and Galitzky, J. (1998) Catecholamine effects on lipolysis dand blood flow in human abdominal and femoral adipose tissue. *J. Appl. Physiol.* **85,** 181–188.

22. Bolinder, J., Ungerstedt, U., and Arner, P. (1992) Microdialysis measurement of the absolute glucose concentration in subcutaneous adipose tissue allowing glucose monitoring in diabetic patients. *Diabetologia* **35,** 1177–1180.

23. Lafontan, M. and Arner, P. (1996) Application of in situ microdialysis to measure metabolic and vascular responses in adipose tissue. *TIPS* **17,** 309–313.

24. Lönnroth, P. (1997) Microdialysis in adipose tissue and skeletal muscle. *Horm. Metab. Res.* **29,** 344–346.

25. Jansson, P.-A., Smith, U., and Lönnroth, P. (1990) Interstitial glycerol concentration measured by microdialysis in two subcutaneous regions in humans. *Am. J. Physiol.* **258,** E918–E922.

26. Jansson, P.-A., Smith, U., and Lönnroth, P. (1990) Evidence for lactate production by human adipose tissue in vivo. *Diabetologia.* **33,** 253–256.

27. Jansson, P.-A., Larsson, A., Smith, U., and Lönnroth, P. (1994) Lactate release from the subcutaneous tissue in lean and obese men. *J. Clin. Invest.* **93,** 240–246.

28. Lönnroth, P., Jansson, P.-A., Fredholm, B. B., and Smith, U. (1989) Microdialysis of intercellular adenosine concentration in subcutaneous tissue in humans. *Am. J. Physiol.* **256,** E250–E255.

29. Lönnroth, P., Jansson, P.-A., and Smith, U. (1987) A microdialysis method allowing characterization of intracellular water space in humans. *Am. J. Physiol.* **253,** E228–E231.

30. Darimont, C., Saint-Marc, P., Ailhaud, G., and Négrel, R. (1996) Modulation of vascular tone and glycerol levels measured by *in situ* microdialysis in rat adipose tissue. *Am. J. Physiol.* **271,** E631–E635.

# 26

## Immunoassay Measurements of Leptin

### Martin Bidlingmaier and Matthias Tschoep

## 1. Introduction

Since the discovery of the circulating peptide hormone, leptin, the recognition of adipose tissue (AT) as an endocrine organ has added a new perspective to obesity research. Numerous studies have shown the involvement of leptin (expressed predominantly in AT, placenta, and stomach) in the physiology of hematopoiesis, immune function, angiogenesis, fertility and the regulation of body wt *(1)*.

However, despite high expectations for using this hormone as a therapeutic agent, leptin has been demonstrated to be inefficacious for the treatment of human obesity. Nevertheless, a new era of understanding the pathophysiology of obesity and regulation of energy homeostasis has arisen from the observation that AT secretes a 16 kDa hormone that informs the hypothalamic regulatory system about the level of peripheral energy stores *(2)*.

In the past 5 yr, hundreds of published clinical and experimental studies, have been based on the quantification of leptin in body fluids, tissues, or cell culture media. Although the quantification of leptin mRNA levels is widely used, immunoassay measurement of leptin levels in body fluids is the most frequently used method that has permitted the clinical studies on leptin to be feasible.

Despite of the frequency and comparatively long history of immunoassays, the generation and interpretation of immunoassay data is frequently underestimated. This chapter provides a short overview of immunoassay principles and some practical tips regarding troubleshooting during leptin measurements. The concentration of hormones in fluids (serum, plasma, cerebrospinal fluid, conditioned media) are analyzed by various immunoassay techniques.

In general, these methods require specific antibodies (Abs) (poly- or monoclonal) to recognize an analyte. A standard preparation of the hormone (used for calibration) and a detection system are utilized to determine concentrations.

From: *Methods in Molecular Biology, vol. 155: Adipose Tissue Protocols*
Edited by: G. Ailhaud © Humana Press Inc., Totowa, NJ

The detection system can utilize radioactive labeling, fluorescence, luminescence, or a colorimetric reaction, which are catalyzed by enzymes. These detection systems have advantages and disadvantages, and in most cases the method is chosen based on equipment availability. In terms of the assay sensitivity, the Ab affinity is much more important than the signal detection system. When multiple detection systems are available, fluorescence immunoassays (FIAs) offer the advantage of stable tracer, which can be used for several years, while radioactive tracers have to be regenerated frequently, which leads to varying signal intensities. Moreover, FIAs use tracers that are comparable in size to the analyte; radiolabeled molecules, such as $I^{125}$, are significantly larger.

Concerning the choice of the best suitable Ab, polyclonal Abs have the ability to recognize a complete spectrum of peptide hormone isoforms in a sample, however, undesirable molecules, such as degradation products, are also detected. This confounding factor can lead to a potential overestimation of the immunofunctional potency of the circulating hormone, in this case, leptin. On the other hand, generating polyclonal Abs is easy and fast, which is of great significance when rapid quantification of newly discovered molecules is desirable.

Monoclonal antibodies (mAbs) only react with distinct, defined epitopes, allowing for precise and reliable measurement of peptide hormones, such as leptin. However, insufficient information about the epitope map may lead to an underestimation of the total amount of biologically active hormone. On the other hand, the recognition of specific epitopes with mAbs allows for a more precise description of what an Ab is measuring. Because mAb generation and characterization are relatively complicated, time-consuming, and risky, their application in established immunoassays is not as common as one would predict from their superior properties.

The two most common types of immunoassay currently in use are the competitive and the sandwich immunoassay. In a competitive assay, the hormone in the sample or standard competes with a labeled hormone fraction (tracer) for Ab binding (fixed amount). This competition leads to a negative relationship between hormone concentration and signal intensity. Higher levels of endogenous, unlabeled hormone in a sample, decreases the amount of tracer that is bound by the Ab. A double-Ab system can be used to separate free and bound antigen. Depending on which tracer-labeling reagent is used, competitive assays are named radioimmunoassay (RIA), fluorescence immunoassay (FIA), luminescence immunoassay (LIA), and enzyme immunoassay (EIA).

In contrast, a sandwich type immunoassay generates a signal, proportional to the amount of antigen in the sample. This assay requires two Abs, which recognize different epitopes. In other words, the analyte is sandwiched between two Abs, one labeled with a radioactive marker (immunoradiometric assay), fluorescent marker (immunofluorometric assay), chemoluminiscent marker (immunoluminometric assay), or enzyme marker (enzyme-linked immunosorbent assay [ELISA]). These sandwich immunoassays are less susceptible to bias caused by circulating analyte fragments, because only those molecules possessing two specific and intact epitopes are measured.

For assay calibration, a standard preparation of the analyte is required. Variability in the standard preparation results in differences in the concentrations obtained. Moreover, different sources of organ-derived (extracted, affinity-enriched) standards are a major source of the heterogeneity in assay results. This caveat can lead to contrasting and contradictory data in clinical and experimental studies.

When recombinant standards are available, differences in manufacturing processes of these preparations (e.g., variant refolding, incomplete glycosylation) can cause discrepancies. However, in the case of leptin, comparable biological and immunogenetic activities for different recombinant leptin preparations have already been demonstrated *(3)*.

To overcome the problem of variabilities in standard preparations, immunoassays must be calibrated against an international reference preparation. Unfortunately, such a reference preparation for leptin is not available. Consequently, the National Institute for Biological Standards and Control (a World Health Organization international laboratory) is developing such a reference preparation *(4)*.

Both competitive and sandwich-type immunoassays for measuring human, rat, and mouse leptin are commercially available. For analyzing samples containing very low concentrations of leptin (i.e., cerebrospinal fluid and conditioned media), a high-sensitivity method is required, with a lower detection limit of at least 0.1 μg/L (in humans). For less-sensitive methods, the addition of a concentration or extraction step prior to the immunoassay procedure may be necessary, but this would be time-consuming and difficult, and can be a source of artifacts and decreased reproducibility. Therefore, the use of high-affinity Abs is recommended. The theoretical sensitivity, or lower detection limit, is calculated by the interpolation of the mean + 2 SD from at least 20 replicates of the 0-standard. A more realistic estimation of assay sensitivity would be to calculate the lower limit of quantification: This involves measuring the amount of analyte that can be measured with an intra-assay variation below 15% (as calculated from a 10-fold determination in one assay). In most cases, the realistic limit is above the theoretical limit.

The standard or sample matrix can generate significant performance and reproducibility problems. For example, measuring leptin in conditioned medium requires diluting the calibrator in native medium: for measurements in serum, the ideal standard matrix would be leptin-free serum from the relevant species. Alternatively, serum from a species showing very low crossreactivity with the Ab, might be used. Manufacturer's instructions, provided with an assay kit, should always include information about the standard matrix, as well as the applicability of plasma or serum samples. A telephone call to the manufacturer's technical services can prevent the repetition of time-consuming experiments, not to mention the generation of misleading data.

Regarding the measurement of serum leptin, physiology can interfere with generation and accurate interpretation of immunoassay data *(5)*. Besides the urgent need for an international reference preparation for recombinant human leptin, the authors would like to briefly emphasize two further issues:

1. Serum leptin circulates primarily bound to a leptin-binding protein, which is probably a dimer consisting of two extracellular domains of the long form of the leptin receptor molecule. Obese individuals with high leptin levels exhibit a relatively lower percentage of bound, and a higher percentage of free, leptin, compared to normal subjects. Information concerning the specificity of commercially available assays for bound or free leptin are rarely given. In addition, the biological activity of the bound and free fractions of circulating leptin remains unclear. The most insightful data would be generated by a parallel measurement of total leptin and free leptin, or, alternatively, by direct measurement of the circulating complexes.
2. Leptin secretion is pulsatile (approx 32 peaks/24 h), and exhibits a circadian rhythm (peak at night); thus, singular leptin measurements within small groups of individuals are not reliable. Therefore, it is recommended that leptin profiles are obtained using larger populations, taking into consideration other interfering factors, such as food intake.

## 2. Materials

### 2.1. Ordering a Commercial Assay Kit

1. Check availability of the relevant detection system.
2. Ask the customer service/company representative about the labeling and date of expiration of the tracer when ordering RIA kits.
3. Verify that the Ab recognizes the correct hormones of the correct species, e.g., the Ab from a rat leptin RIA that shows crossreactivity with mouse leptin of less than 20% cannot be recommended for the quantification of mouse leptin.
4. Double-check with your in-house radio safety officer that your research facility is in possession of the mandatory approval for the use of the radioactive isotope you are ordering. Receive approval from your radio safety officer to obtain the relevant isotopes.
5. Find the person at your facility who receives radioactive immunoassay kits (and, in case of arrival during a weekend, where the kit could be stored safely and at proper cooling conditions). Ensure that personnel receiving RIA can safely and properly store the kits.
6. Ensure that the size and/or amount of assays ordered are sufficient for the number of samples you plan to measure. For example: 100 Ab-coated vials are included in an assay kit, and four vials are needed for the control samples (duplicates

should be included in the assay kit). Between 10 and 16 vials are needed for standard concentrations (which are also included; performance again in duplicates), and four vials are necessary for total counts and nonspecific binding (NSB) measurements. Accordingly, 38–41 samples could be analyzed in duplicate with this kit. Samples from the same study should always be measured with the same kit, the same Ab and the same batch of tracer, if possible.

7. Commercially available kits should provide sufficient information about appropriate sensitivity, specificity (crossreactivity), linearity of dilution, recovery, and assay precision (interassay and intraassay). If a noncommercial immunoassay is used, determine that the crossreactivity with other molecules is minimal, intra- and interassay variance do not exceed 12–15%, and sufficient amounts of tested controls are available.

## 2.2. Materials Often Required but not Regularly Supplied

1. Polystyrene test tubes/96-well microtiter plates.
2. Test tube racks.
3. Precision and repeating pipets.
4. Sponge rack for decantation, or a vacuum suction device.
5. Centrifuge/microcentrifuge.
6. Microtiter plate shaker.

## 2.3. Preparation of Specimen

1. Leptin specimens should be stored frozen at –80°C. Storage at 2–8°C for longer than 12 h should be avoided, as should repeated freezing and thawing of samples.
2. Hemolyzed or lipemic samples should not be analyzed.
3. Frozen samples should be equilibrated to room temperature (RT) before analysis, then mixed thoroughly by gentle inversion or vortexing prior to use.

# 3. Methods
## 3.1. Competitive Assays (see Notes 1–4)
### 3.1.1. Standard Assay Procedure

For competitive RIA, please read and follow specific instructions given by the manufacturer.

1. Label and arrange tubes for total counts, NSB, standards, controls, and unknowns (duplicates).
2. Pipet standards, controls, or unknowns into the appropriate tubes. Add 0-standard to the NSB counts. (Lyophilized leptin standard may dissolve slowly. If gentle mixing yields in complete resuspension, transient increase in pH may be necessary.)
3. Add tracer to each tube.
4. Add antiserum to each tube, except NSB- and total-counts tubes.
5. Vortex gently.
6. Incubate at RT (overnight incubations are acceptable).
7. Add precipitating reagent to all tubes (except the total-count tubes), and vortex.

8. Incubate at RT
9. Centrifuge sample for 20 min at 2000$g$.
10. Decant samples, except the total counts (do not disturb the pellet), by simultaneous inversion (eventually with a sponge rack) into a radioactive waste receptacle. Drain on absorbent material, and gently blot tubes, to remove any droplets adhering to the rim, before returning them to an upright position. Failure to blot tubes adequately may result in poor replication and invalid results.
11. Quantify signal intensity of samples in counter.

### 3.1.2. Calculation of Results

Software for calculating the final results should be installed on the detection system. Normally, a log–linear curve fit is used. The basic calculation procedure typically includes the following steps:

1. Calculation of the mean cpm for each standard, control, and unknown.
2. Subtract the mean cpm of the NSB tubes from all standard/control/unknown samples to obtain normalized counts.
3. Calculate the normalized counts for each standard/control/unknown sample as a percentage of the normalized counts of the zero standard ($B/B_0$).
4. Plot the concentrations of the leptin standards against the corresponding $B/B_0$ values. Draw a standard curve through the mean of the standard points.
5. Read the concentration of each control/unknown sample corresponding to its $B/B_0$ in comparison to the standard curve.

## 3.2. Sandwich Immunoassays (see Notes 1–4)

### 3.2.1. Standard Assay Procedure

For sandwich-type ELISA, please read and follow specific instructions given by the manufacturer.

1. The assay procedure is performed on a 96-well microtiter plate, in a simultaneous incubation format. For some assays, a sequential procedure will be necessary.
2. Prepare a 96-well microtiter plate coated with the appropriate antileptin Ab.
3. Pipet standards, controls, or unknowns, in duplicate, to the appropriate wells. To minimize carryover, standards should be pipeted, starting with the blank and continuing with increasing concentrations of the leptin standard.
4. Add appropriate, enzyme-labeled second Ab.
5. Incubate for 2 h at RT on a plate-shaker.
6. Wash plate with washing buffer.
7. Add the substrate for the colorimetric reaction.
8. Incubate for at least 30 min at 37°C (longer incubations may be performed at RT).
9. Add the stopping solution.
10. Read the absorbance (optical density [OD]) at the appropriate wavelength. The 0 ng/mL standard must be set as blank. To correct for differences in the OD between the wells of the microtiter plate, a background wavelength correction should be used (usually at 600 or 620 nm).

### 3.2.2. Calculation of Results

Software for calculating the final results should be installed on the detection system. Normally, a log-linear curve fit is used. The basic calculation procedure typically includes the following steps:

1. Calculating the mean absorbance (OD) for each standard, control, and unknown sample, after correcting for background.
2. Plot a curve count of the leptin standards against the leptin concentration. Draw a standard curve through the mean of the standard points.
3. Determine the mean from the duplicates of each control or unknown sample, by matching their mean OD readings with the corresponding leptin concentration.

### 3.2.3. Interpretation of Data

1. Measured concentrations should plot within the linear phase of the standard graph. When performing an ELISA/EIA, the OD values obtained must be within the linear range of the photometer. Most photometers are limited to an OD of up to 2–3; thus, interpretation of higher values are not accurate. If the OD exceeds the limit of the photometer, either the amount of substrate or the incubation time must be reduced.
2. Any sample reading higher than the highest standard, or outside the linear range of the standard curve, should be diluted appropriately with the 0-standard, and reassayed.
3. Any sample reading lower than the lowest standard should be reported as such.
4. Duplicates for standards, controls, or unknowns may not differ more than 15% from each other. If this is the case, the assay may have to be repeated.
5. Verify that the controls are within the appropriate range.
6. Are the results plausible? Do obese individuals have generally higher leptin levels than lean individuals?

## 4. Notes

1. Background problems. High background may lead to a poor, lower detection limit, and may make it difficult to use the standard curve. The background may be caused by either NSB of the Abs or a contamination of a reagent or measuring device by the marker used in your assay system. The first possibility should be ruled out by using an appropriate Ab with a high specificity for the analyte. In addition, the background could be lowered by adding a substance that blocks NSB to the assay buffer (e.g., BSA and/or bovine γ-globulins). In most cases, contamination causes problems not only in a single assay, but in all assays using the contaminated reagent (e.g., washing buffer) or measuring device. In RIAs, the racks for counting should always be checked. One major source of contamination could be the preparation of reagents containing high concentrations of the labeling substance (e.g., preparation of stock solutions, tracer labeling procedures). These steps should be carried out away from the workplace where the assay is performed. Regardless of the marker used, contaminated pipets could be another source of background.
2. Low signal. In this case, the marker reagent should be controlled. Depending on the isotope used, radioactive tracer substances may have short half-lives. Within

enzyme-catalyzed colorimetric reactions, the enzyme may not work, because of an interfering substance, such as sodium azide or hypochlorous acid. The laboratory water used to prepare buffers can be a source for such substances.

3. Poor duplicates may result from nonhomogenous material in the tubes/microtiter plates, causing differences in background signal,e.g., scratches on the bottom of a plate may affect the absorbance. More often, the washing steps cause problems. For example, the size of the pellet in a RIA procedure may vary because of uneven decantation. Furthermore, if an automatic washer for microtiter plates is used, blocked tubes often lead to uneven washing of the wells. Finally, all pipeting steps may result in poor duplicates, either because of different amounts pipeted, or carryover from wells/tubes containing higher concentrations of the analyte.

4. Implausible high or low results. Aside from the possibility of wrongly labeled vials and and technical problems, such as carryover from other wells (*see* **Note 3**), there are other causes of implausible results. All assays should be tested for high-dose hook effect, which indicates that very high concentrations of the analyte may be disturbing Ab binding, resulting in implausibly low results. Reassaying the sample after serial dilution may solve this problem. Many immunoassays are using Abs raised in mice. Human anti-mouse Abs (HAMA) found in some human serum samples can interfere with Ab binding and, thus, result in implausible high or low assay results. As a rare case for implausible low results, a mutation of the analyte not crossreactin with the Abs used in the assay may be present in the samples from certain patients.

## Acknowledgments

The authors would like to thank Todd Suter for experienced advice and language editing of the manuscript, as well as R. Schwaiger, S. Freihoefer, and C. J. Strasburger for advice and helpful disscussions.

## References

1. Friedman, J. M. and Halaas, J. L. (1998) Leptin and the regulation of body weight in mammals. *Nature* **395,** 763–770.
2. Jequier, E. and Tappy, L. (1999) Regulation of body weight in humans. *Physiol. Rev.* **79(2),** 451–480.
3. Varnerin, J. P., Smith, T., Rosenblum, C. I., Vongs, A., Murphy, B. A., Nunes, C., et al. (1998) Production of leptin in *Escherichia coli*: a comparison of methods. *Protein Expres. Purif.* **14(3),** 335–342.
4. Robinson, K. (1997) WHO standard for leptin, in *Leptin: The Voice of the Adipose Tissue* (Blum, W. F., Kiess, W., and Rascher, W., eds.), Barth-Heidelberg, Leipzig, Germany, pp. 309–314.
5. Blum, W. F. (1997) Round table discussion on standardization of leptin assays, in *Leptin: The Voice of the Adipose Tissue* (Blum, W. F., Kiess, W., and Rascher, W., eds.), Barth-Heidelberg, Leipzig, Germany, pp. 315–317.

# Index